DRUG-INDUCED DEMENTIA

A PERFECT CRIME

Books by Grace E. Jackson, MD

Rethinking Psychiatric Drugs: A Guide for Informed Consent

Contributing Author

Making or Breaking Children's Lives

Rethinking ADHD: International Perspectives

DRUG-INDUCED DEMENTIA

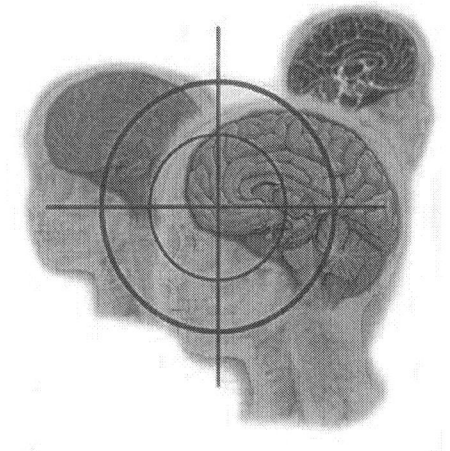

A PERFECT CRIME

GRACE E. JACKSON, MD

AuthorHouse™
1663 Liberty Drive
Bloomington, IN 47403
www.authorhouse.com
Phone: 1-800-839-8640

© 2009 Grace E. Jackson, MD. All Rights Reserved.

No part of this book may be reproduced, stored in a retrieval system, or transmitted by any means without the written permission of the author.

First published by AuthorHouse 5/28/09

ISBN: 978-1-4389-7231-2 (sc)
ISBN: 978-1-4389-7232-9 (e)

Library of Congress Control Number: 2009905329

Printed in the United States of America
Bloomington, Indiana

This book is printed on acid-free paper.

DEDICATION

For Jim Gottstein, Esq.

…founder of the Law Project for Psychiatric Rights
…tireless opponent of the Pharmacaust

And the Lord said to Cain: "Where is your brother?"
And he said: "I know not … Am I my brother's keeper?"
And He said: "What have you done?
The voice of your brother's blood cries unto me from the ground."

Genesis, Chapter 4, verses 9-10

CONTENTS

Prologue		ix – xix
Chapter One	Dementia	1-11
Chapter Two	Mechanisms	12-55
Chapter Three	Epidemiology	56-93
Chapter Four	Antidepressants	94-140
Chapter Five	Antipsychotics	141-181
Chapter Six	Anxiolytics (Anti-Anxiety Drugs)	182-207
Chapter Seven	Mood Stabilizers	208-258
Chapter Eight	Stimulants	259-316
Epilogue		317-322
Appendices		323-351
Endnotes		352-440

Prologue

Case #1

The story begins with an unidentified female whose psychiatric history commenced in her teens.[1] Diagnosed with bipolar disorder (manic depression) at that time, she subsequently experienced treatment with numerous psychiatric medications and numerous hospitalizations over the next 30 years.

By the age of 44, this woman had developed chronic lung and heart disease (including COPD, unstable high blood pressure, and congestive heart failure), persistent auditory hallucinations, and an inability for self-care. The unexpected death of her mother – her primary caregiver – triggered an emotional crisis which involved worsening psychotic symptoms and involuntary commitment to a Virginia facility later that year. After a one-month course of treatment with three antipsychotic drugs (fluphenazine, quetiapine, risperidone) and an antiepileptic drug (carbamazepine), the woman was transferred to a State institution for continuing care.

Medications were changed in the new facility. The revised regimen consisted of one anticonvulsant drug (valproate, aka Depakene) and the continuation of one antipsychotic medication (risperidone). On this combination, the patient soon became agitated and confused.

On the 10th day of valproate, the woman displayed symptoms of an unobserved, seizure-like event: slow eye blinking, foam in the mouth, and incontinence of urine. She was taken to the nearest emergency room where a blood test revealed evidence of drug intoxication (drug level of valproate = 141 ug/mL; normal < 100 ug/mL). She was diagnosed with a **pseudoseizure** – a term which means a psychologically caused or possibly "faked" convulsion, and she was sent back to the psychiatric hospital where her drug therapies remained essentially unchanged (only a slight reduction in the dose of valproate).

On the 11th day of valproate, the patient suffered a generalized, tonic-clonic seizure. This event was witnessed by a roommate. She was again transported to the nearest emergency room where she was admitted for observation. On this occasion, she was evaluated by specialists from four different fields: emergency medicine, internal medicine, cardiology, and neurology. Partly due to the patient's continuing agitation, these assessments were inconclusive. After two days, she was discharged back to the psychiatric hospital with the following diagnosis:

"two episodes of unresponsiveness, of unclear etiology."

Over the ensuing eight days, she displayed disorganized thought, fluctuating confusion, disorientation, and new-onset visual hallucinations (e.g., seeing devils and angels on her shoulders).

Prologue

On day #21 of valproate – after two seizures, two emergency room visits, and one overnight stay on an inpatient medical ward – a physician decided to order the very first blood test of ammonia. (This is a byproduct of protein metabolism which becomes abnormally elevated in at least 50% of all valproate patients). This lab finding revealed an extraordinarily high level of 445 ug/dL – more than five times normal (8.5-85 ug/dL). This was also consistent with the diagnosis of **valproate-induced hyperammonemic encephalopathy or VHE** – a potentially life-ending, drug-induced inflammation of the brain which accounted for this woman's seizures, confusion, and psychiatric deterioration.

The patient spent two more months in the hospital on a treatment regimen which consisted of risperidone plus a replacement anticonvulsant. Although her disorientation, confusion, and visual hallucinations resolved rapidly with the normalization of serum (and brain) levels of ammonia, she remained disorganized in her speech and thinking, and she continued to display a labile mood.

Case #2

A 61-year old woman with a 30-year history of mixed anxiety and "rapidly cycling" mood reported two weeks of progressive depressive symptoms to her social worker.[2] She was admitted to the local inpatient psychiatry ward for the stabilization of poor sleep and social withdrawal. At the time of admission, her diagnoses included bipolar II disorder, several anxiety conditions, early onset Alzheimer's disease, hypothyroidism, and gastroesophageal reflux (heartburn) for which the following treatments were concurrently and actively prescribed:

for anxiety:	clonazepam	1.5 mg three times a day
for depression:	tranylcypromine	30 mg each morning and noon
	dextroamphetamine	15 mg each morning and noon
for "rapid cycling":	lamotrigine	100 mg twice a day
	risperidone	0.5 mg morning/evening; 2 mg at bedtime
for insomnia:	quetiapine	300 mg at bedtime
	trazodone	50 mg at bedtime
	zolpidem	10 mg at bedtime
for dementia:	rivastigmine	3 mg twice a day
	memantine	10 mg twice a day
for hypothyroidism	levothyroxine	125 ug each morning
for reflux	lansoprazole	30 mg each morning

Prologue

Soon after admission, the patient began to complain of chest tightness. Although this had been reported during previous hospitalizations (and attributed at that time to anxiety), the most recent events were qualitatively different. First, the chest tightening was noted to occur with each dose of rivastigmine. Second, the tightness was accompanied by shortness of breath. Third, the chest discomfort was accompanied by repeated episodes of retrocollis (aka, cervical dystonia: an abnormal, sustained contraction of the neck muscles that result in the backward displacement of the head):

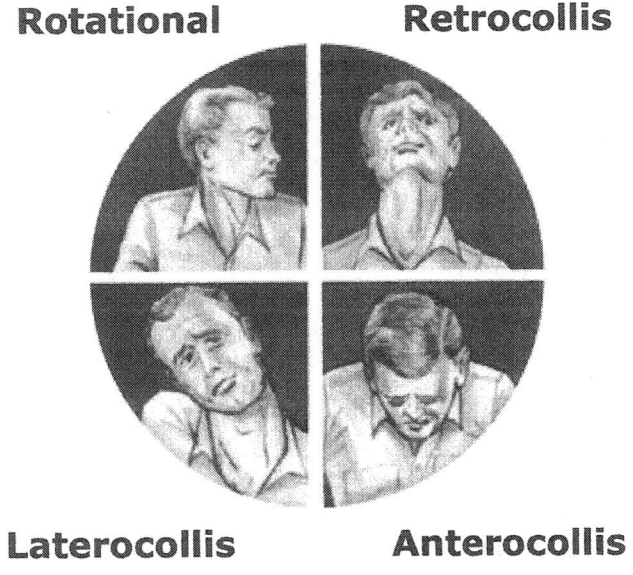

Source: National Spasmodic Torticollis Association

Emergency treatment was initiated with the anticholinergic drug, benztropine (Cogentin). This resolved the neck spasm. Within 30 minutes, the patient's chest tightness also began to abate. An internal medicine specialist was consulted to rule out any cardiovascular cause of the patient's symptoms. At the recommendation of this provider, dextroamphetamine and rivastigmine were discontinued "due to concerns about a potential interaction with the patient's antipsychotic drugs."

Two days later, a different physician resumed treatment with rivastigmine in an effort to improve the patient's memory. Within two hours of the regular morning dose of this anti-dementia drug, the dystonia recurred. Cogentin was again administered with good results, and the order for rivastigmine was discontinued for a second time. Throughout the remaining course of this woman's two-week hospitalization, she experienced no recurrence of the muscle contractures. Although slight adjustments were made to her polypharmacy regimen (tranylcypromine was stopped, risperidone bedtime dose was reduced from 2 mg to 1 mg), she was discharged back to the community on seven different neuropsychiatric drugs.

Prologue

What lessons do these cases hold for physicians and their patients? *To their credit*, the writers of these reports identified several key features of drug toxicity that are not generally emphasized in medical school, residency programs, or continuing medical education curricula. It is *not* common for physicians to hear about the *seizure-provoking effects* of the anticonvulsant, valproate. Similarly, it is *not* common for physicians to hear about the *neurotoxicities* of rivastigmine (Exelon) – an anti-dementia drug which can replicate many effects of poisoning by insecticides or nerve gas. However, as important as these real-life accounts may have been in terms of alerting others to the drug-induced phenomena which the authors emphasized (encephalopathy and dystonia), they were equally important in terms of revealing how much medical professionals seemed *not* to know. Specifically, these episodes vividly displayed a profession-wide failure to appreciate the **numerous ways in which psychiatric drugs disable the human brain**:

Case #1   failure to consider **NCSE** (non-convulsive status epilepticus) as the cause of the patient's seizures, confusion, agitation, and worsening psychosis [3-10]

Case #2   failure to consider **Respiratory Dyskinesia, Tardive Akathisia, and Drug-Induced Hypothyroidism** as a cause of chest tightness, shortness of breath, anxiety, and unstable mood [11-25]

Both Cases   failure to consider **Tardive Dementia** as a cause of chronic psychosis (Case #1), chronic mood lability (Case #2), and early-onset Alzheimer's disease (Case #2)

Drug-Induced Dementia

Governmental regulators are not concerned about it. Doctors are not trained about it. And patients are not warned about it.

It is only at sporadic intervals in the history of medicine that astute observers have attempted to alert their peers about it. For example, in the 1980s, a group of physicians in North Carolina, a visiting scientist in Tel Aviv, and a clinician in Vancouver, British Columbia all contributed papers to the medical literature, recommending the adoption of the term ***Tardive Dysmentia*** as a name for the delayed-onset (tardive) deterioration experienced by patients under the influence of antipsychotic drugs.[26-28] These professionals shared their experiences and concerns about the emergence of progressive mood instability, cognitive deficits, and/or disfiguring movement abnormalities due to the chronic effects of dopamine-blocking agents.

Even the American Psychiatric Association has reserved a term for it. Although the construct is seldom mentioned or clinically used, it is significant that the official handbook of mental illnesses – the *Diagnostic and Statistical Manual of Mental*

Prologue

Disorders (aka, *DSM*) – recognizes the persistent cognitive limitations which are induced or enhanced by *drugs of any kind*:

> "Features that are associated with **Substance-Induced Persisting Dementia** are those associated with dementias generally....This disorder usually has an insidious onset and slow progression, typically during a period when the person qualifies for a Substance Dependence diagnosis. The deficits are usually permanent and may worsen even if the substance use stops, although some cases do show improvement...
>
> Substance-Induced Persisting Dementia can occur in association with the following classes of substances: alcohol, inhalants, sedatives, hypnotics and anxiolytics; or **other or unknown substances**... [emphases added]" [29]
>
> – *Diagnostic and Statistical Manual of Mental Disorders, 4th Edition, Text Revision* (2000)

However, for reasons that one can only presume (legal, economic, psychological, linguistic), the terms Tardive Dysmentia and Substance-Induced Persisting Dementia never gained ground. What did gain ground, though, was the scope of this problem in the context of ageing populations and societies which became increasingly dependent upon pharmaceuticals. Over the ensuing years, what has emerged is **a *hidden epidemic* of *drug-induced dementia*.** Silent, ignored, and deadly, it has become *psychiatry's perfect crime*.

The Perfect Crime: *Murder on the Orient Express* + *Michael Clayton*

<u>Murder on the Orient Express: The Challenge of Multiple Suspects</u>

In the Agatha Christie novel and film, *Murder on the Orient Express*, a group of co-conspirators board the famous trans-European railway with the goal of killing a fellow passenger. As the plot unfolds, Christie's famous protagonist – a Belgian sleuth by the name of Hercule Poirot – must sort through a collection of confusing and seemingly excessive clues. Although the story ultimately leads to two possible endings, one of them is germane to the field of psychiatry. The problem for Poirot is not the fact that he lacks suspects for the murder. Rather, the problem is that he discovers evidence and motives which implicate not one, but twelve different individuals who have all participated in the crime.

In clinical practice, physicians face a similar Poirot-like challenge in terms of sorting through an assortment of clues and suspects. ***In this case, the "crime" is drug-induced dementia (with or without immediate death)***. The "clues" include a variety of neuropsychiatric symptoms: movement abnormalities, seizures, mania, depression, anxiety, hallucinations, delusions, and cognitive decline. The "suspects" include a vast array of potential weapons (antidepressants, antipsychotics, anticonvulsants, sedatives/hypnotics, and stimulants) and the agents who mandate or encourage their use: judges, lawyers, educators, journalists, "key opinion leaders," and many more.

Michael Clayton – "Make It Look Like An Accident"

In the Academy Award nominated movie, *Michael Clayton*, one of the heroes of the film is eliminated by hit men in an effort to protect the commercial interests of a chemical manufacturer. Notwithstanding the substantive relevance of the plot to the theme of this book, the point to be made here is how the pivotal assassination in the film is cleverly conceived and executed so that it ***looks like an accident***.

First, the hero is knocked unconscious with a stun gun. Next, pills are forced down the back of his throat in order to create the *illusion* of intentional drug overdose. Finally, the victim is injected between the web of two toes with an unidentified, untraceable, and rapidly lethal substance (presumably, potassium or another electrolyte which results in a brief convulsion, the cessation of breathing, and sudden cardiac death). The hit men achieve the perfect crime because – to the unwary characters in the screenplay – the murder is initially misinterpreted as a suicide.

What possible relevance could these movie events have for physicians and patients?

The answer to this question lies outside of medicine. Within the profession of *toxicology*, scientists study the "mechanisms and modes of action by which chemicals produce adverse effects in biological systems." [30] When a substance is absorbed and distributed in the human body, toxicologists refer to the location where it produces damage as the *target organ*. In sum, *toxicology – as a field of scholarly and applied endeavor – is dedicated to the identification, prevention, and mitigation of **target organ toxicity***.[31] Unfortunately for patients, though, the profession of medicine is not. This discrepancy – especially **when toxicity concerns the intended target of treatment** – is the basis of how and why the effects of America's health care system are, at this moment, frequently analogous to a murder mystery.

Target organ toxicity is often undetected when there are too many suspects.

To return to the case involving rivastigmine (Case #2), the patient was receiving *ten different neuropsychiatric drugs* at the time of hospitalization. Ironically, it was this combination of agents which created confusion, resulting in the delayed (and only partial) recognition of medications as the cause of the patient's distress. For example, neither lamotrigine, nor memantine, nor dextroamphetamine, nor the two antidepressant drugs were ever suspected as contributing or determining causes of dystonia, because the treating physicians focused only upon the cholinesterase inhibitor (rivastigmine) and two antipsychotic drugs.[32-33]

Prologue

Target organ toxicity is not suspected by physicians when an event "looks like an accident."

Throughout their training and throughout the practice of their various clinical specialties, physicians are seldom (if ever) exposed to the concept of Target Organ Toxicity. Thus, whenever a patient dies from an event which is exceedingly common in the general population (such as pneumonia, heart disease, or cancer) it is highly unlikely that a physician will ever once consider the role of medication as a cause or contributor to that event. In other words, common diseases provide a perfect alibi for pharmaceuticals, because the background prevalence of a given problem "masks" or "hides" the contributions which are made by prescription chemicals.

Related to this phenomenon is the fact that physicians are neither trained, nor encouraged, to identify the Target Organ Toxicities associated with the standard treatments of their vocation. For a cardiologist, Target Organ Toxicity involves the heart, arteries, and veins. For a gastroenterologist, Target Organ Toxicity involves the gut. For neurologists and psychiatrists, Target Organ Toxicity involves the brain, the spinal cord, and the peripheral nerves. Although health care professionals are certainly provided with long laundry lists of adverse effects in medical school, and although they are expected to memorize these effects for examinations, the *applied experience* of the medical school graduate is far different.

In reality, few clinicians ever contemplate the poisonous effects which their treatments exert upon their intended, target organs. For example, in cardiology this applies to an anti-arrhythmic or lipid-lowering drug which damages the heart (blame it on the underlying heart disease); in gastroenterology, to an intestinal drug which harms the gut (blame it on the patient's ulcers or colitis); in oncology, to a chemotherapy or radiation therapy which creates a tumor or blood malignancy (blame it on the original cancer). Additionally, in neurology, this involves an anticonvulsant or anti-dementia drug which harms the brain (blame it on the epilepsy or advancing age), etc.

Target organ toxicity is ignored by governmental regulators, such as the leaders and staff members of the Food and Drug Administration. This happens for at least two reasons.

First, there is currently no law or policy which mandates the testing of new or existing chemicals for Target Organ Toxicity.

For psychiatrists and their patients, this means that none of the agents in use today have ever been tested *methodically* to establish their *neurotoxic* effects. If such a standard did exist, it would clearly require much different procedures than the "hit or miss" detection strategy which now applies. For example, it would require the postmortem inspection of animals (and humans) using electron microscopy to confirm the qualitative features and temporal progression of drug-induced damage and cell death. It would require the use of unambiguously relevant doses in lab animals, including the aggressive pursuit of *low-dose toxicities* and the investigation of non-linear (non-monotonic) dose-response relationships. It would require the study of toxicities associated with *chronic* and *intermittent* exposures. And it would require the consideration of chemical imprinting – the potential for agents which act upon the central nervous system to change gene expression and to re-wire the brain, thereby altering circuits and electrochemical activity in ways that result in *delayed and potentially irreversible* toxicities.

Second, to the extent that Target Organ Toxicity ever crosses the radar screen of drug regulators, drug makers, or "key opinion leaders" in medical education, the current procedure for identifying this problem is systematically flawed. (In fact, the current structure of the American medical and legal systems is designed to ignore, deny, and conceal this very problem.)

Like the crucial murder scene in *Michael Clayton*, the FDA's drug-review system, its leadership, and its expert consultants make many drug-induced events "look like an accident." For example, it was the FDA's failed drug-review system which resulted in a 20-year delay in identifying the problem of antipsychotic-induced tardive dyskinesia (a delayed and usually permanent disability which results in the involuntary movement of muscles in the face, limbs, and/or trunk); and a 10- to 60-year delay in recognizing the lethality of these drugs when given to older patients with dementia, as well as the capacity of these drugs to induce diabetes in all age groups.

How has the FDA accomplished such extraordinary errors of judgment?
The answer requires a brief review of medical history in order to appreciate how physicians (or at least, how governmental regulators) have decided to resolve the complicated issue of disease causation.

Determining Causation in Medicine

The determination of causation within medicine has always been critical for the prevention and treatment of disease.[34] In Ancient Greece, all disease was attributed to imbalances among four key bodily fluids: blood, phlegm, yellow bile, and black bile. Beginning in the 1800s, the *germ theory* held that infectious microorganisms were a necessary cause of disease. According to this notion – which grew out of the invention of the light microscope, and Robert Koch's identification of the bacterium which causes tuberculosis – the cause of any illness was believed to be certain when and if it satisfied four criteria (Koch's postulates): 1) the causal agent could be isolated from the affected host; 2) the causal agent could then be grown in pure culture; 3) the causal agent could be transferred to another host; 4) the causal agent would reproduce the original illness in that new host.

Prologue

Ultimately, Koch's criteria proved to be imperfect predictors of disease causation, due to the complex interactions which exist between microorganisms (causal agents) and hosts. As it became apparent that the ability of any infectious agent to produce disease depended not only upon the agent, but equally upon the immunity (defenses) of the host, Koch's criteria of disease causation had to be revised.

The Hill Criteria

In 1965, Sir Austin Bradford Hill – the British statistician whom some regard as the "father of epidemiology" – proposed a set of criteria which were intended to *guide but not restrict physicians* in evaluating evidence for disease causation.[35] In appreciating the difficulties which confront clinicians in distinguishing between *association* and *causality*, he proposed the consideration of nine different variables. It is these variables which have come to be used by the Food and Drug Administration (and others) in determining the safety hazards of pharmaceuticals and medical devices. [Theoretically, these are the factors which might serve as the foundation of official determinations of Target Organ Toxicity: is a psychiatric or neurological drug the cause of brain dysfunction or brain disease?]

Hill Criteria	**Today's Interpretation**
strength of association	How many people were harmed (absolute risk)? How many more people were harmed by the drug, relative to the non-exposed (relative risk)? Is the effect statistically significant (did it occur by chance)?
consistency of the association	Has the toxic effect been replicated in more than one study, in different age groups or species, or in different study designs (case reports, cohort studies, randomized controlled trials, surveys, registries, etc.)
specificity of the association	Was the drug "uniquely toxic" in terms of causing an otherwise rare or unusual problem? Was the observed effect the specific (sole) toxicity of a given drug?
timing (temporality) of the association	Did the observed effect emerge only after exposure to the given drug? Did the adverse effect occur in multiple patients within the same time frame (temporal clustering)?

presence of a dose-response relationship	Has the drug hazard been linked to a dose threshold? Have studies shown that higher doses or longer durations of exposure increase the risk or the severity of the effect?
biological "likelihood" (plausibility) of the association	Have scientists determined a plausible mechanism which would account for the adverse effect?
coherence of the association	Was the toxic effect consistent with the principles and findings of biology, toxicology, and neuroscience?
experimentation	Has the effect been tested experimentally to see if repeated use and withdrawal predict or prevent the hazard? (challenge/dechallenge/rechallenge, as exemplified in the rivastigmine case)
reasoning by analogy	Has the toxic effect been observed in conjunction with other drugs that are similar in chemical structure or in their physiological mechanisms of action?

Systematic Errors in the Evaluation of Drug Safety

Unfortunately for the safety of patients, the application and interpretation of the Hill criteria have been inappropriately promoted and distorted by attorneys, scientists, and clinicians who represent the pharmaceutical industry, and by judges who are unfamiliar with biochemistry and neuroscience. These distortions have been particularly egregious in the aftermath of the Supreme Court's decision in *Daubert vs. Merrell Dow Pharmaceuticals* – a 1993 verdict which enshrined federal judges as the sole arbiters of the kinds of scientific evidence and professional opinions which may be considered in deliberations of drug safety.[36]

Tragically, the court system's and the FDA's perpetual misconstruction of the Hill criteria has introduced a systematic bias *against* the identification of Target Organ Toxicity in the case of psychiatric (and neurologic) drugs. In other words, for patients exposed to neuropsychiatric pharmaceuticals, the ***flawed application of the Hill criteria has become a serious obstacle to the detection of treatment-related, brain damaging effects***. (For an in-depth analysis of this problem, see **Appendix A**.)

Prologue

No More Crimes, No More Accidents . . .

Where does this leave physicians who – like Detective Poirot – face the formidable challenge of apprehending the pertinent suspects in drug-induced disability and drug-induced death? And where does this leave patients, who have the right to expect that their treatments will not result in dire outcomes (such as dementia or death) which others will be allowed to construe as accidents?

It leaves them both in need of a resource which will review relevant aspects of the Hill criteria, while respecting the limitations of those criteria and the complexity of the human brain. It leaves them in need of a resource which will integrate research findings from basic biology (animal experimentation), clinical science (neuroimaging, pathology), and epidemiology (observational studies of diverse populations who use psychiatric drugs). It leaves them in need of a resource which will explain the reality and the significance of one of the most serious hazards which can occur during or after exposure to psychiatric drug therapies: ***Drug-Induced Dementia***.

The goal of this book is to serve as that resource. The following chapters explain *what dementia is* (Chapters 1 and 2); describe *how psychiatric drugs cause or enhance this problem* (Chapter 2); explore the *prevalence and patterns of drug-induced dementia* in real populations (Chapter 3); and present *the scientific evidence for this devastating drug effect* (Chapter 4: antidepressants; Chapter 5: antipsychotics; Chapter 6: anxiolytics; Chapter 7: mood stabilizers; and Chapter 8: stimulants). The book concludes with a brief discussion of possible approaches which policy makers, health care professionals, and health care consumers might consider in an effort to protect themselves and others from the causes and consequences of drug-induced dementia…psychiatry's perfect crime.

Chapter One

Dementia

*Do Doctors Know It When They See It ... and When They See It,
Do They Know What to Do?*

In a famous case which appeared before the U.S. Supreme Court in 1964, Justice Potter Stewart was called upon to deliberate the limits of free speech. In clarifying the legal threshold of obscenity, he opined:

> "I shall not today attempt to further define the kinds of material I understand to be embraced within that shorthand description; and perhaps I could never succeed in intelligibly doing so. ***But I know it when I see it...***"[1]

Although these words were written within the context of Constitutional jurisprudence, rather than within the context of healing, they reflected the twin challenges of phenomenology (describing what something is) and epistemology (knowing how to know what something is) – challenges which are common to medicine and law. Specifically, there is a striking parallel in the way that American society has empowered both professions to serve as arbiters of propriety and human normalcy.

In no areas of medicine has this expectation been more clearly realized than within the fields of neurology and psychiatry. Curiously, despite the proliferation of knowledge and technologies in the Postmodern era, it has been difficult for physicians to fulfill these expectations. An excellent illustration is evident in the phenomenon of dementia.

"Every definition is dangerous...."

- Desiderius Erasmus (1465-1536)

Deriving from the Latin term for "senselessness" or "insanity" (demens/dementis ="out of one's mind") the word *dementia* entered the parlance of Western medicine in the 17th century.[2-3] Since that time, the concept has evolved through many variations, but a **lack of measurable, objective determinants** which can be used in **living patients** has contributed to an atmosphere of clinical confusion which continues to this day.

For example, in its 1990 classification of illnesses known as the *ICD-10* (*International Classification of Diseases, 10th Edition*), the World Health Organization described dementia as:

> "...a syndrome due to disease of the brain, **usually of a chronic or progressive nature** in which there is disturbance of multiple higher cortical functions including memory, thinking, orientation, comprehension, calculation, language, and judgment. **Consciousness is not clouded**. The impairments of cognitive function are commonly accompanied, and occasionally preceded by deterioration in emotional control, social behavior or motivation." [4]

According to the *Diagnostic and Statistical Manual of Mental Disorders*, prepared at various intervals by the American Psychiatric Association, dementia has been characterized as a mental disorder which involves:

DSM-III (1980)   a loss of intellectual abilities of sufficient severity to interfere with social or occupational functioning; **memory impairment**; and at least one of the following: impairment of abstract thinking, impaired judgment, other disturbance of higher cortical functioning (aphasia, apraxia, agnosia), or change in personality [5]

DSM-IIIR (1987)   **impairment of short- and long-term memory**, and at least one of the following: impaired abstract thinking, impaired judgment; other disturbances of higher cortical functioning (aphasia, apraxia, agnosia); personality change – which significantly interfere with work or usual social activities or relationships with others [6]

DSM-IV (1994)   the development of multiple cognitive deficits manifested by both **memory impairment [short-term or long-term]** and one or more of the following: aphasia (language disturbances), apraxia (impaired ability to carry out motor activities despite intact motor function); agnosia (failure to recognize or identify objects despite intact sensory function); disturbance in executive functioning (i.e., planning, organizing, sequencing, abstracting) – all of which cause significant impairment in social or occupational functioning and represent a significant decline from a previous level of functioning [7]

[emphases in bold added by author]

Regrettably, the information in these and other sources contradicts the reality of many patients. For example, the ***ICD's exclusion of clouded consciousness*** has ignored the fact that changes in arousal level occur in several dementias, such as Lewy body disease, Creutzfeldt-Jakob disease, and hydrocephalus.[8-10] Similarly, the ***DSM's inordinate emphasis upon memory loss*** has ignored the fact that memory deficits are commonly absent in the early stages of frontotemporal and subcortical dementias, including Pick's disease, Huntington's chorea, and Parkinson's disease.[11-12] Furthermore, depending upon the cause of symptoms, and the quality and timing of interventions, it is clinically nihilistic and scientifically false to suggest that dementia is a universally, inevitably chronic and progressive condition which is limited to the elderly.[13-14]

Sachdev's Challenge

In light of these limitations and ambiguities, some audacious clinicians have recently challenged the ethics and utility of the standing dogmas. In a provocative essay[15] which appeared in the spring of 2000, Dr. Perminder Sachdev suggested that it might even be time to *retire dementia from the medical lexicon* for the following reasons:

the lack of a clear, operationalized (i.e., measurable) definition:

> As there are no biological "markers" which can be measured at this time in a *living* patient to determine many of the subtypes of dementia (such as Parkinson's disease), the diagnosis has depended upon checklists of symptoms which have been inconsistently and ambiguously defined. This has resulted in serious discrepancies both between and among raters, even when they have used the same diagnostic tool.

> For example, in a Canadian investigation of more than 1,800 subjects, the use of the ICD-10 criteria revealed a dementia prevalence of just 3%.[16] In contrast, when the same team of raters applied the criteria of various editions of the *DSM (DSM-III, DSM-IIIR,* and *DSM-IV)*, the prevalence of dementia in the same patient group was higher by *4- to 10-fold* (29%, 17%, and 14%, respectively).

the negative clinical effects arising from diagnostic uncertainty:

> Currently, dementia is a diagnosis of exclusion. Either a patient is found be suffering from a general medical condition which accounts for his or her cognitive dysfunction (secondary dementia), or the patient is presumed to be suffering from a primary brain disease (primary dementia). At this time, however, there are no unique markers or tests which can be used in a living patient to confirm a

primary dementia. Without an objective method to determine the presence or absence of changes which reflect "normal" versus "abnormal" ageing, patients have been (and continue to be) over-diagnosed, underdiagnosed, and misdiagnosed. This has resulted in delayed and/or inappropriate care.

the pejorative connotations of the current terminology:

In light of the public and professional biases which have come to associate dementia with irreversible deterioration, the very use of this diagnostic label has resulted in a variety of unintentional but harmful effects. For some patients, the term has resulted in a self-fulfilling prophecy about inactivity, functional limitations, and the inability for independent living. For physicians and for the legal system, the term has resulted in a set of beliefs about progressive debilitation which has contributed to the chronic isolation and demise of many people.

The Dementia Industry Juggernaut – Are Doctors Doing the Right Thing?

In the nine years since Sachdev's opinion piece was printed, the medical community has clearly *not* erased dementia from the dictionary. Neither has it erased the attendant problems (semantic ambiguity, clinical harm, and stigma). In fact, what the health care profession has created is a robust dementia industry which includes **new drug products**, new screening tools, new clinics, new journals, and new consumer advocacy organizations – all clamoring for the use of treatments which lack clinically meaningful, beneficial effects.

A superficial inspection of advertisements and medical education materials might lead one to conclude that these developments must be benefiting many patients and their caregivers. Unfortunately, the so-called "anti-dementia" drugs have been largely *unsuccessful* for the majority of their recipients.[17-21]

For example, in an extensive review of the medical literature which appeared in the *Annals of Internal Medicine* in March 2008, Canadian investigators concluded that while the cholinesterase inhibitors and memantine (the latter, a glutamate-inhibiting drug) had produced ***statistically*** *significant* changes on cognitive rating scales in more than 50 unique studies, the small size of these changes had been *unlikely* to produce ***clinically*** *significant results.*[22]

The question then emerges: if the new drugs do not reverse dementia, can they at least *prevent* it?

One unfounded claim that has recently gained momentum is the notion that early intervention with anti-Alzheimer's drugs is needed for the newly created disease entity known as *mild cognitive impairment* or MCI (aka, **pre-dementia**).

In a recent analysis which pooled data from 15 studies, each of which spanned five years or more, investigators in the U.K. found that the overwhelming majority of unmedicated MCI patients (approximately 70%) experienced not cognitive decline, but rather, *stable or improved abilities.*[23] **In other words, for reasons unknown, the MCI in these individuals did not become dementia.**

Similarly, in a systematic review of six placebo-controlled trials which focused upon MCI, an international team of researchers showed that the benefits of the new "anti-Alzheimer's drugs" were essentially indistinguishable from placebo in terms of preventing or delaying dementia (13 to 25% of medicated patients progressing to dementia, vs. 18 to 28% on placebo).[24]

None of this information should be surprising to health care providers, if one considers the following facts:

the therapeutic value of the anti-dementia drugs remains unclear

The randomized controlled trial data which led to the approval of the cholinesterase inhibitors (anti-Alzheimer's drugs) were of poor quality; demonstrated minimal benefits relative to placebo (response rates of only 20 to 50%; response magnitudes of questionable significance); and presented little scientific justification for the use of these agents in Alzheimer's and other dementias (i.e., unclear benefit in those who had already lost substantial numbers of cholinergic neurons).[25] Post-marketing studies have demonstrated or suggested the harmfulness of these agents for patients with pre-existing deficiencies of butyrylcholinesterase, seizure disorders, and subcortical disease.[26-31]

the long-term safety of the anti-dementia drugs remains unexplored

Doctors presumably forget, and the public presumably does not know, that the biological mechanisms of most anti-dementia drugs are *identical* to pesticides and nerve agents. Common to all these substances is the inhibition of an enzyme which breaks down acetylcholine (acetylcholinesterase). For example, this shared enzyme effect accounts for the most frequent adverse reactions, especially in elderly patients: nausea, vomiting, and diarrhea. Although toxicologists have characterized some of the long-term and delayed brain hazards of pesticides and weapons of war, they have not established the safety limits of analogous medicinal chemicals.

While this topic may not be of concern to patients with end-stage or severe dementia, it should be of great concern to those individuals who are candidates for extended therapy. Specifically, there is no research evidence at this time which confirms that these medications will *not* replicate the delayed or chronic effects of low dose organophosphate poisoning (e.g., by inhibiting Neuropathy Target Esterase).[32-33] Moreover, animal studies suggest that the neuronal adaptations which occur with continued anti-cholinesteras therapy – such as the downregulation of muscarinic receptors, or the downregulation of acetylcholine synthesis and release – would eventually oppose any early benefits of these drugs.[34-38]

the causes and treatments of Alzheimer's disease remain contentious

Although the prevailing consensus in neurology has long implied that genes (apoE genotype) and abnormal protein deposits (tau tangles and beta-amyloid plaques) are the *proximal cause* of a discrete entity known as Alzheimer's disease, it is no longer clear that these are necessary or sufficient *determinants* of that condition. A consistent paradox has been the failure of tau and beta-amyloid levels to predict the severity of cognitive impairment, or to distinguish Alzheimer's disease from other dementias or normal ageing.[39-41]

Furthermore, it is notable that the theoretical and applied research of some scientists has presented a cogent challenge to the biological foundations of the anti-dementia drug industry by demonstrating the toxicity of certain combined therapies, and by proving the beneficial role of brain cholinesterases in *preventing* neurodegeneration.[42-47] These and other findings cast doubt upon the dogma which emphasizes plaques, tangles, and the use of drugs which may boost cholinergic transmission excessively, because they suggest that the true causes and mitigants of Alzheimer's disease lie elsewhere.[48-49]

The Problem of Performativity

As the practice of American medicine has come to focus upon the *chemical suppression of symptoms*, rather than the *prevention and eradication of root cause*, textbooks and curricula have increasingly avoided a *rational* approach to many conditions. This has been especially true for ambiguous syndromes – like dementia – where an excessive focus upon **phenomenology** (signs and symptoms) and **pathology** (anatomical effects of disease) has deflected attention away from **pathophysiology** (causal mechanisms of disease).

One origin of these developments within neurology has been a longstanding tradition of **eponymous nosology**, meaning the classification of diseases according to the names of their discoverers. For example, it was eponymous nosology which resulted in the syndromes named for Parkinson, Huntington, Pick, Creutzfeldt-Jakob, Alzheimer, Wernicke-Korsakoff, Kluver-Bucy, and many more. While these schemes may have been helpful to doctors who lived *before* the emergence of molecular biology and advanced microscopy, they are a liability for today's practitioners and patients if or when they result in a narrow focus upon phenomenology (checklists of symptoms + matching pills).

Similarly, advances in biochemistry and recombinant gene techniques have given rise to a contemporary obsession with subcellular pathology, according to which disease entities are being reclassified in terms of dysfunctional cell parts (the mitochondria, endoplasmic reticula, cytoskeleton, lysosomes) and component proteins (e.g., abnormally folded deposits of tau, amyloid, alpha-synuclein, ubiquitin).

There is a fitting term which describes this bevy of interesting but arguably futile activity. As contrived by the late French philosopher, Jean-Francois Lyotard, the concept of ***performativity*** refers to the way in which knowledge has been legitimized in the Postmodern era:

> "…the goal is no longer truth, but performativity – that is, the best possible input/output equation…Scientists, technicians, and instruments are purchased not to find truth, but to augment power…"[50]

Within the conduct of medical practice, medical education, and medical research, it is performativity which has constricted the quest for knowledge and the authentic pursuit of healing. Rather than serving as a branch of science which seeks to understand and reverse the fundamental causes of disease, allopathic medicine in America has come to focus upon the efficient performance of power-enhancing functions. These functions include conformity with clinical practice guidelines (CPGs), treatment algorithms, consensus statements, pay-for-performance (P4P) mandates, and other manifestations of commercially biased Group Think, in order to obtain or maintain academic tenure, political appointments, research funding, and the maximization of corporate profits and investment returns. Applying this back to the theme of dementia, it is *performativity* which accounts for the delusion that physicians have been handling a "well defined disease" (such as MCI or Alzheimer's) safely and effectively, despite abundant evidence to the contrary.

The Global Epidemic of Dementia

According to the projections of international research teams, the planet stands on the brink of a dementia epidemic. In the year 2000, the global *prevalence of dementia* (# of existing cases) was 25 million. However, under the influence of declining birth rates, expanding longevity, and the changing structure of populations around the world, this number is expected to *increase by more than four-fold* in the next forty years:

Global Dementia Projections [51-53]			
	2000	2030	2050
# ≥ 65 years old	420 million	973 million	~ 2 billion
# ≥ 65 years old with dementia	25 million	63 million	114 million

Within the United States, dementia is also expected to grow exponentially as members of the "Baby Boom" generation (those born between 1945 and 1964) move into their twilight years.[54] By combining projections from a recent study on Alzheimer's disease with other common research findings (i.e., that Alzheimer's accounts for approximately 70% of all dementias),[55] the following changes in dementia prevalence are expected to occur:

U.S. Dementia Projections [56-58]			
	2000	2020	2040
Total Population	281 million	341 million	406 million
≥ 65 years old with dementia	7.6 million	8.9 million	18.3 million
% of U.S. population with dementia	**2.3%**	**2.6%**	**4.5%**
Medicare Expenditures for Dementia			
	2005	2010	2015
	$91 billion	$160 billion	$189 billion

These numbers have important implications for patients, caregivers, communities, economists, and governments. However, before policy makers can react to these predictions responsibly, they will need to consider the fact that these numbers more than likely *underestimate* future events. First, existing projections have been based upon faulty data. The problem stems from the use of datasets which have inevitably suppressed the number of dementia cases, due to diagnostic uncertainty, misdiagnosis, or a failure to record dementia as a cause of death. Second, it is notable that many epidemiologists have used a methodology which *assumes an unchanging prevalence of dementia by age bracket*. In several studies, researchers have applied the findings of previous investigations (such as the meta-analysis of Fratiglioni and Rocca, 2001)[59] in which the dementia rate was found to double every four to five years:

Prevalence of Dementia by Age Group	
60 to 64	1%
65 to 69	1.5%
70 to 74	3%
75 to 79	6%
80 to 84	13%
85 to 89	24%
90 to 94	34%
≥ 95	45%

While it is clear that the number of people entering each age bracket will increase in concert with the "graying" of the population, it is dangerous to assume that the *incidence* of disease (the # of new cases every five years) will remain constant. In other words, it remains uncertain that the number of dementia cases will rise solely because more people will be living long enough to experience an *unchanging risk* (extended longevity), rather than because the *risk factors themselves will have increased.* This latter situation appears likely in light of the proliferation of precursors to dementia: cancer and its treatments, diabetes, vascular disease, and chronic exposure to environmental and medical toxins. Third, it is significant that existing projections have neglected the phenomenon of *early-onset dementia* which begins in young or middle age. For industrialized and developing countries which are now experiencing profound reversals in their economic dependency ratios (EDRs), a dementia epidemic which strikes at the core of the tax-paying labor force is the last thing which politicians, economists, and health care systems can afford to ignore.

United States Economic Dependency Ratios (EDRs) [60]
[EDR = Size of Population ≥ 65 Relative to Size of Population Aged 15 to 64]

2005	2010	2015	2020
20.9	22.2	25.1	29.1

The Importance of Defying Performativity

In the U.S. Surgeon General's 1999 report on the subject of mental health, Dr. David Satcher and his research team made the following remarks about the care of older adults:

> "…the past 15 to 20 years have been marked by rapid growth in the number of clinical, research, and training centers dedicated to the…needs of older people. Risk factors [for dementia and other illnesses] include co-occurring, or comorbid, general medical conditions, the high numbers of medications many older individuals take, and psychosocial stressors…These are cause for concern…but they also point the way to possible new preventive interventions. The goal of such prevention strategies may be to limit disability or to postpone or even eliminate the need to institutionalize an ill person." [61]

Ten years have elapsed since Dr. Satcher's words were printed, but the costs and consequences of dementia have expanded. Ultimately, if there is to be real progress in reducing the global burden of this condition, it will be necessary for physicians to think critically and behave ethically in defiance of performativity. This can only happen by acknowledging and correcting the unnecessary harmfulness of policies which prioritize, or coerce, the use of dementia-inducing, dementia-enhancing, and dementia-sustaining drugs.

Summary:

Dementia is a wastebasket term which refers to a constellation of cognitive symptoms (such as impaired memory, attention, judgment, or planning) serious enough to disrupt social, occupational, and independent functions. Imprecision and inconsistencies in the definition of dementia have resulted in professional turmoil. In the year 2000, Dr. Perminder Sachdev consequently suggested the retirement of the term 'dementia' due to the combined problems of semantic ambiguity, clinical harm, and stigma. However, the word dementia has *not* been erased from the medical dictionary. Rather, an entire dementia industry has emerged, emphasizing the detection and treatment of controversial disease entities and the use of marginally beneficial drug therapies whose long-term risks remain unexplored.

One way of understanding the evolution of the dementia industry is by appreciating its place within the context of history. According to the late philosopher, Jean-Francois Lyotard, knowledge in the Postmodern era is legitimized not by the quest for truth, but rather by the quest for power. The term which he created for this phenomenon was *performativity*.

Returning to theme of dementia, performativity has resulted in a clinical focus upon phenomenology (checklists of symptoms) and pathology (anatomical effects of disease), rather than pathophysiology (causal mechanisms of disease). A preference for the illusion of controlling symptoms has deflected attention away from the identification, prevention, and mitigation of the processes which are already well known to result in dementia.

As the world's populations enter their twilight years and longevity continues to grow, epidemiologists project that the worldwide prevalence of dementia *among the elderly* will expand in the next four decades by at least *four-fold*. Within the United States, the prevalence of elderly dementia is expected to increase two-fold. Although alarming, even these statistics quite likely underestimate the scope of the coming crisis for several reasons. They do not adjust for the under-detection and under-reporting of dementia, they omit dementia rates among the young and the middle-aged, and they assume that the *incidence* of dementia will remain unchanged.

Present and impending costs associated with dementia will exert financial and social pressures upon communities around the globe, particularly within countries which are experiencing a rapid climb in their EDRs (economic dependency ratios). Due to contracting birth rates which began in the 1960s, the burden of the dementia epidemic will fall upon the shoulders of a declining pool of caregivers and laborers under the age of 65. Furthermore, the risk of dementia among these subgroups of the population is a potential reality which epidemiologists and policy makers have long ignored.

In his 1999 report on the subject of mental health, U.S. Surgeon General, Dr. David Satcher, emphasized the importance of recognizing and mitigating risk factors for dementia. Among the risk factors which he overtly identified were general medical conditions and "the high numbers" of prescription drugs consumed by older adults.

Judging from the trends in the use of pharmaceuticals which occurred in the intervening years, it appears that performativity has had the upper hand. In a survey of individuals conducted in 2005, researchers found that 90% of Americans, aged 57 to 85, were using at least one medication regularly.[62] Approximately 30% of this same group reported the concurrent use of five different prescription pharmaceuticals. Meanwhile, the burden of dementia in 2005 had escalated to include 6 million patients and medical expenditures which exceeded 90 billion dollars from Medicare (the federal government's insurance program) alone.[63] (This amounted to 30% of all Medicare outflow, or approximately 1% of the entire gross domestic product that year.)[64]

In the face of these developments, one wonders how the dementia epidemic might be contained. If there is to be a quest for true prevention or true healing – rather than the continued quest for the augmentation of corporate and personal power – it will be necessary for physicians to redirect their attention to the *root causes* of dementia. This will require critical thinking and ethical action. It must also include the recognition of the research evidence which demonstrates the dementia-inducing properties of psychiatric drugs.

Chapter Two

Mechanisms of Dementia

This book is dedicated to the problem of dementia and to the evidence which shows that psychiatric medications both cause and sustain it. However, before one can appreciate the link between psychopharmaceuticals and the cognitive changes which accompany drug-induced degeneration, it is important to understand a few fundamental features of brain biology. From a developmental and functional perspective, scientists have historically described brain structure and activities with reference to three key parts:

name		location	functions
the cortex	=	outer (or top) layer	"human" functions planning, intending, meaning
the subcortex	=	middle layer	"animal" functions appetite, sex, emotions
the brainstem	=	base layer	"vegetative" functions sleep/wake, breathing, heart beat

For the non-biologist, a helpful metaphor might be the popular candy lollipop with the tootsie roll (chocolate) center, otherwise known as the Tootsie Pop:

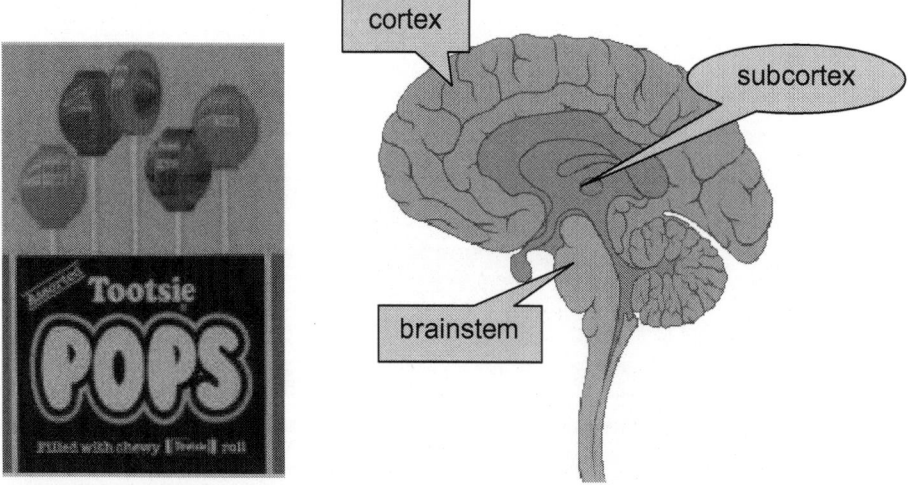

Tootsie Pops reproduced with permission of Tootsie Roll Industries. Brain from Wikimedia Commons.

        candy coating   =   cortex
        tootsie roll center   =   subcortex
        lollipop stick   =   brainstem

Mechanisms

If one examines the various regions of the brain (the candy coating, the tootsie roll middle, or the stick) under a microscope, one will find two major kinds of component cells. These are called *neurons* and *glia* (shown below), and it is the coordinated interaction of these units which determine the health of the brain.

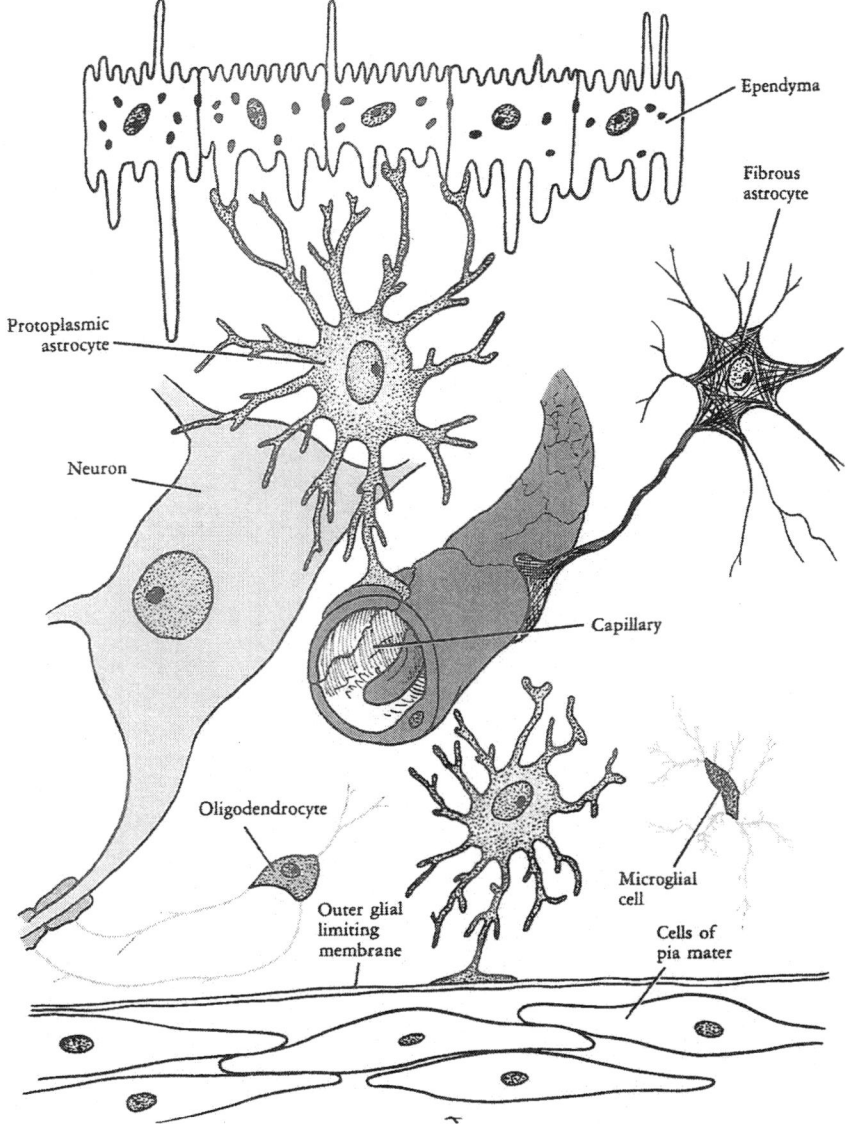

FIGURE 3-1 A diagrammatic representation of the arrangement of different types of neuroglial cells.

Source: R.S. Snell. *Clinical Neuroanatomy for Medical Students, 4th Ed.* (Philadelphia: Lippincott-Raven Publishers, 1997), 78. Reproduced with permission of Wolters Kluwer and LWW.

neurons – These are the nerve cells which make, absorb, store, and release chemicals known as neurotransmitters. Neurons are comprised of three major parts: a <u>soma</u> (cell body) which contains the cell nucleus; short projections known as <u>dendrites</u>; and an extension called an <u>axon</u> (the latter, often wrapped in a layer of fatty tissue called myelin).

For the past century, neurons have dominated the theory and practice of Western medicine and pharmacology, largely because of the observations of a Spanish anatomist named Santiago Ramon y Cajal. In the late 19th century, Cajal proposed that the neurons were the fundamental actors of the brain. This view, known as the <u>*neuron doctrine*</u>, initially contributed to many advances in neuroscience. It also contributed to a rigid belief system whose tenets are no longer true (such as: the belief that neurons are the "master cells" which control the electrical and chemical events of the brain; and the belief that action potentials and synaptic transmitter release are the most critical events in the brain).

glial cells – According to Cajal's neuron doctrine, glial cells have been regarded as second class citizens that exist solely to support or serve neurons. This perspective has slowly been invalidated. In fact, glial cells participate in cell-to-cell communication to such a large degree that the terms "neurotransmission" and "neurochemicals" are no longer entirely correct. Rather, it is far more accurate to refer to the structures and activities of the brain in terms of an integrated web of *neuroglial* transmission and *neuroglial* chemicals. To date, *four different kinds of glial cells* have been identified:

astrocytes	star-shaped cells which participate in neuroglial transmission; provide structural support; perform nutritional and detoxifying functions; phagocytize (eat up) dead or waste material; and stabilize the *blood-brain barrier* (the capillary defense system in the brain which prohibits invasion by potentially harmful intruders)
ependymal cells	cube-shaped cells which produce, secrete, and absorb fluid in brain cavities (ventricles, cisterns)
oligodendrocytes	small cells which produce myelin, a fatty-layer of insulation which surrounds axons; myelin facilitates the speed of electrical and chemical transmission (cell-to-cell communication) in the brain
microglia	the smallest of the glia, these are defense cells which are inactive under normal conditions; microglia expand in number and move throughout the brain in response to injury or infection; like astrocytes, they consume damaged or dead material

Beginning in the late 1600s, practitioners of *allopathic medicine began identifying *specific subtypes of dementia*. As theories and technologies evolved over the centuries, these conditions came to be described in a variety of ways:

>according to differences in signs and symptoms (phenomenology)
>according to differences in anatomical abnormalities (pathology), and
>according to the biological mechanisms which cause disease (pathophysiology)

Unfortunately, there has never been a systematic method for classifying dementia, and many of the conditions which were historically considered to be unique entities (such as Parkinson's disease vs. Lewy body dementia; or Alzheimer's disease vs. vascular dementia) have turned out to share many of the same pathologies. Furthermore, the latest tools and theories of molecular biology have resulted in even greater confusion, because developments in genetics and biochemistry have led scientists to focus upon smaller and smaller units of pathology.

One can appreciate this reality by noting the proliferation of journal articles and textbook chapters devoted to the subject of *proteinopathies*: *diseases,* including the dementias, which are allegedly caused by abnormal amounts, abnormal locations, or abnormal 3D arrangements (folding) of specific molecules (tau, beta-amyloid, S100B, alpha-synuclein, ubiquitin, parkin, huntingtin).[1] These discoveries are all quite interesting. However, unless science can prove that these naughty proteins are the *cause*, rather than the *effect* of conditions which give rise to suffering and disability, it is difficult to comprehend how patients will benefit from the massive project which now seeks to redefine every condition in neurology according to the *parts* of *cell parts*.

In an effort to make sense out of the chaos of dementia nosology, this chapter will summarize four traditional systems of disease classification (Signs and Symptoms, Brain Location, Intracranial Origin, Brain Pathology). The chapter will conclude with a detailed examination of a fifth approach to dementia (Essential Cause), focusing upon the ways in which psychiatric medications replicate all of the disease mechanisms which are proven causes of brain cell damage and death.

---

* allopathic – pertaining to a therapeutic system in which a disease is treated by
  producing a second condition that is incompatible with or antagonistic to the first [2]

Scheme I   Signs and Symptoms

According to this classification scheme, the major dementias are identified with reference to specific clusters of signs and symptoms.

Dementias According to Signs and Symptoms	
Alzheimer's disease	memory loss; declining executive functions and language skills; confusion; psychosis; inability to recognize self or others; inability to communicate
Creutzfeldt-Jakob disease	insomnia; impaired coordination and/or muscle jerks; depression and personality changes; impaired memory, judgment, and thinking; paranoid delusions and/or hallucinations
frontotemporal lobe dementia/Pick's disease	problems with executive functions; personality changes (social withdrawal, lethargy, or disinhibition); impaired language skills (understanding or speaking)
Lewy body dementia	impaired thinking, memory, and language skills; early onset of psychotic symptoms (such as visual hallucinations); varying levels of attention and alertness; motor problems similar to Parkinson's (stiffness, gait disturbance, tendency to fall)
normal pressure hydrocephalus	apathy, inattention, social withdrawal; abnormal gait; impaired memory; confusion; incontinence (classic triad: "wet, wobbly, wacky")
Parkinson's disease	resting tremor, difficulty initiating movement, shuffling gait; impaired cognition; decreased motivation and mood
vascular dementia	"stepwise" progression of memory problems and decline of executive functions (planning, organizing, judging); changes in personality; difficulty with motor skills

The problem with this approach to classification is the tremendous overlap in the signs and symptoms of various neurodegenerative conditions.[3-7] Furthermore, as many drug therapies contribute to the onset or worsening of cognitive and behavioral features, it becomes increasingly difficult to distinguish these various dementias from one another when patients are followed over sustained periods of time.

Scheme II   Brain Location – Cortical or Subcortical?

According to this classification scheme, dementia is identified with reference to the primary location of cell degeneration or cell loss. This method (common until the late 1990s) asks the question: is the dementia primarily cortical or subcortical, or is the dementia associated with damage primarily within the front, middle, or back of the brain (frontal, parietal, temporal lobes, etc.):

Cortical	Subcortical
Alzheimer's disease	Huntington's chorea
frontotemporal dementia	Parkinson's disease
Pick's disease	

The problem with this approach is the fact that it ignores the inseparability of all brain regions when it comes to understanding complex functions. For example, although Parkinson's disease has long been considered the prototype for subcortical dementia, recent research evidence has demonstrated that this condition includes degenerative changes in the cortex.[8] Other neurological conditions are similarly inconsistent with the cortical versus subcortical dichotomy, since they also involve damage to multiple brain regions. These include the dementias associated with chronic alcoholism, Lewy body disease, corticobasal degeneration, multiple system atrophy, and progressive supranuclear palsy.[9]

Scheme III   Intracranial Origin – Vascular or Non-Vascular?

According to this classification scheme, dementias are identified with reference to intracranial origin: either vascular (blood vessels) or parenchymal (brain cells). Vascular dementias refer to brain cell damage which has been caused by disrupted cerebral blood flow. This can occur via blood clots, bleeding, or inflammation. The disrupted delivery of oxygen, glucose (sugar), and other nutrients then results in the damage or death of brain cells.

Although Thomas Willis was the first physician to recognize vascular or "post-stroke" dementia (and remarkably, did so in the late 1600s), it fell to his successors to adopt this entity as a method of dementia classification. Subsequently, textbooks of neurology came to emphasize dementias caused primarily by cerebrovascular disease (large and small strokes, isolated or episodic) versus those which were presumed to originate in the brain tissue itself (non-vascular or parenchymal disease):

Vascular	Non-Vascular (Parenchymal)
Binswanger's disease = dementia related to small vessel disease resulting in scar tissue (lacunes); primarily affects white matter in subcortical regions	everything else
multi-infarct dementia = dementia arising from repeated large vessel events (major strokes); usually due to reduced blood flow (ischemia) or bleeds (hemorrhage)	
other vascular dementias = dementia arising from silent infarcts, transient ischemic attacks (TIAs), high blood pressure, heart disease, diabetes, vasculitis, hereditary vascular disease (CADASIL, CARASIL, CAA, et al)	

Definitions: [10]

CADASIL – cerebral autosomal dominant arteriopathy with subcortical infarcts and leukoencephalopathy

CARASIL – cerebral autosomal recessive arteriopathy with subcortical infarcts and leukoencephalopathy

CAA – cerebral amyloid angiopathy

hemorrhage – escape of blood from the vessels (i.e., bleeding)

infarct – localized area of dead tissue due to interruption of arterial blood flow or venous drainage; *silent* infarcts refer to asymptomatic events

ischemia – deficiency of blood due to constriction or obstruction of blood flow

vasculitis – inflammation of blood vessels (can be caused by infection or autoimmune condition, such as Lyme disease or systemic lupus erythematosus)

The problem with this scheme is the tremendous overlap which exists between vascular and parenchymal (tissue-based) pathologies. Increasingly, the autopsies of many patients are providing evidence of a high rate of mixed dementias (vascular + non-vascular disease). Furthermore, it must be appreciated that one of the defining features of Alzheimer's disease is the accumulation of amyloid protein within the cells which line the arteries and veins of the brain (so-called *amyloid angiopathy*). For this reason alone, it is incorrect to speak of Alzheimer's disease as a strictly parenchymal condition.

Scheme IV   Brain Pathology – Proteinopathy or Non-Proteinopathy?

According to this classification scheme, dementias are identified with reference to defective cell type (neurons or glia; dopamine/serotonin/GABA); defective organelles (mitochondria, endoplasmic reticula, cytoskeleton, lysosomes); or allegedly defective molecules (i.e., ***abnormal accumulations of proteins, such as ubiquitin, S100B, alpha-synuclein***).[11-12]

Dementias According to Abnormal Proteins	
(aka, The New Proteinopathies)	
protein	subtypes of dementia
alpha-synuclein	found in Alzheimer's disease, Creutzfeldt-Jakob disease, Down syndrome, Gaucher disease, Lewy body dementia, multiple system atrophy (aka, Shy-Drager disease), pantothenate kinase-associated neurodegeneration or PKAN (aka, Hallervorden-Spatz disease), Parkinson's disease
beta-amyloid	found in Alzheimer's disease, Creutzfeldt-Jakob disease, Down syndrome, Lewy body dementia, normal ageing, Parkinson (Guamanian) dementia complex
tau	found in Alzheimer's disease, corticobasal degeneration, Creutzfeldt-Jakob disease, frontotemporal dementia and parkinsonism linked to chromosome 17 (FTDP-17), Lewy body variant of Alzheimer's, PKAN, Pick's disease, progressive supranuclear palsy
ubiquitin	found in Alzheimer's disease, corticobasal degeneration, frontotemporal/Pick's, Lewy body dementia, multiple system atrophy, progressive supranuclear palsy

Regrettably, the molecular classification of dementia (shown above) is one of the clearest examples of performativity in medicine. Despite the fact that research now proves the impossibility of using any one of these proteins to reliably distinguish specific types of neurodegeneration – since the same proteins keep turning up in multiple types of dementia – scientists are scrambling to redefine these conditions as "primarily" due to amyloid, or "primarily" due to tau (etc.). While all of this activity and speculation is doubtless productive for research labs, academic institutions, and pharmaceutical companies, it is difficult to see how any of this will be helpful to patients unless or until scientists and clinicians direct their attention to *essential causes* of dementia.

Scheme V   Essential Cause – Primary or Secondary Disease?

According to this classification scheme [the suggestion of this writer, in an effort to bring order to the usual hodgepodge], physicians are obliged to identify and respond to *essential causes of dementia*.[13-14] This consists of four major steps.

First, there is a search to determine the **origin of dementia with reference to the brain**. Does the essential cause *originate in a disturbance which lies within the brain* (**PRIMARY DEMENTIA**) or does it *originate in a disturbance in the *soma outside the brain* (**SECONDARY DEMENTIA**).

Second, there is a search to identify temporal factors. Is the dementia associated with a **new (acute) and possibly temporary illness**, or is it associated with a **pre-existing chronic disease**?

Third, there is a search to identify the cause as **genetic (internal) or acquired (external)**. Here, it is critical to know if the dementia originates in a toxic environmental exposure, or if it represents a first-generation or transgenerational inherited effect.

Fourth, there is a search to identify **key mechanisms of disease causation** in order to recognize processes which might be reversed, prevented, or modified.

Putting all of these elements together, the classification of dementia by **Essential Cause** can be charted in the following way:

---

* soma – a term which refers to the entire body; however, soma will be used here and elsewhere to emphasize the organs which lie beyond the brain (i.e., neck, torso, limbs)

Dementias According to Essential Cause

Primary   vs   Secondary

ORIGINATING IN THE BRAIN   **ORIGINATING IN THE SOMA**

Acute or Chronic
Genetic or Acquired

Key Mechanisms of Causation

Autoimmune
    multiple sclerosis

Electrophysiological
    seizures

Idiopathic (*cause unknown*)
    neurodegenerative conditions
    normal pressure hydrocephalus

Infectious (encephalitis, meningitis)
    [1]CJD, neurosyphilis, Lyme

Mechanical (head trauma)
    bomb blasts, contact sports

Metabolic/Toxic-Metabolic
    Endocrine/[2]Hormonal
    Mitochondrial
    Storage disorders

Neoplastic/Paraneoplastic (abnormal growths)
    primary tumors of the brain
    metastatic spread to the brain

Vascular
    strokes, bleeds, vasculitis

Autoimmune
    lupus, Sjogren's

Electrophysiological
    cardiac dysrhythmias

Idiopathic (*cause unknown*)
    essential hypertension

Infectious
    pneumonia, hepatitis

Mechanical (somatic trauma)
    car accidents

Metabolic/Toxic-Metabolic
    Endocrine/[2]Hormonal
    Mitochondrial
    [3]Nutritional
    Storage disorders

Neoplastic/Paraneoplastic (abnormal growths)
    tumors of non-brain

Vascular
    heart disease, vasculitis,
    bleeds, clots, emboli

---

[1] CJD = Creutzfeldt-Jakob disease (may be idiopathic, infectious, genetic, or toxic)
[2] Hormonal = includes neuroglial chemicals (acetylcholine, dopamine, etc.)
[3] Nutritional = includes glucose, oxygen, vitamins/minerals

The classification scheme which seeks to identify Essential Cause provides a systematic approach to the diagnosis and treatment of dementia. A comprehensive review of the key mechanisms would easily fill another book. Indeed, these mechanisms are the very core of entire branches of medicine, such as consult-liaison psychiatry and behavioral neurology. The point to be made in this chapter is simply this: ***psychiatric drugs are a common – but frequently unrecognized -- source of primary and secondary dementias.***

How and why medications function this way requires an understanding of the biological processes which damage brain cells. These processes will be described in the following sections of this chapter. The five major classes of psychopharmaceuticals will be identified using the following key:

ADs = antidepressants
    1st generation = monoamine oxidase inhibitors (e.g., isocarboxazid, tranylcypromine); tricyclic antidepressants, such as: imipramine, clomipramine, amitriptyline, nortriptyline

    2nd generation = serotonin reuptake inhibitors and other new agents such as: fluoxetine, fluvoxamine, paroxetine, sertraline, bupropion, duloxetine, milnacipran, venlafaxine

APs = antipsychotics
    1st generation = chlorpromazine, clozapine, haloperidol, thioridazine
    2nd generation = olanzapine, quetiapine, risperidone, ziprasidone
    3rd generation = aripiprazole

AAs = anxiolytic agents or anti-anxiety drugs: specifically, benzodiazepines
    such as: alprazolam, clonazepam, diazepam, oxazepam.

MS = mood stabilizers (lithium and antiepileptic drugs)
    such as: carbamazepine, oxcarbazepine, topiramate, valproate

Stimulants = ADHD medications
    such as: amphetamine, *atomoxetine, **bupropion, dextroamphetamine, methamphetamine, methylphenidate

---

* atomoxetine – not marketed in the United States as a stimulant, but clearly identified as such by the World Health Organization's Anatomic, Therapeutic, and Chemical (ATC) drug classification system

** bupropion – a dopamine reuptake inhibitor and another "covert" stimulant which has been marketed in the U.S.A. as an antidepressant. Structurally related to amfepramone (diethylcathinone), bupropion is a drug which stimulates the sympathetic nervous system and suppresses appetite.

Psychiatric Drug Dementia Mechanism #1   Autoimmune

Autoimmune diseases refer to conditions in the brain or soma (non-brain) which involve the activation of white blood cells and the production of inflammatory or cytotoxic proteins. All of these reactive components of the body's natural defense system can turn against host tissues, resulting in damage to otherwise healthy organs.

Primary dementia:
Examples of autoimmune diseases which originate inside the brain include multiple sclerosis (a condition which results in the loss of myelin both within and beyond the brain); primary angiitis of the central nervous system (PACNS = an inflammatory condition which harms the blood vessels of the brain and spinal cord); and paraneoplastic limbic encephalitis (PLE = caused by cancer-induced proteins which attack the subcortical structures of the brain).

Secondary dementia:
Examples of autoimmune diseases which originate outside the brain include systemic lupus erythematosus, Sjogren syndrome, diabetes, and thyroid disease (e.g., Hashimoto's thyroiditis). Connective tissue disorders – such as lupus and Sjogren syndrome – are particularly well known to cause central nervous system effects. Damage may be mediated directly, via components of the immune system (circulating immune complexes, anti-neuronal antibodies, cytokines, and anti-phospholipid antibodies) or indirectly, via seizures or atypical strokes.

What the psychiatric drugs do:
Many psychiatric medications induce a syndrome which is immunologically identical to systemic lupus.[15-20] By triggering an abnormal activation of the body's defenses, drugs replicate two lupus-related phenomena: 1) *lupus cerebritis* (inflammation of the brain) and 2) *lupus vasculitis* (inflammation of the blood vessels inside the brain).[21-22] How often these effects result in disability or death remains unknown. What is concerning, though, is a recent proliferation of case reports in which clinicians have described *cerebral vasculitis* (not lupus, per se) in child and adult recipients of methylphenidate and other stimulant drugs.[23-26]

	ADs	APs	AAs	MS	Stims
[1] drug-induced lupus		x		x	
cerebral vasculitis					x

[1] most commonly seen with anticonvulsants and 1st generation antipsychotics, including clozapine

Note: Clozapine was invented in 1959, tested in Europe in the 1960s, and approved by the FDA in 1989. It has been repeatedly mischaracterized as a second-generation neuroleptic in the U.S.A.

Psychiatric Drug Dementia Mechanism #2   Electrophysiological

Electrophysiological disorders refer to conditions which involve the disruption of electrical events. Similar to the workings of a car battery, the cells of the brain and heart transmit messages which are both chemical and electrical. This conduction of energy is particularly vulnerable to interference.

Primary dementia:
Within the brain, electrophysiological abnormalities include seizure disorders (aka, epilepsies) and sleep cycle disturbances (e.g., REM sleep suppression). The harmfulness of these conditions arises from the fact that repeated seizures and REM suppression both result in damage to the hippocampus of the temporal lobe. If these effects are severe, they impair learning and memory.

Secondary dementia:
Within the soma, electrophysiological abnormalities involve cardiac dysrhythmias (abnormal or dysregulated beating of the heart). If these abnormalities occur with sufficient intensity or regularity, they can impair the delivery of oxygen, sugar, and other nutrients to the brain.

What the psychiatric drugs do:
Many psychiatric drugs lower the seizure threshold (provoking or predisposing to seizures), suppress REM sleep, and induce potentially life-ending dysrhythmias of the heart.[27-34] [Note: One might reasonably expect electrophysiological crises with intentional or accidental drug *overdose*. The point to be made here is that *electrophysiological toxicity occurs even under the influence of ordinary or so-called therapeutic doses of psychopharmaceuticals*].

	ADs	APs	AAs	MS	Stims
↓ seizure threshold	[1]x	[1]x			[1]x
REM sleep suppression	x	[2]x	x	x	x
cardiac dysrhythmias	x	x		[3]x	x

[1] bupropion, clozapine, tricyclic antidepressants
[2] clozapine, haloperidol, risperidone
[3] carbamazepine, lithium

Psychiatric Drug Dementia Mechanism #3   **Idiopathic**

Idiopathic disorders refer to conditions whose fundamental cause remains unknown.

Primary dementia:
With the exception of those illnesses which are inherited (genetically transmitted), idiopathic conditions of the brain include the primary neurodegenerative dementias: Alzheimer's disease, Lewy body dementia, Parkinson's disease, Pick's disease. Other idiopathic conditions include normal pressure hydrocephalus – a potentially reversible phenomenon which involves the excessive accumulation of cerebrospinal fluid within the brain; pseudotumor cerebri (aka, idiopathic intracranial hypertension) – a condition which involves excessive pressure inside the skull, mimicking the symptoms of a brain tumor; and the sporadic form of Creutzfeldt-Jakob disease – a condition which results in the diffuse and often rapid destruction of brain tissue.

[Note: It is possible that the term idiopathic may eventually fall into disuse when discussing many of the aforementioned conditions. Some scientists speculate that environmental exposures account for many of these conditions. Genetic contributions and mitochondrial defects conceivably account for many more. See *metabolic/toxic-metabolic* section, below].

Secondary dementia:
Idiopathic conditions within the soma include high blood pressure (essential hypertension), bone marrow abnormalities (idiopathic anemia and platelet deficiencies) and diseases of the lungs (interstitial fibrosis, interstitial pneumonia).

What the psychiatric drugs do:
Many psychiatric drugs replicate or enhance the features of these idiopathic conditions. For example, antipsychotic drugs precipitate the pathologies which are associated with Alzheimer's disease and Parkinson's disease. Mood stabilizers, such as valproate and lithium, induce hydrocephalus, pseudotumor cerebri, and Creutzfeldt-Jakob disease.

The idiopathic somatic conditions (high blood pressure, anemia, platelet dysfunction, lung disease) are replicated by many psychopharmaceuticals.[35-40] Stimulants, for example, are so-named precisely because of their capacity to stimulate the sympathetic branch of the peripheral nervous system, resulting in chronic elevations of blood pressure. Interestingly, researchers have only recently documented the capacity of one antipsychotic drug (clozapine) to provoke similar hypertensive effects. Although this area of investigation is not widely known, it provides yet another mechanism through which psychiatric medications contribute to heart disease and strokes.

[see *infectious* and *metabolic/metabolic-toxic* sections, below]

	ADs	APs	AAs	MS	Stims
Alzheimer's pathology		x	¹x		
Creutzfeldt-Jakob				x	
hydrocephalus			¹x	x	
Parkinson's	x	x		x	x
pseudotumor cerebri				x	
bone marrow suppression	x	x	x	x	
high blood pressure	²x	³x		³x	x

¹ chronic exposure to benzodiazepines has been linked to enlargement of the ventricles, suggestive of hydrocephalus; also to cell membrane features which occur in Alzheimer's disease

² notably, bupropion, duloxetine, and venlafaxine. [Although bupropion is marketed in the United States as an antidepressant, it is technically a stimulant because of its physiological and chemical properties.]

³ if high blood pressure occurs, it is usually part of the metabolic syndrome

Psychiatric Drug Dementia Mechanism #4   Infectious

Infectious diseases refer to bloodborne or tissue-based illnesses which are caused by the toxic effects of invading pathogens, such as bacteria, viruses, fungi, or protozoa. According to the relatively modern theory of "prion" disease, infections may also arise from exposure to *protein particles* rather than discrete microorganisms which carry their own genetic material. For example, the neurodegenerative condition known as Creutzfeldt-Jakob disease is allegedly caused by the proliferation of brain prions following exposure to "infected" food, "infected" medical instruments, or "infected" body parts (e.g., corneal transplants). However, it should be noted that some scientists believe the prion theory is only partly correct. Their views are based upon the detection of competing or complementary agents which appear to account for some patients' symptoms (e.g., spiroplasma and CJD).[41] Similar concerns have been raised with respect to Human Immunodeficiency Virus (HIV) and AIDS (Acquired Immunodeficiency Syndrome), particularly in light of the failure of research scientists to isolate, culture, transmit, and inspect the causal microorganism using unambiguous techniques of microbiology and electron microscopy.[42-46]

Primary dementia:
Within the brain, infectious causes of dementia include viruses (cytomegalovirus, hepatitis, herpes simplex, JC virus, measles, polio, West Nile); bacteria (borreliosis = Lyme disease, *Streptococcus*, syphilis, tuberculosis); fungi (*Aspergillosis, Cryptococcus, Coccidiomycosis, Pneumocystis jiroveci*) and protozoa (babesiosis, toxoplasmosis). One of the challenges for patients and clinicians alike is the capacity of certain infectious agents – such as those which cause Lyme disease, measles, shingles, and polio – to become latent. After initial infection, these agents can become episodically reactivated, resulting in delayed or long-lasting autoimmune changes that harm the brain. Another infection-related cause of primary dementia is the toxicity associated with many drug therapies, such as AZT, other nucleoside analogues used for HIV, interferon-alpha, and steroids.[47-52]

Secondary dementia:
The same microorganisms which give rise to various forms of brain disease (encephalitis = diffuse inflammation of the brain; meningitis = inflammation of the tissue layers which surround the brain; abscess = walled-off pocket of infectious organisms and immune cells) can also cause various somatic illnesses. These infectious conditions include pneumonia (an inflammatory disease of the lungs which is accompanied by the accumulation and solidification of fluid), myocarditis (inflammation of the heart muscle), endocarditis (inflammation of the heart valves), hepatitis (inflammation of the liver), and infections of the urinary tract. Any of these conditions can harm the brain via immunological changes (such as cytokine release) or metabolic disturbances (e.g., impaired oxygenation, impaired blood flow, or impaired clearance of toxins).

What the psychiatric drugs do:
Psychiatric medications contribute to infectious processes when they suppress the natural defenses of the host.[53-63] Especially common among members of the antipsychotic drug class, these immunosuppressant toxicities include the reduction of white blood cell populations, such as neutrophils and T-lymphocytes; or the disruption of white blood cell function. In the worst case scenarios, these effects result in the deaths of patients from organ failure or diffuse septicemia (bloodborne toxins). Psychiatric drugs also weaken the blood-brain barrier (BBB).[64-72] This is the protective layer of capillary cells and adjoining astrocytes which normally prevents intruders from crossing out of the brain's blood vessels and into the nearby brain tissue (parenchyma). By impairing the integrity of this barrier, many psychiatric drugs may facilitate the entry of infectious microorganisms. Alternatively, disruption of the BBB may contribute to an *autoimmune disorder* by permitting T-lymphocytes or cytokines to attack brain cells. (Interestingly, the latter process has been proposed as the most likely cause of multiple sclerosis or MS: a condition which destroys the fatty tissue layer around neurons in the *central and **peripheral nervous systems).

---

* central nervous system – brain and the spinal cord
** peripheral nervous system – nerves supplying the limbs, face, internal organs, etc.

	ADs	APs	AAs	MS	Stims
[1]blood-brain barrier defect	x	x		x	x
immunosuppression	x	x	x	x	

[1] blood-brain barrier dysfunction has been documented in response to tricyclic antidepressants, chlorpromazine, haloperidol, topiramate, valproate, and amphetamine

Psychiatric Drug Dementia Mechanism #5   **Mechanical**

Mechanical damage to the brain occurs via transportation accidents, sports injuries, falls, assaults, and war (direct impact of shrapnel, percussion injury from shock waves of blasts). As patients accumulate these injuries via repetitive events (as in boxers or quarterbacks), or as patients experience the normal physiological changes which accompany ageing, the delayed or lasting effects of mechanical trauma facilitate the onset or intensity of dementia.

Primary dementia:
Mechanical trauma damages neurons and glia by compressing brain tissue against bone; by compressing brain tissue beneath clumps of blood or fluid (e.g., due to subdural hematoma, or due to the obstruction of ventricles, venous sinuses, or granulation tissue); by starving the brain of nutrients and oxygen due to loss of blood (subarachnoid hemorrhage, intracranial hemorrhage, aneurysm); or by tearing intracranial structures (such as the pituitary gland) which perform essential life-sustaining functions.

Secondary dementia:
Within the soma, mechanical trauma can result in serious damage to vital organs and skeletal muscles. Depending upon the intensity and location of these injuries, cerebral blood flow and nutrition can be severely compromised. Furthermore, if the liver and kidneys are severely harmed, toxins and toxic metabolites may accumulate within the central nervous system, resulting in the dysfunction or death of brain cells.

What the psychiatric drugs do:
Although psychiatric drugs do not strike the brain like a physical blow, they produce changes which can lead to mechanical trauma and impede recovery from it. First, psychopharmaceuticals can cause new or repeated episodes of trauma due to falls or accidents. These events arise from drug-induced changes in arousal; brain-based or heart-based syncope (fainting); seizures; muscular weakness; postural instability; or unsteady gait.[73-91] Second, psychopharmaceuticals retard or prevent recovery from past or recent head injury by disturbing blood flow and/or hemostasis (clotting); by disrupting endocrine functions; and by damaging brain cells directly via other metabolic effects.[92-94]

	ADs	APs	AAs	MS	Stims
↓ arousal/alertness	x	x	x	x	
endocrine disruption	x	x	x	x	x
fainting (syncope)	x	x		x	
imbalance/instability	x	x	x	x	

Background Information

Traumatic Brain Injuries or **TBIs**, are associated with high rates of disability and death. Within the United States alone, there are more than one million emergency room visits for TBI each year – most, associated with motor vehicle accidents, contact sports, and falls. (A fairly new source, not reflected in these statistics, is the population of civilians and military members who experience the effects of explosions in acts of terror or war.) Approximately 25% of these subjects die in the acute phase of injury and recovery.[95] However, for those who survive the immediate trauma, *15 to 40% experience a potentially life-threatening endocrine disturbance which may persist for three years or more.*[96-97]

First recognized by E. Cyran in 1918, a phenomenon known as **post-TBI hypopituitarism** involves the imbalance of one or more hormones which is caused by damage to the *pituitary*. Hanging from the hypothalamus of the brain by a sliver of tissue, known as the infundibulum (stalk), the pituitary is a gland which participates in the regulation of virtually all life-sustaining functions.

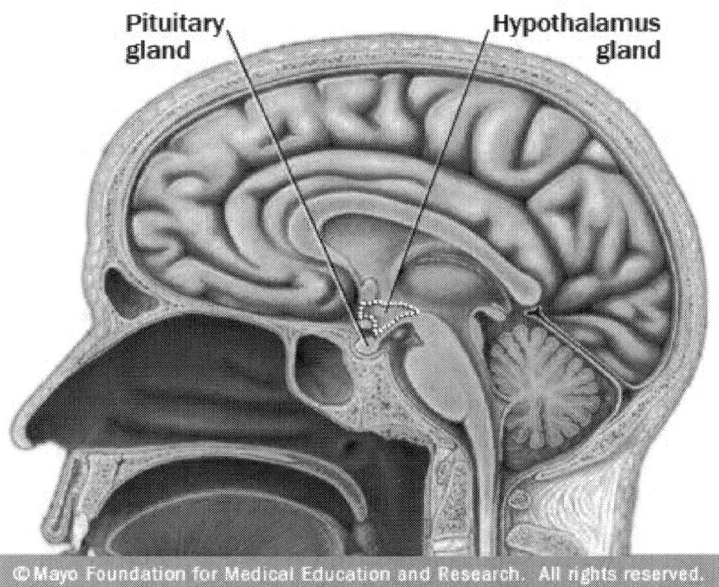

Copyrighted and used with permission of Mayo Foundation for Medical Education and Research, all rights reserved.

Because of its location deep within the skull, and because of the location of its blood supply, the pituitary is particularly vulnerable to damage from rotational forces or shearing; from laceration (the actual cutting of the stalk); or from ischemic insult after trauma due to a variety of secondary events: low blood pressure, anemia, high intracranial pressure, bleeds or blood clots, and brain swelling (edema).

Why is post-TBI pituitary damage an important topic for this book?

First, hypopituitarism is a potential source of dementia and death which is commonly misdiagnosed or ignored.[98-100] Second, the treatment regimens which TBI patients receive in the context of chronic rehabilitation frequently prioritize the use of psychiatric drugs for emotional and neurological symptoms. Third, the use of psychopharmaceuticals is a probable (but unrecognized) cause of prolonged disability and premature death in the brain-injured population, precisely because these substances cause or enhance pituitary dysfunction.

Hypopituitarism as a Source of Dementia

When the pituitary gland is damaged as a direct or indirect consequence of traumatic brain injury, a series of hormonal fluctuations occur. These imbalances involve chemicals which are produced by the *hypothalamus* of the brain (so-called releasing hormones), and/or chemicals which are made and released by the pituitary and its target organs (e.g., the thyroid, adrenal glands, gonads, and kidneys):

hormones		modulatory functions
growth	(GH, IGF-1)	skeletal and brain growth, cardiac function
prolactin	(PRL)	lactation, reproduction, coagulation
sex hormones	(estrogen, testosterone)	fertility, mood, skeletal integrity
steroids	(ACTH, cortisol)	blood pressure, insulin/glucose
thyroid	(TSH, T4)	neurodevelopment, mood, metabolism
vasopressin	(aka, antidiuretic hormone)	fluid balance

Mechanisms

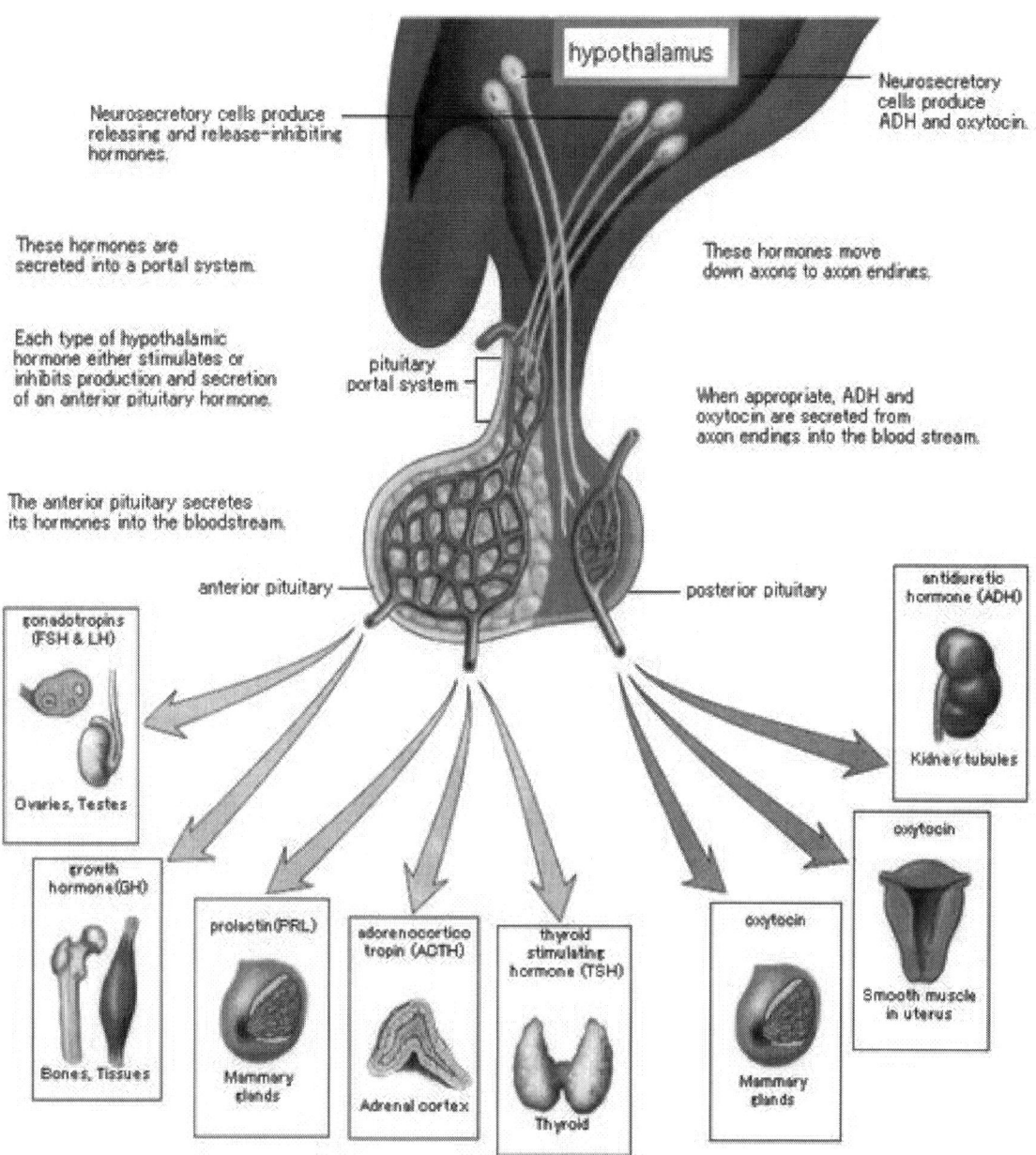

Reproduced with permission of Scholarpedia.

Deficiencies and excesses of hormones can impede the recovery of previously injured brain tissue, destroy neurons and glia that were initially unharmed, undermine the quality of life, and elevate the risk of premature death by 50 to 200%.[101-105] Relevant to the theme of this book, any and all of the aforementioned disturbances (low sex hormones, low growth hormone, low thyroid hormone, low cortisol) can cause or enhance dementia by disrupting neuroglial transmission or by killing brain cells through direct and indirect means (via ischemic stroke, hypoxia due to low blood pressure, low or high blood sugar, excessive insulin levels). Furthermore, numerous studies – both retrospective and prospective – have demonstrated that all survivors of head trauma are at risk for these endocrine abnormalities, irrespective of age, gender, or the severity of original brain injury.

Treatment Regimens in TBI: An Unrecognized Source of Dementia and Death

Currently, the treatment guidelines and recommendations of professional associations (the American Psychiatric Association), governmental organizations (the National Institute of Mental Health, the Department of Defense, the Department of Veterans Affairs/Veterans Administration), and consumer advocacy groups (Brain Injury Association of America, Brain Trauma Foundation) emphasize the use of psychiatric drugs for the long-term rehabilitation of traumatic brain injury.[106-115] Nothing could be worse for these patients, because psychiatric medications – *regardless of class* – are essentially and inherently *neuroendocrine disruptors which damage the central and peripheral nervous systems* [see *metabolic/toxic-metabolic* section, below]. **Furthermore, the contribution of psychopharmaceuticals to sudden and premature death – via growth hormone deficiency, cortisol deficiency, and/or excessive levels of prolactin – is habitually ignored by treating clinicians and even by medical examiners at autopsy**, particularly as these drug-related toxicities are generally not investigated after death. In sum, the tragedy of existing treatment schemes is the systematic exacerbation of Traumatic Brain Injury with **Chemical Brain Injury**, thus facilitating long-term disability and early death.[116-120]

Mechanisms

Psychiatric Drug Dementia Mechanism #6   Metabolic/Toxic-Metabolic

Metabolic diseases affect single or multiple organs of the body. When they are caused by mutations in the DNA (genetic defects), they are inherited as diseases which usually emerge in childhood. When they are acquired by exposures to toxins in the environment, or in the course of medical treatment, they can occur at any age. Common to all of these conditions, though, is the disruption of fundamental cell properties due to the impairment of energy production and other life-supporting processes.

Primary dementia:
Metabolic dementias include those which arise from inherited endocrine disorders (i.e., hormonal abnormalities which *originate* in the brain); mitochondrial disorders (i.e., diseases which are caused by inherited dysfunction of mitochondria – the energy-producing units found in most cells); and storage disorders (i.e., genetically acquired enzyme deficiencies which result in the abnormal accumulation of sugars, proteins, or fats). When the origin of these conditions is readily identifiable in the environment, these illnesses are collectively referred to as ***toxic-metabolic dementias***. Causes include heavy metals (aluminum, arsenic, lead, mercury); noxious gases (e.g., ammonia, carbon monoxide, chlorine); organic chemicals (PCBs, styrene, toluene); and medications (see below).

Secondary dementia:
Metabolic and toxic-metabolic diseases of the soma have the same causes as those which originate in the brain. However, in this case, the dementia-producing effects are secondary to cell dysfunction which *begins in organs outside the brain.* For example, endocrine, mitochondrial, or storage disorders which affect the liver and kidneys can result in the accumulation of toxic metabolites, such as uric acid or ammonia. High levels of these waste products travel through the bloodstream, subsequently disrupting the structure and activity of the brain. Somatic nutritional deficiencies also threaten the survival of neurons and glia by depriving them of metabolic requirements, such as glucose or essential vitamins and minerals.

What the psychiatric drugs do:
Psychiatric drugs are a common but generally unrecognized source of toxic-metabolic disease. These medications induce or enhance dementia precisely because they impair the endocrine system, suppress mitochondrial function, undermine nutrition, and replicate many features of inborn (inherited) storage disorders.

Endocrine Disruption

All of the psychiatric medications on the market today disrupt the function of one or more endocrine organs. These disruptions are particularly dangerous for children (who require hormones for the proper function and development of the growing brain) and for individuals at risk for dementia from pre-existing endocrine disease (such as diabetes, or pituitary dysfunction following brain injury). Furthermore, the chronic use of psychiatric medication, as promoted and often mandated in the United States today, *results in adaptive changes in the brain which oppose the intended and immediate drug effects.* It is these adaptations which textbooks of pharmacology, medical school and postgraduate curricula, and Continuing Medical Education courses (CME programs) are especially prone to overlook.

Few physicians are trained to appreciate the fact that **the *brain itself is the most important endocrine organ in the body*.** As a consequence, neuroglial chemicals – such as dopamine and serotonin – exert effects which influence the growth, repair, and ordinary function of tissues that are located *within* **and** *beyond* the brain. Regrettably, the brain's long-term adaptations to exogenous substances are usually in the direction of electrochemical "shut down," or in the direction of changes which result in the exacerbation of initial symptoms.

For example, through a complex series of changes involving cell receptors and enzymes in the neurons and glia, antidepressants trigger an enduring sensitivity of specific circuits within the brains of many patients. Using dietary methods to temporarily provoke transient but severe dips in the blood and brain levels of different chemicals, research teams have repeatedly made the following observations: [121-124]

> serotonin drugs create a lasting vulnerability to depressed mood
> via the serotonin system
>
> norepinephrine drugs create a lasting vulnerability to depressed mood
> via the norepinephrine system.

Dozens of studies, spanning more than 30 years of research, have demonstrated this finding. Notably, formerly depressed individuals who have received treatment with psychotherapy (but who have avoided pharmaceuticals) have *not* displayed this reaction when subjected to the same methods of dietary manipulation (monoamine depletion).[125-127]

SSRI responder (past or current use)	**deplete serotonin** deplete norepinephrine	>> >>	**depression returns** no symptoms return
NRI responder (past or current use)	**deplete norepinephrine** deplete serotonin	>> >>	**depression returns** no symptoms return
healthy subject	deplete serotonin deplete norepinephrine	>> >>	no depression appears no depression appears
drug-naive depressed (never used antidepressant)	deplete serotonin deplete norepinephrine	>> >>	no worsening no worsening
psychotherapy responder	deplete serotonin deplete norepinephrine	>> >>	no depression returns no depression returns

SSRI = selective serotonin reuptake inhibitor (e.g., fluoxetine)
NRI = norepinephrine reuptake inhibitor (e.g., desipramine)

This chart reflects the fact that approximately 60 to 80% of formerly medicated patients experience a rapid return of depressive symptoms which corresponds to the target of past treatment (serotonin drugs persistently perturb the serotonin circuits; norepinephrine drugs persistently perturb the norepinephrine circuits).

The biological basis of these persistent toxicities has been confirmed in numerous experiments. In studies involving mice, rats, and monkeys, neuroscientists have regularly discovered that chronic exposures to antidepressants reduce or deplete brain chemicals.[128-146] *Contrary to prevailing beliefs and published deceptions, the treatments which are widely prescribed to patients in an effort to relieve depression appear to reduce, rather than enhance, the synthesis, release, and/or target sensitivity of the neurotransmitters which are presumed to be responsible for healthy mood.*

Most importantly, drug-induced disruptions in brain processes have been demonstrated in the context of chronic treatment for all psychiatric drugs. While many of these adaptations have been verified using neuroimaging techniques in human subjects, the clearest evidence has been drawn from research involving the measurement of electrical and chemical changes in experimental animals:

Drug Class	Adaptations
anticholinergic drugs/ anticholinergic properties[147-160]	increased release of acetylcholine in cortex, hippocampus, and striatum (presumably, due to blockade of pre-synaptic muscarinic autoreceptors)

Note: As psychiatric patients consume anticholinergic drugs or antipsychotic drugs which possess strong anticholinergic properties, their bodies respond by making and releasing *more acetylcholine*. This adaptation replicates the effects of organophosphate poisoning (whether by nerve gas, by insecticide, or by anti-Alzheimer's pharmaceuticals) by ***over-stimulating acetylcholine circuits of the brain***.

This explains why consumers of anti-Alzheimer's drugs develop dystonic features (abnormal, involuntary and sustained contractions of muscles) identical to those which emerge among consumers of antipsychotics. It also accounts for the overlapping features of two potentially lethal conditions: ***neuroleptic malignant syndrome*** or NMS (confusion or coma, *autonomic instability, muscle rigidity and breakdown, fever) and ***organophosphate poisoning*** or OP (acute: loss of consciousness, autonomic instability, paralysis, tremors, seizures; intermediate: weakness of respiratory and limb muscles, dystonic posturing, cranial nerve palsies; delayed or persistent: polyneuropathy, dementia, dystonia, psychosis, anxiety, depression, aggression).

It is for this reason that physicians should avoid polypharmacy regimens which combine antipsychotic, anticholinergic, and anti-Alzheimer's drugs. Even in neurological patients who suffer from Parkinson's disease, there is good reason to rethink the use of popular drug "cocktails." Recent research has linked the use of anticholinergic drugs in Parkinsonian patients with increased levels of amyloid plaques and neurofibrillary tangles, the defining pathologies of Alzheimer's disease.

* autonomic instability – profuse sweating, high blood pressure (OP-nicotinic), low blood pressure (NMS, OP-muscarinic), respiratory distress, drooling, urinary or fecal incontinence, ↑ heart rate (NMS or OP-nicotinic), ↓ heart rate (OP-muscarinic)

Drug Class	Adaptations
anticonvulsants[161-165]	increased glutamate levels ("pro-seizure") effect); changes in $GABA_A$ receptor composition; diminished synthesis of GABA (vigabatrin); decreased sensitivity of sodium channels (loss of use-dependent blockade); downregulation of calcium channels (gabapentin) and adenosine receptors (carbamazepine); elevated levels of glycoprotein B, resulting in diminished entry of drugs into brain & enhanced efflux of drugs out of brain via blood-brain-barrier
antidepressants	decreased intracellular levels and synthesis of neurotransmitters; decreased turnover and evoked (pulsatile) release of neuroglial chemicals; decreased density of monoamine transporters; reductions in density and/or sensitivity of postsynaptic receptors
anxiolytics (benzodiazepines)[166-169]	decreased # of GABA and benzodiazepine binding sites; decreased sensitivity of GABA receptors; functional uncoupling of benzodiazepine recognition site and $GABA_A$ receptor due to changes in subunit composition, phosphorylation, and/or conformation (low affinity vs. high affinity) of $GABA_A$ receptor; decreased levels of GABA (brainstem); increased activity of glutamate system (particularly, in limbic structures)
lithium[170-172]	enhanced reactivity of cerebellar neurons to norepinephrine; decreased # of benzodiazepine receptors and decreased GABA levels in frontal cortex; downregulation of Nurr 1 (a nuclear transcription factor) in hippocampus [associated with impaired spatial learning]
neuroleptics (antipsychotics)[173-178]	increased # and sensitivity of D_2 receptors; decreased synthesis of dopamine; decreased firing of dopamine neurons due to depolarization blockade (the latter, a common but not universal finding in lab animals)

Drug Class	Adaptations
stimulants[179-189]	increased sensitivity of D_2 receptors (↑ high affinity state); decreased synthesis and release of dopamine; decreased firing of neuronal subgroups within striatum and brainstem; decreased dopamine transporter density
Note:	The long-term adaptations to stimulants reflect many changes in brain anatomy and function identical to those which accompany chronic dopamine blockade (antipsychotic therapy). This may account for the tics and dyskinesias which emerge in a considerable number of habitual stimulant consumers. It also explains why combination therapy with neuroleptics and stimulants presents an increased risk for Parkinson's disease, neuroleptic malignant syndrome, and various tardive phenomena.

Similar disruptions occur with respect to each and every endocrine organ *outside the central nervous system*. These disturbances include dysfunction of the anterior pituitary gland, the thyroid gland, the adrenal gland, the kidneys, and the pancreas.[190-196] All of these effects contribute to cognitive deterioration and dementia, because all of these endocrine organs make and release chemicals which are essential for the development and maintenance of brain cell activity.

Mitochondrial Dysfunction
The term ***mitochondrial dementia*** has recently been proposed by some researchers as a name which describes the cognitive deterioration caused by malfunctioning mitochondria (the energy-making units of cells). There are two possible origins of this phenomenon: 1) *genetically transmitted disease*; and 2) *environmentally acquired disease*.[197-200]

Genetically Transmitted Disease
Neurologists have long recognized the existence of genetically transmitted diseases which arise from mutations of mitochondrial or nuclear DNA. In order to appreciate the significance of these conditions, one must first understand the location and function of the pertinent cell parts.

Mechanisms

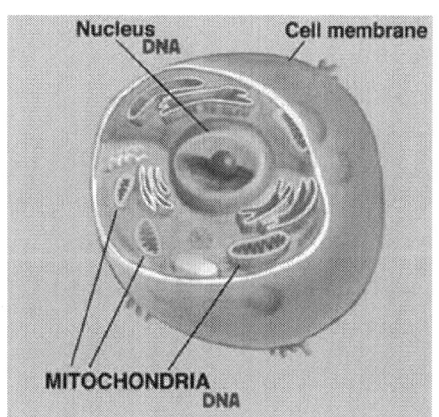

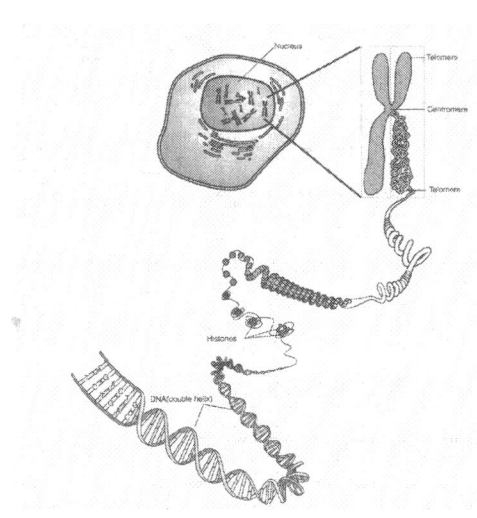

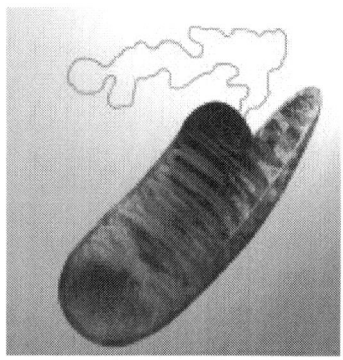

Top right: **A human cell** consists of multiple components or organelles ("tiny organs"). With the exception of mature red blood cells (no nucleus or mitochondria) and platelets (no nucleus), cells within the human body contain a single nucleus and multiple copies of mitochondria. Both the nucleus and the mitochondria are home to genetic material known as deoxyribonucleic acid, or DNA. The purpose of this of substance is to serve as an architectural blueprint which directs the synthesis of almost everything that a cell requires to sustain its functions. [Image reprinted with permission of Answers in Genesis. Artist: Dan Lietha]

Lower left: **Nuclear DNA** is contained within x-like structures known as chromosomes. When this material is uncoiled into its simplest form, it assumes the appearance of a ladder-like, double strand. The function of nuclear DNA is to serve as a template for the production of chemicals (proteins, fats, sugars) which are used throughout the cell. [Source of image: National Institutes of Heath, www.genome.gov, public domain.]

Bottom right: **Mitochondrial DNA** (mtDNA) is found only within the mitochondria. Unlike the DNA which is housed within each nucleus, mtDNA assumes the form of a single-strand loop. The purpose of mtDNA is to provide a template for the synthesis of chemicals which contribute to mitochondrial activity and structure. [Image reprinted with permission of Russell Knightley media, rkm.com.au]

Genetically transmitted mitochondrial disorders arise *either* from mutations in nuclear DNA (which codes for the majority of mitochondrial proteins) or from defects in mtDNA.

Environmentally Acquired Disease

Alternatively, mitochondrial disorders can also be *acquired*. Mitochondrial disorders of this kind are precipitated by exposure to numerous toxins in the environment, in the food supply, and in the form of medical therapies.

Why is Mitochondrial Dysfunction Being Emphasized in this Book?

The reason why mitochondrial dysfunction receives emphasis here is because of its debilitating effects and potential lethality. Particularly for high-energy consuming organs, like the heart and brain, mitochondrial disturbances are a predictable cause of disease and suffering.

Key Functions of Mitochondria

Source: public domain, licensed under Creative Commons

- Production of energy (ATP) via Cell Respiration + Oxidative Phosphorylation
- Production of heat
- Modulation of cell death functions (apoptosis vs. necrosis vs. autophagy)
- Decoding of mitochondrial genes (e.g., protein synthesis)

Located inside the inner membrane of each mitochondrion is a string of protein units (complexes I through V) known as the **Electron Transport Chain or ETC**:

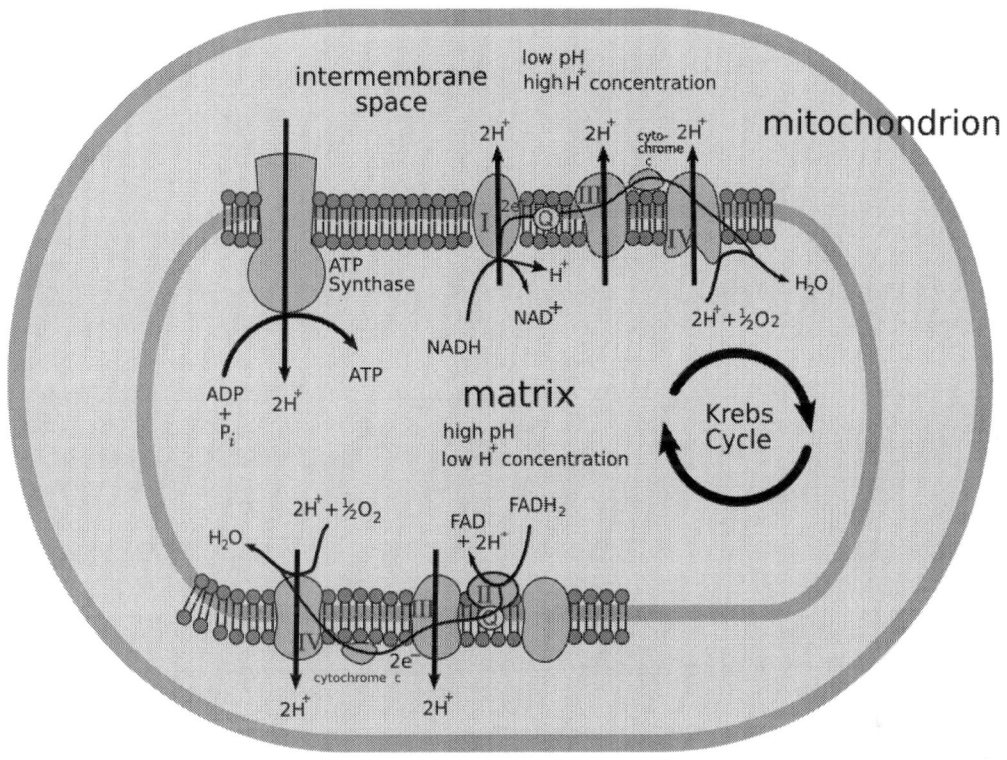

Source: public domain, licensed under Creative Commons

As the cells of the body burn fuel to produce energy, electrons (negatively charged particles) are passed down the ETC until they meet up with oxygen. A complex series of chemical reactions then convert oxygen into water (aka, cell respiration) and synthesize energy in the form of ATP (aka, oxidative phosphorylation).

Psychiatric medications impede the proper functioning of mitochondria in one of two ways: 1) by impairing energy production via the suppression of cell respiration and/or oxidative phosphorylation; and 2) by inducing oxidative stress ("free radical" toxicity).[201]

Energy Production

Some psychopharmaceuticals or drug metabolites suppress the activity of the electron transport chain. Others impair only the process of oxidative phosphorylation. The net result of either effect is the reduction of energy production in the affected cells.[202-217] The importance of this toxicity can be appreciated by considering the fact that this is precisely how and why poisons – such as carbon monoxide, cyanide, and the pesticide known as rotenone – exert their lethal effects.

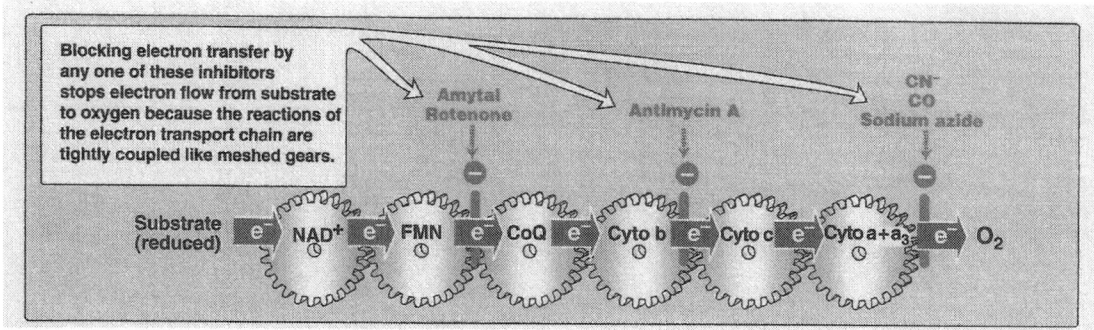

Source: P.C. Champe, R.A. Harvey, and D. R. Ferrier. *Lippincott's Illustrated Reviews: Biochemistry, 3rd Ed.* (Philadelphia: Lippincott Williams and Wilkins, 2005), 76. Reproduced with permission of LWW.

Oxidative Stress

Many psychiatric medications or drug metabolites cause a harmful cell phenomenon known as *oxidative stress*.[218-239] Because the human body uses oxygen to burn fuel and produce energy, an inevitable byproduct of all life-sustaining functions is the production of unstable chemical entities known as *free radicals*. As these molecules react easily with various components of each cell, they are sometimes called *Reactive* Oxygen Species (ROS) or *Reactive* Nitrogen Species (RNS). When ROS and RNS interact with the lipids (fats), sugars, and proteins, they cause damage which can result in cell death.

Psychopharmaceuticals, including the anticonvulsants, antipsychotics, and stimulants, accelerate the production of free radicals. Although the body maintains its own supply of antioxidant proteins which act as a defense mechanism against these substances (for example: catalase, superoxide dismutase, glutathione reductase), these defenses can either be overwhelmed by an excessive number of reactive species, or they can be directly suppressed by medications.

For example, one of the most extensively researched phenomena in toxicology is the oxidative damage induced by *amphetamine and other stimulants*.[240-246] Although a full discussion of this subject lies beyond the scope of this section, scientists have repeatedly demonstrated the mechanisms through which these chemicals incite the degeneration of dopamine and serotonin nerve terminals (neuronal endings).

How or why this occurs is no longer a mystery. Specifically, whether a person consumes illicit street drugs or prescription stimulants, dopamine is released inside and/or outside of the dopamine-producing cells in the brain. This results in the formation of free radicals, such as 6-hydroxy-dopamine, hydrogen peroxide, and quinones. These reactive compounds then damage the fats, sugars, and proteins which are found in all cell membranes and organelles, including the mitochondria and nuclear DNA. Ultimately, oxidative stress results in the destruction of nerve endings (axon terminals) and – in some cases – the death of entire cells.

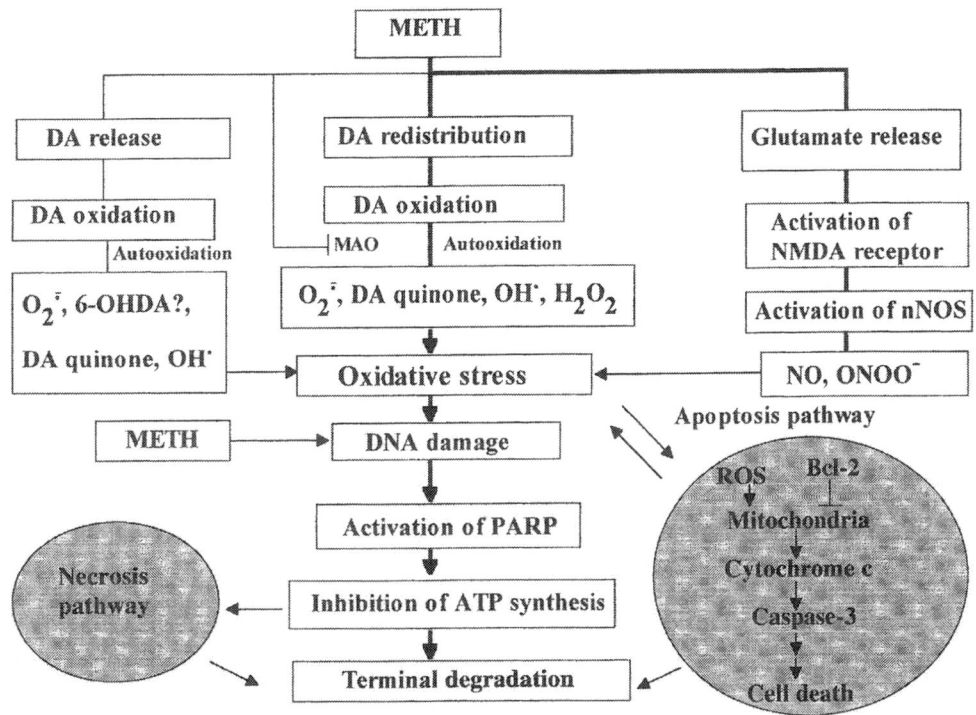

Source: T. Kita, G.C. Wagner, and T. Nakashima, "Current Research on Methamphetamine-Induced Neurotoxicity: Animal Models of Monoamine Disruption," *Journal of Pharmacological Sciences* 92 (2003): 187. Reproduced with permission of the Japanese Society of Neuropsychopharmacology.

This chart demonstrates the complexity of stimulant-induced toxicity. Although the diagram was produced by authors who used it to describe the effects of methamphetamine (METH), the free radical events and resulting cell damage apply equally to other stimulants, including the prescription drugs which are used to modify the symptoms of human inattentiveness or hyperactivity (e.g., ADHD).

Nutritional Deficiencies

Psychiatric medications induce a variety of metabolic abnormalities by impeding the absorption or the action of essential nutrients. For example, anticonvulsant drugs – such as valproate and carbamazepine – often reduce plasma and brain levels of folate (vitamin B9), cyanocobalamin (vitamin B12), and/or pyridoxine (vitamin B6). Because these vitamins are critical co-factors in the chemical reactions which control the synthesis of methionine and homocysteine (aka, the **one-carbon-transfer cycle**, shown below) disruptions in the absorption of B vitamins can result in excessive homocysteine.[247-249] This condition is associated with small-vessel strokes, seizures, Parkinson's disease, and Alzheimer's disease, presumably due to oxidative stress and the effects of a cytotoxic metabolite (homocysteic acid).

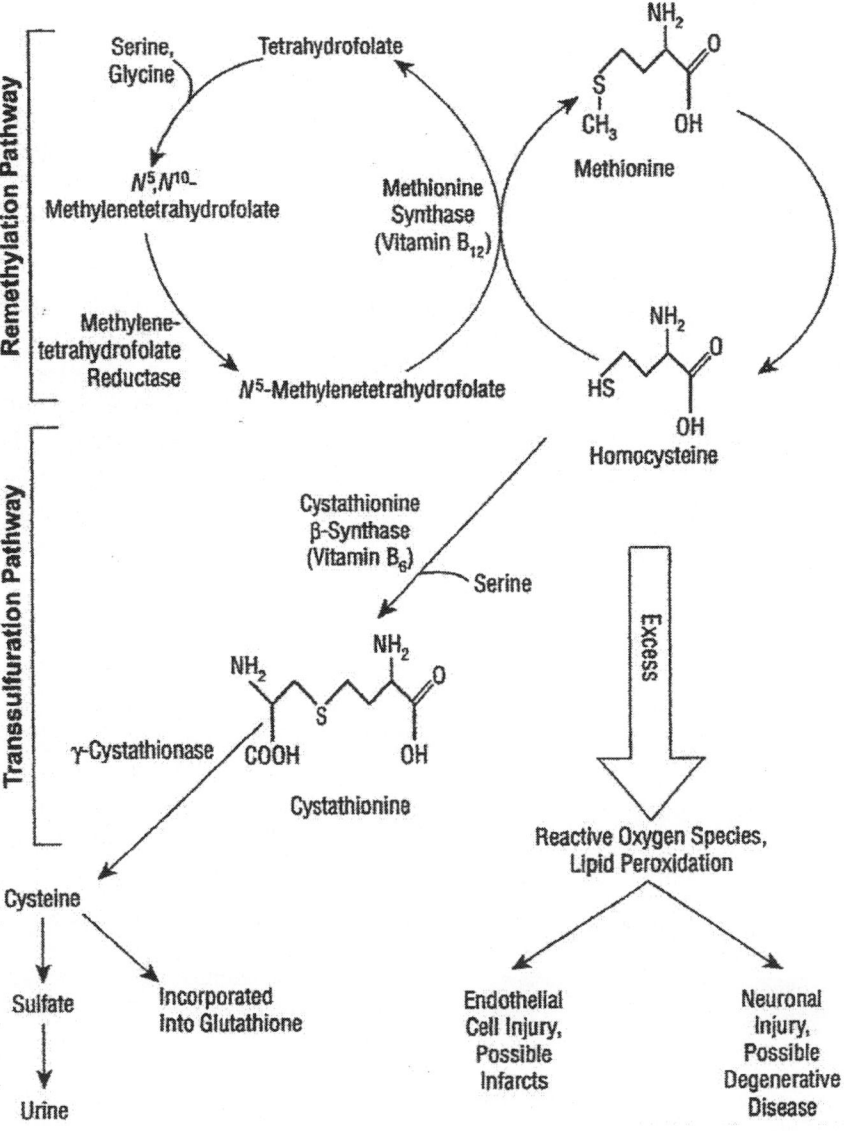

Source: R. Diaz-Arrastia, "Homocysteine and Neurologic Disease," *Archives of Neurology* 57 (2000): 1423. Reproduced with permission. Copyright 2000, American Medical Association. All rights reserved.

Other metabolic disturbances caused by psychiatric drugs include the partial depletion of carnitine or selenium (e.g., valproate, clozapine); the reduction of liver enzymes, such as n-acetyl-glutamate or NAG (e.g., valproate); and the inhibition of carbonic anhydrase (e.g., topiramate).[250-255] These last three effects (carnitine, NAG, carbonic anhydrase) contribute to high concentrations of ammonia in the blood and brain, resulting in further impairment of cell energy production, excitotoxicity (too much glutamate activity), glial swelling, and brain cell death.[256-260]

Storage Disorders

Many psychopharmaceuticals replicate the biological features of inherited conditions known as lysosomal storage disorders or LSDs. Under ordinary conditions, the *lysosomes* of each cell act as recycling and storage sacs which absorb large molecules and process them according to cell needs. Valuable materials are retained. Toxic materials are destroyed. However, in certain genetically transmitted diseases, deficiencies of key enzymes result in the abnormal accumulation of proteins, sugars, or fats.[261] This results in the dysfunction and deformation of lysosomes, and – eventually – dysfunction and deformation of the organs in which they are found (the eyes, kidneys, lungs, liver, heart, and brain). Although textbooks of medicine and pharmacology do not currently describe it, the phenomenon of **drug-induced phospholipidosis (DIPL)** refers to the accumulation of a fatty material known as **phospholipid** within the lysosomes of the body's cells.[262-264] As phospholipids build in number (either because the drugs become attached to them and prevent their degradation; or because the drugs inhibit the chemicals that would ordinarily break them down), they assume the form of stacked layers of membranes called myelin whorls or lamellar bodies (aka, myelin bodies, multi-vesicular bodies, lamellar inclusion bodies, etc.). When these structures are viewed with a high-powered lab instrument (the transmission electron microscope), these myelin bodies resemble a stack of pancakes or the layers of an onion skin:

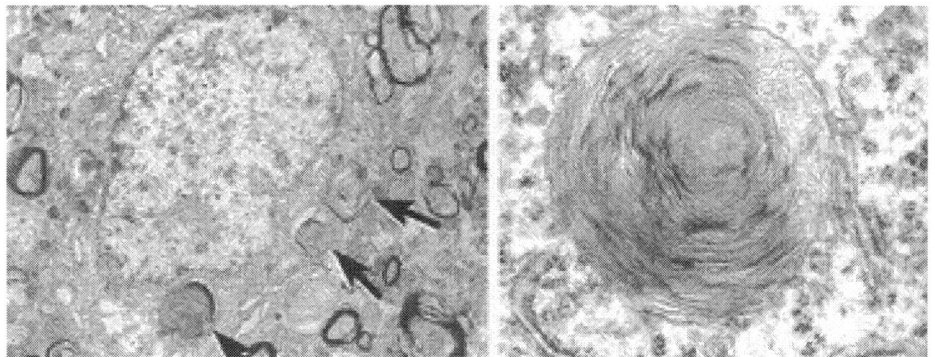

Reproduced with permission of the Japanese Society of Toxicologic Pathology from Nonoyama T, et al. Drug-induced Phospholipidosis – Pathological aspects and its prediction. *Journal of Toxicology and Pathology* 21: 9-24, 2008.

Left: Arrows point to **myelin bodies** in the nerve cell of a rat, following exposure to **fluoxetine** (Prozac). Right: An enlarged myelin body can been seen in this cerebellar neuron, also obtained from a fluoxetine-exposed lab rat. Similar changes have been found in humans who developed lung disease or nerve damage during treatment with fluoxetine or other antidepressant drugs.[265-267]

At the present time, the medical literature reflects great confusion about the *meaning* of **drug induced phospholipidosis (DIPL)**, and whether or not it denotes a true pathology. However, despite the fact that some scientists have consistently denied that DIPL is a necessary or sufficient cause of *any* toxic effect, the FDA has required all drug companies to test new products for this property.[268-269]

Why is the FDA concerned?

When phospholipidosis occurs in the context of genetically inherited conditions – such as Tay-Sachs disease, and Niemann Pick disease (Types A, C, and D) – it is associated with the degeneration of the brain and other organs. Symptoms include dementia, seizures, slurred speech, impaired coordination, loss of muscle tone, and the loss of vision and hearing. Some forms of these diseases are rapidly fatal, and the affected individuals die in infancy. Although the pathophysiology of these conditions has not been fully elucidated, there is no dispute about the harmful nature of these lipid aberrations.

At the current time, more than 50 different pharmaceuticals – including several popular psychiatric drugs – are known to induce the same lysosomal phenomena which are seen in Tay-Sachs, Niemann Pick, and other kinds of storage disease.[270-274] In some cases, clinicians have taken the time to report the appearance of tissue samples obtained via autopsies, biopsies, or bronchial lavage.[275-276] Importantly, these specimens have demonstrated DIPL in the target organs that were associated with symptomatic and sometimes fatal disease.

Psychiatric Drugs Known to Cause Phospholipidosis

antidepressants	antipsychotics	stimulants
amitriptyline	chlorpromazine	fenfluramine
citalopram	clozapine	methylphenidate
clomipramine	haloperidol	phentermine
fluoxetine	thioridazine	
fluvoxamine		
imipramine		
iprindole		
maprotiline		
nortriptyline		
promazine		
sertraline		
zimelidine		

* methylphenidate – myelin whorls have been identified in the sarcoplasmic reticulum, rather than the lysosomes, of human and non-human heart tissue

The clinical significance of DIPL is an area of longstanding and ongoing debate, due to the fact that phospholipidosis can and does occur within the body under some conditions of health. For example, phospholipidosis is a component of normal pulmonary functions when the myelin bodies appear within specific types of lung cells (i.e., the type II pneumocytes which secrete surfactant). However, phospholipidosis can also occur in conjunction with many kinds of pathologies in addition to the aforementioned storage diseases (e.g., acute respiratory distress syndrome, atherosclerosis). Thus, the complex manifestations of DIPL have resulted in several competing and contradictory interpretations.

According to one view, DIPL represents a ***detoxification process to prevent disease***. This hypothesis proposes that DIPL is an adaptation which permits a cell to store excessive levels of a harmful substance (such as a chemical which could trigger oxidative stress). In other words, the lysosome performs a triumphant act by walling off or sequestering a pharmaceutical when it traps the drug within myelin whorls. Under appropriate conditions, the lysosome would then release its contents to the outside world (via a process known as exocytosis), allowing the toxin to leave the body via the bile or urine.

According to a second view, DIPL represents a ***pathophysiological cause of disease***. First, the accumulation of drug products within the lysosome may interfere with the activity of other enzymes in that organelle. This could have negative effects upon the breakdown and recycling of other cell components, sabotaging the activities of the host cell. Second, the accumulation of drug products within the lysosome could result in the weakening or total fragmentation of the lysosome's own outer layer. This could result in the leakage of digestive enzymes (hydrolases) into the cell's interior, causing mitochondrial dysfunction and cell death. (In support of this hypothesis, a recent experiment involving cultured white blood cells demonstrated the capacity of amiodarone and imipramine to induce the formation of myelin bodies, concurrent with apoptotic cell death).[277]

According to a third view, DIPL reflects the ***pathophysiological effect of disease*** (autophagy, heterophagy). When the units of a cell become defective or damaged, they are absorbed and metabolized by their own lysosomes.[278-283] This process is known as autophagy (literally, "eating the self"). Alternatively, lysosomes may also absorb fragments of lipid membranes or lipid particles from neighboring cells which have collapsed around them (a process known as heterophagy). In either event, the myelin bodies which accompany these developments represent the end stages of a concurrent or pre-existing chain of toxic events.

Presently, some *experts in the field of toxicology recognize the harmful nature of phospholipidosis when it occurs in conjunction with *certain medications*, and when it produces *certain kinds of effects*. For example, phospholipidosis is weakly suspected of exerting toxic actions when it accompanies the *lung damage* associated with the cardiac drug, amiodarone.[284] On the other hand, the same American authorities dismiss this effect completely as a determinant of amiodarone-related nerve damage, such as sensorimotor neuropathy (which includes the loss of myelin), optic neuropathy, encephalopathy, movement disorders (Parkinsonism, dyskinesias), and brainstem dysfunction.[285-292] Similarly, phospholipidosis is clearly accepted as *exerting toxic actions* when it accompanies the *renal damage* associated with the antibiotic known as gentamicin.[293] On the other hand, the same observers deny phospholipidosis as a toxic determinant of gentamicin-related encephalopathy, peripheral neuropathy, respiratory insufficiency, or hearing loss.[294-298]

These contradictions may seem trivial to some, but what is most disturbing is the continuing inability or unwillingness of pertinent authorities to acknowledge the breadth and depth of patient suffering which has occurred in the context of DIPL. In these cases, the resulting organ pathologies have most certainly *not been* "clinically insignificant," nor has the treatment-emergent phospholipidosis been reflective of a successful "detoxification" process. It should concern physicians and patients alike that no one has stepped forward to recommend the aggressive investigation of DIPL with respect to psychiatric drugs, particularly in light of the strong case reports which have been published documenting this pathology in the vital organs of patients who were emphatically symptomatic (or dead).

To their credit, officials within the FDA convened a special working group on DIPL in 2004 in an effort to discern – once and for all – the prevalence, implications, and biomarkers of this effect. To their continuing discredit, though, no one yet has mentioned DIPL as a plausible cause of cardiac, pulmonary, and brain disease among consumers of psychopharmaceuticals. This continuing attitude of ambivalence says as much about the ambiguity of medical epistemology (how does one "know" if DIPL is the cause, rather than effect, of bodily dysfunction) as it does about anatomical and medical realities.

Nevertheless, whether or not myelin bodies reflect a detoxification process, a pathological *cause* of disease, or the pathophysiological *effects* of a toxic process, the clinical implications seem clear. Either the body is recognizing and responding to many psychiatric medications as invading toxins and walling them off (possibly leading to mitochondrial dysfunction) or the medications are inducing cellular disintegration (autophagy or heterophagy) or possibly both. For those who will heed the warning signs, myelin whorls appear to be an ominous signal of drug-induced harm.

---

* experts whose work has been partly funded by the Pharmaceutical Manufacturers Association Foundation (PhRMA), Proctor and Gamble Pharmaceuticals, and Rosetta Inpharmatics / Merck and Company, Inc.

	ADs	APs	AAs	MS	Stims
endocrine disruption	x	x	x	x	x
metabolic dysfunction	x	x	x	x	x
nutritional impairment		x		x	
storage disorder (DIPL)	x	x			x

Psychiatric Drug Dementia Mechanism #7 — Neoplastic (& Chemo Brain)

Neoplasms refer to benign or cancerous growths within solid organs or within the cells of the bone marrow. They contribute to dementia via direct and indirect processes.

Primary dementia:
Dementia is directly related to neoplasms when it arises from tumors of the glia or neurons, or from the spread of blood-based cancers to the brain (such as lymphoma). The reason why these entities are damaging to cognitive functions is because they compress healthy brain tissue, impair normal neurotransmission, divert nutrients and blood flow away from non-cancerous cells, induce harmful autoimmune reactions, and contribute to strokes or ischemia via the formation of blood clots (coagulopathies). Furthermore, the interventions which are aimed at malignancies (radiation therapy, chemotherapy) often serve as an additional source of brain cell dysfunction and death.[299]

Secondary dementia:
Dementia is an indirect byproduct of cancer when those pathologies originate in the soma (non-brain). In this scenario, malignancies of solid organs or the bone marrow impair the delivery of nutrients, oxygen, and blood to the brain; and/or impede the clearance of potential brain toxins. Somatic malignancies also stimulate autoimmune toxicities (via the production of proteins known as "paraneoplastic factors" which travel to the brain and harm cells) and induce coagulation defects which increase the risk of heart attack, pulmonary embolism, or stroke. Radiation and chemical therapies used in the treatment of these conditions are often poisonous to the neurons and glia of the brain.

What the psychiatric drugs do:
The medical literature features many accounts of abnormal growths attributable to psychiatric drugs. For example, case reports, cohort studies, and epidemiological investigations have described neoplastic side effects which include leukemia (valproate, lithium); breast cancer (fluoxetine, tricyclic antidepressants, antipsychotics); and colon cancer (antipsychotics).[300-308] Furthermore, benign pituitary tumors have been observed in the context of antipsychotic use (risperidone).[309-310]

More commonly (and with increasing frequency), the various classes of psychopharmaceuticals drugs have been tested and patented as *potential chemotherapies* to restrain the uncontrolled growth of cells within the soma and brain. An unaddressed problem in this new field of research, however, is the potential for medications to exert cell-destroying effects which go beyond their intended targets. This is not a hypothetical or trivial concern.

The term *chemo brain* refers to a spectrum of cognitive deficiencies which arise in the aftermath of treatment for cancer. Ranging in severity from mild to severely disabling, and affecting as many as 75% of chemotherapy recipients, this phenomenon has only recently gained recognition as a "real," rather than psychosomatic, entity. The change in perspective has been caused by new investigations in radiology and molecular biology which have confirmed that the **same agents which kill cancer cells throughout the body (soma) are equally or more harmful to normal tissues of the brain.**[311]

Ironically, at the very moment that oncologists have acknowledged the biological basis of chemo brain within their specialty, along with a duty to avoid or modify the therapies which cause it, psychiatrists have remained oblivious to a parallel development in their own field. An expanding body of animal research and preliminary clinical trials has documented the potential utility of many existing psychopharmaceuticals in the treatment of various neoplastic conditions. However, it should concern oncologists *and* psychiatrists that the very chemicals which are now being used experimentally to curb the proliferation of *cancerous* glia and neurons are also drugs which inhibit the proliferation or survival of *healthy* brain cells.

Although a comprehensive analysis of chemo brain lies beyond the scope of this chapter, the point to be made here is the fact that numerous medications which are used by psychiatrists are potent cell killers. One of the best examples of this is the anticonvulsant drug, valproate, which possesses not one but at least five different anti-cancer properties.[312-327] Small wonder, then, that this drug – like other chemotherapies – causes birth defects, bone marrow suppression, and hair loss in many patients:

Valproate (Depakene/Depakote) - 5 Chemotherapies in 1	
valproate properties	anti-cancer / cell-killing effects
↓ energy production >>	↓ cell respiration & oxidative phosphorylation
↓ methionine (↓ folate, ↓ B6) >>	disrupts one-carbon transfer; ↑ homocysteine
HDAC inhibition >>	turns off gene expression in nucleus
↓ tubulin assembly >>	disrupts cell division (mitosis)
urea cycle disruption >>	increases ammonia

Mechanisms

In actuality, the capacity of psychiatric drugs to kill rapidly proliferating cells (via apoptosis, via autophagy, and via necrosis) originates in the numerous pathologies which have been previously described:

- ➢ the disruption of endocrine functions
- ➢ the reduction of oxygenation and blood flow
- ➢ the depletion of essential nutrients
- ➢ the dysregulation of metabolic cycles (one-carbon transfer, urea, Krebs)
- ➢ the impairment of energy production (mitochondrial toxicity), and
- ➢ the activation of programmed cell death.

To the extent that psychopharmaceuticals destroy cancerous *and* non-cancerous cells alike, and to the extent that these destructive events occur equally within the soma *and* the brain, the chemotherapy actions of psychiatric drugs explain why these agents are a devastating source of immediate and delayed dementia.

	ADs	APs	AAs	MS	Stims
cancer	x	x		x	[1]x
chemo brain	x	x	x	x	

[1] Researchers in Texas recently observed chromosomal abnormalities in the white blood cells of children, following exposure to methylphenidate. Although these findings have not yet been replicated in other patients, DNA damage has subsequently been detected in the blood and brain cells of drug-exposed rats. These discoveries have provoked regulatory concern and multiple follow-up studies in an effort to clarify the risks which this stimulant may pose for humans, particularly with respect to cancer.[328-331]

Psychiatric Drug Dementia Mechanism #8   **Vascular**

Vascular causes of dementia originate in any disease which impairs the oxygenation and circulation of cerebral blood.

Primary dementia:
Within the brain, vascular origins of dementia include inflammation of the cerebral blood vessels (such as primary cerebral angiitis); obstruction of cerebral blood vessels (arteries or veins); or the loss of blood (leakage) from cerebral blood vessels (e.g., subarachnoid hemorrhage, subdural hematoma). All of these conditions starve brain cells of the oxygen, glucose, and essential nutrients which they need for their high-energy functions.

Secondary dementia:
Somatic origins of vascular dementia include diseases of the heart (which restrict the delivery of blood to the brain); diseases of the lungs (which impair the oxygenation of blood); diseases of the bone marrow (e.g., red blood cell or platelet disorders, which alter the transport of oxygen and clotting functions); and various metabolic conditions (e.g., diabetes, growth hormone deficiency, or hyperprolactinemia – which disturb the structural integrity of the body's blood vessels and heart).

What the psychiatric drugs do:
The electrophysiological and metabolic effects of psychiatric drugs accelerate the risks of vascular disease and vascular dementia.[332-364] By disabling the electrical activity of the heart, many psychiatric drugs present a risk for sudden heart block or heart attack. This can result in cardiogenic stroke. Other drug-induced vascular defects include autoimmune inflammation of the arteries or veins (vasculitis); high blood pressure; vascular obstruction (e.g., clots, emboli); and a variety of metabolic and endocrine disturbances, such as those shown below:

Endocrine Disturbance	Vascular Effects [365-379]
adrenal insufficiency	low blood pressure, electrolyte anomalies
diabetes insipidus	electrolyte anomalies (heart block)
diabetes mellitus	endothelial dysfunction, peripheral vascular disease, heart attack, stroke
growth hormone deficiency	endothelial dysfunction, ↓ cardiac output, ↓ ventricular size, ↓ coronary flow reserve
hyperprolactinemia	vasoconstriction, platelet aggregation, cardiac enlargement, heart failure
hyperthyroidism	cardiac dysrhythmias, pulmonary and systemic hypertension, heart failure
hypothyroidism	endothelial dysfunction, high lipids, arterial stiffness, diastolic hypertension, impaired left ventricular filling, increased risk of heart failure

	ADs	APs	AAs	MS	Stims
cardiac dysrhythmias	x	x		x	x
cerebrovascular disease:					
↓ [1]cortisol	x	x	x	x	
diabetes mellitus	[2]x	x		x	
↓ growth hormone	x	x		x	
↑ homocysteine				x	
hyperprolactinemia	x	x	[3]x	[3]x	
hypertension	x	[4]x		[4]x	x
hypothyroidism	x	x	[5]x	x	x
metabolic syndrome		x		x	
stroke	x	x		x	x
vasculitis/vasculopathy	x	x	x	x	x

[1] experimental findings have varied (some, but not all, studies have documented suppression of basal or evoked secretion of cortisol with chronic drug treatment)

[2] most notably, tricyclic antidepressants

[3] possible effect of several anticonvulsants (but rarely seen with valproate or lithium); also observed with ultra-high doses of diazepam

[4] neuroleptics and mood stabilizers can cause high blood pressure as a part of the metabolic syndrome; clozapine has been found to cause hypertension even without the added problems of obesity, insulin resistance, and high lipids

[5] changes in the density of nuclear thyroid hormone receptors have been observed in the brain tissue of experimental animals

Summary:

In its simplest form, the brain can be described as a three-part structure involving a top, a middle, and a base. Similar to the parts of a tootsie-roll lollipop, the brain consists of the cortex (candy coating), the subcortex (chocolate center), and the brainstem (lollipop stick). Within these three main structures, two major cell types can be found: neurons, which have been traditionally enshrined as the masters of the brain; and glia. The latter encompasses four different kinds of cells which participate in all of the brain's electrical and chemical activities, including the formation of cerebrospinal fluid; the synthesis of myelin; the stabilization of the blood-brain barrier; and the destruction of invading pathogens or toxins.

Throughout the history of medicine, there have been many **different schemes** for classifying dementia. This author has taken the liberty of supplying names to four of these systems using the following terms: 1) **Signs and Symptoms**; 2) **Brain Location**; 3) **Intracranial Origin**; and 4) **Brain Pathology**. All of these classification schemes are arguably outdated or irrelevant, either because they include overlapping and contradictory distinctions, or because they refer to dementia in terms (such as protein molecules) which are prone to confuse the cause and effect of disease.

In an effort to circumvent the limitations of these traditional dementia nosologies, a fifth approach was proposed, based upon **Essential Cause**. The potential advantages of this method arise from the systematic determination of fundamental, and potentially reversible, causes of dementia. Examples were given by dividing dementias into two major categories: 1) primary = dementia that originates from a disease process within the brain, and 2) secondary = dementia that originates from a disease process within the soma beyond the brain.

Common to primary and secondary dementias is the existence of **eight essential mechanisms** which can lead to brain cell dysfunction or death: 1) autoimmune processes; 2) electrophysiological processes; 3) idiopathic processes; 4) infectious processes; 5) mechanical trauma; 6) metabolic or toxic-metabolic processes; 7) neoplastic / anti-neoplastic processes; and 8) vascular disease.

Remarkably, all psychiatric drugs replicate the mechanisms of primary and secondary dementia. This chapter reviewed how and why the five major classes of psychopharmaceuticals achieve this harm. A special emphasis was placed upon specific examples of drug toxicity, such as endocrine disruption; DIPL (drug-induced phospholipidosis); and chemo brain. The reason for this particular effort was to highlight toxicities which are not well known by physicians or patients, and which pose significant hazards for special populations (such as children, pregnant or breastfeeding women, and individuals with traumatic brain injury or pre-existing endocrine disease).

Endocrine Disruption
It is significant that the human brain reacts to psychiatric medications with chemical and anatomical adaptations. This chapter reviewed some of the ways in which antidepressants impel long-lasting vulnerabilities to recurrent depression. It also reviewed the processes through which other classes of medication are believed to produce tolerance (diminished benefits) and disability (worsening of symptoms) over time.

DIPL (drug-induced phospholipidosis)
Also significant is the capacity of antidepressants, antipsychotics, and stimulants to incite the formation of myelin whorls within cells. This phenomenon, called Drug-Induced Phospholipidosis (DIPL), replicates the features of certain inherited conditions known as lysosomal storage diseases (LSDs). Although the implications of DIPL remain the object of speculation and debate, not all observers believe in the prevalent attitude of regulatory neutrality and neglect. Whether or not the accumulation of myelin bodies represents a defense mechanism, a cause of cell dysfunction, or the end stages of cell death is unclear. However, all of these possibilities are worrisome because they imply that phospholipidosis signifies the body's recognition of drugs as invading toxins.

Chemo Brain
Because of the fact that psychiatric drugs have become popular among researchers as adjunctive treatments for various types of cancers, it is important for clinicians to stop and think about the unintended consequences of these new uses. In light of the potential for psychopharmaceuticals to kill cells within cancerous *and* non-cancerous tissue, including healthy glia and healthy neurons, it remains essential for the medical community to acknowledge the potential problem of chemo brain and to avoid or withdraw the agents which cause it.

Chapter Three

Epidemiological Studies

Evidence for Drug-Induced Dementia and Death

The term *epidemiology* refers to the study of the distribution and determinants of disease within populations.[1] Epidemiological studies – sometimes referred to as observational studies – report the experience of ordinary patients as they respond to treatments in the "real world." This knowledge is essential for physicians and patients, because the lived experience of those who consume medications for extended intervals – irrespective of age, ethnicity, gender, educational achievement, socioeconomic status, or physical infirmity – provides data which is frequently quite different from the results of the short-term, randomized, placebo-controlled trials (RCTs) upon which drugs are initially approved.

Throughout the history of allopathic medicine, in general – and throughout the course of psychiatry, in particular – physicians have been reluctant to study the long-term consequences of prescription drugs. This has had inimical effects upon the quality and quantity of life for many patients.

The purpose of this chapter is to present a brief overview of the epidemiological evidence which links exposure to psychopharmaceuticals with dementia and death. Although this body of research cannot, and does not, *independently* prove that psychiatric drug therapies are a *cause* of brain degeneration, it is essential for at least two reasons.

First, it provides incontrovertible evidence that the existing drug therapies do not protect patients against dementia (despite the contrary claims of the so-called Key Opinion Leaders in psychiatry). Second, they provide an essential context for interpreting the results of other research, such as animal experiments, neuroimaging studies, and postmortem tissue exams. Ultimately, it is the collective body of all research evidence which must inform the decisions of toxicologists, governmental regulators, and clinicians in assigning a causal role to any medication which results in debilitating and lethal effects.

Epidemiological Studies – A Quick Overview

Epidemiological studies employ a variety of research designs in order to identify the features, risk factors, and/or prevalence of disease within populations. These include case-control, cohort, cross-sectional, and ecological studies.[2-5]

Case-Control Studies
Case-control studies compare outcomes or other features between two groups. First, the study's subjects are selected by identifying the presence (case) or the absence (controls) of a given feature. Information is then obtained with regard to earlier exposures or risk factors, in an effort to determine why the *cases* eventually manifested a particular illness or disease.

Cohort Studies
Cohort studies (aka, case series) track patients over time. A *prospective study (prospective cohort design)* begins by identifying a group of subjects who share a specific feature (e.g., radiologists, nuclear reactor technicians, astronauts) or exposure (e.g., radiation). These subjects are followed forward in order to identify the emergence of a pre-determined or unexpected outcome (e.g., leukemia). In contrast, a *retrospective study (retrospective cohort design)* begins by identifying a group of individuals who share a given outcome (e.g., cancer). The researchers interview pertinent informants or dig back through records and files in order to determine risk factors which might have predicted or modified the medical concern.

Cross-Sectional Studies
In *cross-sectional studies*, researchers identify all of the individuals who satisfy the criteria for a given condition (e.g., diabetes) at a *fixed point in time*. The goal is to provide a snapshot about the prevalence or features of illnesses, risk factors, or treatments as they exist in a population at one discrete moment in time.

Ecological Studies
Ecological or "correlation" studies involve a survey of features across two or more subsets of a population (e.g., groups which vary on the basis of demographics, geography, or history). The purpose of these studies is to generate hypotheses about features which are variably associated with different groups of people and their risks for certain types of disease.

Psychiatric Drugs and Dementia: Statistics Underestimate the Risk

Under ordinary conditions, dementia is a disease of the elderly. Thus, before one can consider the issue of iatrogenic (treatment-caused) *dementia*, one must first consider the problem of *iatrogenic death*. The reason for this will hopefully be obvious to the discerning reader. Before a doctor can detect an age-related disease in an older patient, that patient must live long enough to manifest the *age-related disease*.

Tragically, the lessons of history have shown that psychopharmaceuticals are incompatible with a normal lifespan. Research conducted over the past forty years has revealed that psychiatric medications contribute to higher mortality rates for all age groups, ultimately shortening longevity by 13 to 30 years.[6-7] Although the most common etiologies of drug-induced lethality have been frequently described – such as death from heart attacks, dysrhythmias, blood clots, and pneumonia – one essential cause has been consistently overlooked.

Despite the fact that comas, strokes, and seizures have long been recognized as immediate precursors to death, the ***lethality of dementia has been repeatedly ignored***. For example, in a recent prospective study of 165 Boston-area nursing home residents who suffered from progressive, advanced dementia, the death certificates of more than 1/3 of the decedents made absolutely no reference to neurodegenerative disease.[8] Even when dementia *was* mentioned, it was included as a "contributing cause of death" in only ½ of the cases, and it was seldom listed as an "immediate" cause of death.

While the following sections of this chapter will provide evidence for the role of psychiatric medications in the onset of dementia, it should be appreciated that the lethality of these chemicals has confounded the available research. Given the fact that many patients have died prematurely as a result of the current "gold standards" of psychiatric care, one must consider the probability that the available literature has grossly underestimated the *true* prevalence and severity of drug-inflicted brain disease.

Studies Involving Multiple Classes of Medication

Cooper and Holmes (1998) [9]

Using a case-control study design, researchers in England identified all confirmed cases of dementia occurring among a geographic subset of patients aged 60 or older. Each of these new cases had been recorded in a dementia register between the years 1993 and 1995. To test for any possible link between previous psychiatric history and late-life dementia, information on earlier diagnoses and treatments was obtained from multiple sources. In addition, and wherever possible, diagnostic interviews were conducted with the subjects, their next-of-kin, or their primary caregivers. Each patient in the dementia register was matched by age, gender, and place of residence to one randomly selected control chosen from a list of elderly inhabitants who resided in South London (boroughs of Lambeth, Southwark, and Lewisham).

Findings

A total of 559 subjects were registered in the dementia database. After excluding fifteen individuals whose psychiatric illnesses were possibly coincidental with the onset of cognitive decline, 70 subjects were identified with psychiatric histories. In other words, among the patients reported to the dementia registry between 1993 and 1995, approximately 13% had experienced psychiatric treatment which clearly preceded the onset of dementia. *The overall prevalence of psychiatric drug use was 3 to 4 times higher among those who eventually developed dementia,* **arguing against a neuroprotective role for these chemotherapies**:

	Dementia Register Patients n = 559	Non-Dementia Controls n = 559
interval between first psychiatric contact and diagnosis of dementia	13.2 to 40.2 years	n/a
# receiving psychosurgery	3 (5%)	0
# receiving ECT	17 (30%)	2 (4%)
# using lithium	4 (7%)	1 (2%)
# using neuroleptics	22 (39%)	6 (11%)
# using antidepressants	31 (56%)	12 (22%)
Alzheimer's disease	46 (66%)	
vascular or mixed dementia	10 (14%)	
Lewy body or Parkinson's	5 (7%)	
*other dementia	9 (13%)	

* other dementia = secondary dementia, such as dementia attributed to alcoholism

In Sum:

In a study of older patients selected from a healthcare database in Southern England, ***a past history of psychiatric treatment*** was a risk factor for late-life dementia. These treatments included ECT, psychosurgery, and standard drug therapies.

The prevalence of psychiatric drug use was 3 to 4 times higher among those who developed dementia, relative to age- and gender-matched controls.

The onset of dementia (elapsed time to diagnosis) ranged from 10 to 40 years following the initial exposure to psychiatric treatment.

The association between psychiatric treatment applied to several forms of dementia, including Alzheimer's disease.

Kessing et al (1999) [10]

Drawing upon a comprehensive database which recorded all psychiatric hospitalizations within Denmark, researchers conducted a retrospective analysis of all patients who developed dementia during a follow-up period of more than 20 years (1970 through 1993). The risk of eventual dementia was correlated with discharge diagnoses assigned to each patient during his or her earliest (pre-dementia) admission in the years 1970 through 1973.

Limitations

This study was limited solely to a review of inpatient hospital data. Thus, the actual prevalence of dementia may have been underestimated, due to the exclusion of subjects whose treatment may have been limited to outpatient care.

Findings

Initial diagnosis	Schizophrenia	Bipolar	Unipolar
# of patients	n = 1025	n = 518	n = 3363
mean age at start of study	31.7	42.8	51.3
# developing dementia within ~ 20 yrs	1-2%	6 to 9%	4 to 6%

Unique to this study was the inclusion of fairly young patients whose mean age at the start of the study ranged from 30 to 50. Despite this demographic feature, **one-third of the patients in this study died over the ensuing 20 years**. Of the surviving patients, **two to nine percent were formally diagnosed with dementia.**

When matched by age and gender to non-psychiatric subjects in the general Danish population, **patients who received treatment for schizophrenia and affective disorder (depression and/or manic-depression) were 14 times more likely to develop dementia**. Presumably, the relatively low risk of dementia identified in schizophrenia was an artifact of the diagnostic criteria for that condition. (Since cognitive impairment is considered to be a baseline feature of schizophrenia, physicians are unlikely to recognize treatment-related dementia symptoms as a toxic effect of drug therapy).

Studies Involving Specific Drug Therapies

Antidepressant Drugs

Brandt-Christensen et al (2006) [11]

In this short report, psychiatrists in Denmark described a retrospective, case-control study using their nation's prescription database. The goal of this study was the determination of incidental Parkinson's disease which may have been caused by specific drug therapies. While this study did not directly address the risk of dementia, its relevance to the theme of this book arises from the fact that Parkinson's disease – and its treatments – commonly result in dementia. Furthermore, many patients who initially receive the diagnosis of Parkinson's disease are later rediagnosed with the neuropathologically related condition known as Lewy body dementia [see chapters 2 and 5 for more on Lewy body dementia].

Epidemiology

Experimental Protocol

Cases were selected by identifying all patients aged 30 or older who purchased an antidepressant or lithium between 1995 and 1999. Next, the researchers screened the database to determine how many of these individuals subsequently purchased treatments for Parkinson's disease within the same calendar period. A randomly selected control group was formed by identifying subjects within the same database who had purchased anti-Parkinsonian drug therapies (APDs) in the absence of pre-existing exposure to antidepressants or lithium.

Limitations

This study relied upon the identity of drug treatments (rather than diagnostic codes) as a proxy for new-onset Parkinson's disease. In an effort to avoid an over-estimation of new cases, the researchers excluded some anti-Parkinson's medications which may have been used for alternative indications. For example, they failed to include any subject who purchased an anticholinergic drug [since these agents are also used for cardiac or bladder dysfunction, peptic ulcers, diarrhea, insomnia, and/or respiratory conditions]. They also omitted data from females under the age of 40 who may have purchased certain dopamine agonists (bromocriptine or cabergoline). These chemicals can be prescribed to young women as a treatment for amenorrhea or infertility.

Both of these exclusions must have reduced the power and significance of the research findings. For example, bromocriptine and cabergoline might have been given to some recipients of psychiatric drugs as a treatment for symptoms associated with excessively high levels of prolactin, or as treatments for neuroleptic malignant syndrome. Similarly, the prescription of anticholinergic agents may have occurred on numerous occasions, because these are first-line antidotes for drug-induced restlessness and movement abnormalities (iatrogenic Parkinsonism). Despite these flaws in methodology, however, several outcomes remained detectable and worthy of note.

Findings

Over the five-year period of study, the **risk of developing Parkinson's disease was approximately doubled by exposure to antidepressants or lithium**. Patients who purchased either class of psychiatric medication **also experienced a higher rate of death** during the follow-up period relative to the randomly selected control group:

	Cases n = 388,851		Controls n = 790,817
	antidepressants	lithium	antidepressants or lithium
# purchasing psych drugs	376,780	12,071	0
Relative Risk of buying anti-Parkinson's drug	**1.79**	**1.88**	--
# of subjects purchasing treatment for Parkinson's disease	3954 (1%)	154 (1.3%)	6676 (0.84)
# of subjects dying between 1995 and 1999	56,553 (15%)	1246 (10%)	59,178 (7.6%)

* Relative Risk = calculated using Poisson regression models

A statistical analysis was performed in order to determine the existence of any *temporal* links between the purchase of psychiatric drug therapy and the risk of developing Parkinson's disease (PD). This analysis suggested that the relative risk of PD was highest within the first six months after the purchase of an antidepressant or lithium (***3X higher risk during this interval***). However, the risk of post-medication Parkinson's disease – which might have involved continuous *or* discontinuous drug therapy – remained elevated across the five-year period of follow-up.

Relative Risk of New Onset Parkinson's Disease		
	Antidepressants	Lithium
within 6 months of exposure to psych drug	3.23	3.76
within 6 to 12 months of exposure to psych drug	2.42	1.94
within 12 or more months of exposure to psych drug	1.29	1.82

Chen et al (2008) [12]

Although not directly concerned with dementia, this investigation had important implications for the users of antidepressant drugs. In this study, an international team of researchers performed a case-control analysis using a massive insurance database in the U.S.A. Epidemiological information was drawn from the records of 45 million people enrolled in 70 different health plans, including seven State Medicaid programs.

Experimental Protocol

Based upon a review of insurance health claims, researchers identified all patients with diagnoses of depression recorded during a four-year period (1/1/1998 through 12/31/2002). Next, incident cases of stroke (new-onset stroke) were selected by identifying only those subjects with a first cerebrovascular event (CVE). These episodes had to have occurred *after* the index date of depression. In other words, **the researchers intended to track the incidence of *post-depression strokes*,** rather than the incidence of post-stroke depression.

A control group was formed by assigning to each case five or six individuals who developed depression within the five-year study interval, but who did not experience subsequent cerebrovascular disease.

Pharmacy and institutional datasets were analyzed in order to determine the *nature* and *timing of drug therapies* relative to the emergence of post-depression, cerebrovascular events. Four different categories of drug use were defined: *current* (drug use within 30 days of stroke); *recent* (drug use occurring 1 to 2 months prior to stroke); *past* (drug use occurring 2 to 3 months prior to stroke); and *remote* (drug use occurring more than 3 months before the onset of stroke).

Unique Features of This Study

In contrast to many previous studies of psychiatric illness, drug treatment, and long-term neurological dysfunction, this investigation employed an expansive definition of cerebrovascular disease, *consistent with the broadest spectrum of potential drug effects*. Specifically, cerebrovascular events (CVEs) included episodes arising from hemorrhage (bleed), ischemia (reduced blood flow due to arterial narrowing or obstruction), transient ischemic attacks (short interruptions in blood flow), and "other but ill-defined" CVEs.

Findings

Of the depressed subjects who experienced post-depression cerebrovascular events, more than 40% had received pre-stroke treatment with antidepressants. When the *timing* of drug use was evaluated for this cohort of patients (using a Cox proportional hazards regression analysis), ***current use of antidepressant drug therapy was associated with a 20 to 40% increased risk of stroke***:

Prevalence of Drug Use and Risk of Stroke Due to Current Therapy		
Depression + Stroke n = 1086	Depression, No Stroke n = 6515	Hazards Ratio (odds of stroke with **current** use)

	Depression + Stroke	Depression, No Stroke	Hazards Ratio
female	63%	63%	
% receiving:			
any AD	466 (43%)	3017 (46.3%)	
SSRI	377 (35%)	2255 (35%)	**1.24** (24% increased risk)
TCA	188 (17%)	849 (13%)	**1.34** (34% increased risk)
Other AD	257 (24%)	1360 (21%)	**1.43** (43% increased risk)

AD =	antidepressant drug
SSRI =	selective serotonin reuptake inhibitor (such as fluoxetine)
TCA =	tricyclic antidepressant (such as imipramine)
Other AD =	other antidepressants (such as venlafaxine, bupropion, trazodone)

Across the entire database of depressed patients identified during this interval (1998 to 2002), a striking ***13.4% of all antidepressant recipients experienced a cerebrovascular event***. Although this was not a statistic which the researchers ever mentioned in their published report, a careful inspection of their raw data shows this to be true:

Total Number of Depression Diagnoses	7601
# with Depression + Stroke	1086
% receiving antidepressants	42.9% = 466
# with Depression Without Stroke	6515
% receiving antidepressants	46.3% = 3017
Total Number of Antidepressant Recipients	3483 (= 466 + 3017)
Percent of Antidepressant Users Experiencing New Stroke	**466/3483 = 13.4%**

For the vast majority of these subjects, antidepressant exposure occurred (and/or ended) within 30 days of stroke:

Use Patterns of Antidepressants by Patients Experiencing Stroke			
	SSRI	TCA	Other AD
current users	352	150	237
recent users	4	3	2
past users	1	5	2
remote users	20	0	16

Interestingly, almost 3/4 of the cerebrovascular episodes (71.5%) occurred among subjects who were younger than 65:

Age of Patients Experiencing Depression Followed by Stroke	
≤ 34	5.1%
35-49	23.7%
50-64	42.7%
65-79	17.5%
≥ 80	11.1%

A unique finding of this particular investigation was a high rate of *atypical* cerebrovascular episodes. [This was particularly important in terms of explaining why other research teams have often failed to corroborate similar risks. Quite simply, they have not searched for all of them.] The reason why this particular study design was critical is because *atypical strokes* and *stroke-like* episodes may be the most likely form of cerebrovascular disease provoked by antidepressants and other psychotropic drugs, based upon the cytotoxic effects of these agents. Unfortunately, atypical forms of stroke and the putative mechanisms which cause them (e.g., mitochondrial dysfunction, autoimmune disturbances, and the direct disruption of neuronal and glial cell processes) appear to be a reality which neither the neuropsychiatric community, nor the toxicology community, nor governmental regulators are presently motivated to address.

Prevalence of Strokes During or After Antidepressant Therapy		
hemorrhagic stroke		8.5%
ischemic stroke		15.2%
other CVEs		**76.3%**
TIA	31.2%	
acute, ill-defined CVE	50.9%	
other and ill-defined CVE	17.9%	

In Sum:

The study by Chen et al (2008) appears to be the first case-controlled study to examine the risk of stroke associated with antidepressant use in a large population of depressed subjects.

Using a broad definition to identify cerebrovascular events, these researchers found that recent (within 30 days) or continuing antidepressant drug therapy was associated with a ***20 to 40% increased risk of stroke***.

Among the depressed subjects who received treatment with antidepressant drugs, ***13% experienced a cerebrovascular event*** within the five-year period of study. The overwhelming majority of these episodes (> 70%) occurred ***prior to the age of 65***.

The clinical relevance of the Chen study arises not only from the potentially disabling consequences of strokes in the immediate aftermath of each event, but also from the fact that 10 to 50% of stroke survivors generally experience some form of dementia.

Kessing et al (2009) [13]

Using a Danish national registry to identify all antidepressant purchases made by patients aged 40 and older between the years 1995 and 2005, Kessing et al performed a retrospective, case-control analysis with the goal of identifying the association between antidepressant drug use and the risk of subsequent (post-drug) dementia.

Experimental Protocol

A random sample consisting of 30% of the Danish population (alive on January 1, 1995) was identified in the national medical registry. From this sample, cases were defined as all persons aged 40 or older who purchased antidepressants at least once in the study period (1/1/1995 through 12/31/2005). Antidepressant purchases were detected by reviewing a national prescription database which included information from inpatient, outpatient, general practice, and specialty providers.

The outcomes of concern were main or secondary diagnoses of dementia following hospitalization or outpatient contact. Excluded from consideration were any subjects whose dementia diagnosis occurred *prior* to the start of an antidepressant. (Data from years prior to 1995 were used to exclude pre-existing dementia.)

Antidepressants were classified into three categories: SSRIs (selective serotonin reuptake inhibitors); newer, non-SSRIs (such as bupropion or venlafaxine); and older antidepressant drugs (primarily tricyclics, such as imipramine or amitriptyline).

Findings

Among the cohort exposed to antidepressants, ***approximately 4 to 6% of these individuals developed dementia within the ensuing ten years***. In contrast, the background incidence of dementia among the non-exposed was approximately 2%:

	Exposed to Antidepressants			**Non-Exposed**
	SSRI	non-SSRI	older AD	(no antidepressant)
# exposed	539,685	162,897	235,877	779,831
age (median)	57.5	53.1	57.2	49.2
total dementia	**30,167 (5.6%)**	**7,110 (4.4%)**	**9,971 (4.2%)**	**12,563 (1.6%)**
Alzheimer's	6,643 (1%)	1,529 (1%)	2,019 (1%)	2,305 (0.3%)
other dementia	23,524 (4%)	5,581 (3%)	7,952 (3%)	10,258 (1%)
dead	124,333 (23%)	22,988 (14%)	57,649 (24%)	119,472 (15%)
AD = antidepressant				

After adjusting for age, gender, calendar period, and the purchase of agents from more than one class (SSRI, non-SSRI, or older), the *relative risk of new dementia was 2 to 5 times higher for the antidepressant users* in comparison to community controls:

	SSRIs	Newer, non-SSRIs	Older Drugs
Relative Risk of Dementia	**3.36**	**4.74**	**1.77**

The researchers also compared outcomes in terms of "number of new dementia cases per 10,000 person years." For example, 10 individuals aged 20 would contribute 200 person years. When the *incidence of dementia* was calculated in this way, antidepressant use was associated with **4- to 5-fold higher rate of neurodegeneration**, relative to those who remained antidepressant drug-free:

# of dementia cases per 10,000 person years				
	SSRI	non-SSRI	older AD	non-exposed
total dementia	93.1	67.8	63.4	19.2
Alzheimer's disease	20.5	14.6	12.8	3.5
other dementia	72.6	53.3	50.5	15.6

In an effort to track the temporal and dose-related aspects of this effect, the research team calculated the relative risks of developing dementia according to incremental changes in drug load. Although they observed a gradual decline in risk as the total number of prescriptions rose above nine, this observation did not control for the possible influence of variable attrition. Furthermore, the risks of dementia for the users of antidepressants from all three categories remained persistently higher than the background rate of dementia in the general population.

In Sum:

In a nationwide, ten-year study (1995 to 2005) of patients aged 40 and older, individuals in Denmark who received antidepressants experienced a *2- to 5-fold increased risk of new-onset dementia*, relative to the non-drug exposed.

Although the relative risk of dementia appeared to be greatest in the early stages of therapy, this risk remained elevated with continued use.

It is possible that the dementia rates in this study underestimated the true prevalence of drug-induced brain dysfunction, due to the unrecognized impact of patient attrition. *More than 20% of those who used SSRIs or older drugs died during the follow-up interval*, but it is not clear how many of these subjects died while suffering from symptoms of dementia.

Antipsychotic Drugs (Neuroleptics)

The National Library of Medicine's main search engine, known as PubMed, contains no references to epidemiological studies which focus upon dementia *incidence* (# of new cases per time unit) among users of neuroleptic drugs. The reason for this oversight, while not entirely forgivable, is at least understandable if one considers the context of medical history.

When, in the late 19th century, psychiatrist Emil Kraepelin coined the term ***dementia praecox*** (meaning: *precocious* or *premature* dementia) to refer to the syndrome which was later renamed ***schizophrenia***, he led his contemporaries and successors to believe that psychosis involved an inherently dementing process.[14] Given this fateful beginning, it was perhaps unsurprising that epidemiologists and clinicians became dismissive of the possibility that antipsychotic drugs (or other interventions) could be the cause of cognitive disturbances in any patient, since they were predisposed to blame any such deficits upon an imagined brain disease.

Tragically, it has taken more than 50 years for authorities to reverse a portion of this legacy. Not until 2005 did the Food and Drug Administration (FDA) request new Black Box Warnings on the labels of the so-called "new and improved" neuroleptics, based upon clinical trial results which had implicated the ***second-generation agents** in the deaths of elderly patients with dementia.[15-16] Just three years later, federal regulators were compelled to expand the range of these Black Box Warnings in response to additional research which confirmed the lethality of the ****first-generation drugs**, as well.[17]

Were he still alive, one cannot help but wonder how Kraepelin would react to these developments. On the one hand, U.S. government officials have openly acknowledged the inadvisability of using antipsychotics in older patients who suffer from *dementia*. On the other hand, the primary, FDA-approved indication for the use of these drugs remains the condition which Kraepelin himself associated with dementia (*dementia praecox*). Consistency demands either a reconceptualization of psychotic syndromes, to suggest that they are *not* akin to dementia – or a reconceptualization of the populations for whom these drugs are ill-advised. (Perhaps the FDA is clinging to the hope that these drugs are only lethal when used as sedatives for "senile" rather than "precocious" dementia.)

---

* second generation – allegedly "atypical" neuroleptic drugs, such as risperidone, olanzapine, ziprasidone, quetiapine, and sertindole

** first generation – so-called "typical" or "conventional" neuroleptic drugs, such as chlorpromazine, thioridazine, and haloperidol

Regardless of how one chooses to resolve these contradictions, it is the opinion of this writer that *the lethality of neuroleptics cannot be overstated*. Perhaps others are beginning to share this view. For example, it is notable that the deadly nature of neuroleptics has caught the attention of researchers associated with the Veterans Administration.[18] In a case-control analysis of more than 10,600 patients aged 65 years and above, all of whom received outpatient treatment for dementia in 2002 and 2003, *the initiation of any antipsychotic medication was associated with the deaths of 20 to 30% of the drug recipients within just one year*. By contrast, demented veterans who avoided neuroleptics (old or new) experienced a death rate of 15%.

Similar results have been documented in the United Kingdom among patients diagnosed with probable Alzheimer's disease.[19] In the longest study of its kind, researchers in that nation have documented their observations of a *prolonged lethality risk* associated with both old and new neuroleptic drugs. Conducted between 2001 and 2004, the investigation followed a cohort of nursing home and residential care patients who initially agreed to participate in a randomized, placebo-controlled trial spanning one year. At the end of the first study period, subjects were given the chance to continue in an extension phase of the trial. This time, half of the participants (n = 83) were *randomly assigned* to receive continuous antipsychotic therapy with one of five medications (thioridazine, trifluoperazine, haloperidol, chlorpromazine, or risperidone). The remaining subjects (n = 82) were assigned to placebo. *At the end of four total years, only 25% of the antipsychotic treatment group remained alive*. In contrast, the survival rate for the placebo group was twice as high (50%).

Clearly, there is now a growing awareness about the dangers of neuroleptics for older patients who have already been diagnosed with Alzheimer's disease or other forms of dementia. However, there remain far too few discussions about the *mechanisms* of drug-induced lethality, and even less curiosity about the prevalence and timing of **neuroleptic-induced dementia** – particularly, as it occurs prematurely and even among the very young.

In contrast to a few daring reports in the 1980s, the medical literature has fallen silent on the topic of tardive (delayed-onset) dementia.[20-23] In fact, there are no epidemiological studies which one can cite in an effort to describe the prevalence of this phenomenon. The most that an observant reader can glean from extant publications is a vast body of research evidence which links neuroleptics *indirectly* to dementia, in terms of their connection to key risk factors for neurovascular and neurodegenerative disease. Regrettably, even this evidence is severely limited, since it ignores the direct brain toxicity of neuroleptics (as subsequent chapters will reveal).

Dementia Precursors Induced by Neuroleptic Drugs

Precursor	% Affected	Risk/Prevalence of Dementia
diabetes [24-30]	≥ 10 to 40%	diabetes ↑ risk of vascular dementia 2X to 3X diabetes ↑ risk of Alzheimer's disease by 2X
*EPS/Parkinson's [31-35]	≥ 10 to 40%	25 to 80%
stroke [36-41]	unknown	drugs ↑ risk of stroke by 2X to 3X among survivors of stroke, 10-30% develop dementia within 1 year; 50% develop dementia within 25 yrs
tardive dyskinesia [42-47]	≥ 5 to 50%	≥ 60%

* EPS = extrapyramidal symptoms refer to movement abnormalities due to dysfunction or anatomical damage within specific pathways of the brain and spinal cord; EPS include the features of Parkinson's disease (abnormal gait, pill-rolling tremor, masked facial expression, and blunted emotion)

Anxiolytics (Anti-Anxiety Drugs)

Concern about the long-term consequences of anti-anxiety drugs (aka, "minor tranquilizers") originated in the 1950s when it was discovered that barbiturates and meprobamate (Miltown) were fueling a silent epidemic of addiction. However, with the invention and proliferation of *benzodiazepines* (such as Librium and Valium) in the 1960s, many professionals believed that the problem of prescription-drug dependence had been successfully resolved. Their enthusiasm was misguided and premature.

The chemical properties of the benzodiazepines *should* have alerted physicians to the likelihood of this outcome. Like their predecessors, the new medications also replicated many biological effects of alcohol. Unfortunately, the medical profession was slow to contemplate the toxic parallels between all of these substances.

Despite the fact that the dementia-inducing potential of *excessive alcohol* has subsequently been appreciated by clinicians, the existence of analogous dementias due to alcohol substitutes (prescription anxiolytics and other *****GABA**-boosting drugs) has been largely ignored.

For many consumers of benzodiazepines, this double standard has had disastrous effects. ***It is in the context of this deplorable record that the recent observations of several research teams have been audacious and instructive.***

---

* GABA – an acronym which refers to the amino acid **G**amma-**A**mino-**B**utyric **A**cid. Physicians are trained to think of GABA as the major inhibitory chemical in the brain, but this concept is over-simplified. [It is the receptors and intracellular enzyme cascades which determine whether or not a neurotransmitter results in lower or higher levels of brain activity – not the identity or quantity of chemicals within a synapse.] Regardless, the point to be made here is the fact that the GABA pathways and GABA receptors represent a common target of alcohol, anesthetics, sleep aids (hypnotics), and several prescription antidepressants which are currently marketed and promoted as non-habit forming inhibitors of serotonin reuptake.[48-52]

[In fact, researchers in Turkey have performed a number of experiments in lab animals which have demonstrated the capacity of serotonin reuptake inhibitors (SRIs) to *substitute for ethanol* in reducing the signs of alcohol withdrawal. This unheralded discovery strongly implies that GABA is involved in antidepressant dependence and discontinuation syndrome, as well as the behavioral disinhibition or driving impairment which many patients experience while under the influence of these prescription drugs.]

Epidemiology

The EVA and PAQUID Studies
Nantes and Bordeaux, France

Understanding the negative consequences of benzodiazepines for various aspects of neurobehavioral functioning – such as attention, memory, and the performance of motor skills – and appreciating the potential public health impact of such effects upon the elderly, two teams of investigators in Western France performed large, *prospective*, *population-based* studies in the early 1990s. Both of these projects focused upon the association between benzodiazepines and cognitive decline.

Paterniti et al (2002) – The EVA Study [53]

Experimental Protocol

The Epidemiology of Vascular Aging Study (EVA) identified 1389 volunteers, selected from the electoral rolls of Nantes in Western France. The cohort consisted of individuals who were born between 1922 and 1932 (aged 60 to 70 during the study).

The project began in the years 1991 through 1993 with the collection of baseline data from each subject. Participants were evaluated by physical examination, face-to-face interview, and neuropsychological testing. Each individual was asked questions pertaining to education, occupation, personal habits (smoking, current alcohol use), medical history, and *all prescription medications used within the previous month*. Baseline ratings of depression and/or anxiety were established using two screening tools (the Center for Epidemiological Studies Depression Scale, or CES-D; and the Spielberger Inventory Trait for anxiety). Finally, a trained psychologist administered a battery of 10 different psychometric tests in order to assess cognitive and psychomotor functioning.

At 2- and 4-year intervals, cognitive testing, diagnostic interviews, and the CES-D (depression screen) were repeated. *Cases* were identified on the basis of benzodiazepine use occurring within 30 days of one or more assessment interviews:

Episodic	referring to benzodiazepine use used within the past month at only one interview
Recurrent	referring to benzodiazepine use within the past month at two different interviews
Chronic	referring to benzodiazepine use within the past month at all three assessment interviews (baseline, 2 years, and 4 years)

Controls were assigned by identifying all other subjects who denied the recent or continuing use of benzodiazepines.

At the end of the evaluation period, cases and controls were compared using statistical methods (logistic regression modeling). After adjusting for the presence of several potential moderating variables (confounders), the researchers calculated the odds ratios for cognitive decline based upon the intensity of drug use.

Limitations

Researchers failed to distinguish between sporadic versus continuous medication consumption, failed to collect data on the identity of each benzodiazepine across all four years, and failed to specify the time interval between cognitive testing and the most recent drug use. No information was collected with respect to cumulative exposure (dose x time), and no effort was made to confirm the intensity or regularity of use. Intermittent users of benzodiazepines may have been misclassified as "non-users" (controls) due to the restrictive case definition: "use within the past month." However, this should not have impeded the detection of effects related to uninterrupted drug use.

Strengths

This study excluded institutionalized individuals and focused solely upon those who were not demented prior to the use of benzodiazepines. The rate of follow-up was quite high (85%), making it unlikely that results reflected an attrition bias. Cognitive functions were assessed by interview and psychometric testing, and outcomes were unlikely to have been skewed on the basis of psychopathology or alcohol consumption. (The number of heavy drinkers was small, and daily alcohol consumption was lower in the users of benzodiazepines than in non-users.) Similarly, outcomes were unlikely to have been influenced by disturbances of mood or anxiety. Study results remained the same when the researchers controlled for each of these factors. Multiple domains of functioning were assessed. Important findings were documented with the following instruments:

Test	Cognitive Domains
MMSE = Mini-Mental Status Exam	orientation, short-term and immediate memory, ability to follow instructions, motor coordination, receptive and expressive language skills, reading, writing
DSS = Digit Symbol Substitution Test	sustained attention, motor speed, logical reasoning
Trail Making Test – Part B	abstraction, attention, motor speed, visual search

Findings

Among patients aged 60 to 70, the chronic use of benzodiazepines over a four-year period was associated with ***two-fold higher risk of cognitive decline***. On three different tests of neuropsychological functioning (Mini-Mental Status Exam, Digit Symbol Substitution Test, and the Trail Making Test – Part B), ***statistically significant deterioration was observed in approximately ¼ to ½ of the chronic users***:

	Benzodiazepine Users n = 264	**Non-Users** n = 912
Episodic users	114 (43%)	
Recurrent users	70 (27%)	
Chronic users	80 (30%)	
Among Chronic Users		**Odds Ratio for Decline**
Mini Mental Status Exam	26% declined	1.9
Digit Symbol Substitution	35% declined	2.5
Trail Making Test - Part B	45% declined	2.3

After adjusting the statistical test (multivariate analysis) for the effects of several medical conditions (coronary heart disease, high blood pressure, diabetes, and hyperlipidemia), these results remained the same. Even when the odds ratios were recalculated for the cohort after excluding 39 subjects with baseline deficits (MMSE < 24), similar findings were obtained.

In a further effort to minimize the influence of potential confounding variables, the research team matched a smaller subset of cases (n = 63) with controls (n = 126) on the basis of gender, anxiety (Spielberger Trait at 2-year interview), and exposure to non-benzodiazepine drugs. After controlling for differences in baseline cognitive score, a statistical analysis of this dataset revealed that the ***odds of cognitive deterioration for chronic benzodiazepine users remained statistically significant***, and remained ***two times higher*** than controls:

	Adjusted Case-Control Comparisons **Baseline Scores Versus 4-Year Follow-Up**		
	Chronic User n = 63	Control n = 126	Odds Ratio of Decline for Drug Users
MMSE	22% declined	15% declined	OR = 2
DSS test	37% declined	25% declined	OR = 2.2
Trail Making Test	43% declined	30% declined	OR = 2.2

Lagnaoui et al (2002) – The PAQUID Study [54]

The PAQUID Study (Personnes Agees Quid) was a prospective, population-based survey of a sample of individuals in the Bordeaux region of France. Focusing upon non-institutionalized individuals aged 65 years and older, this investigation monitored the participants for **eight years** with the goal of determining the risk factors, incidence, and features of dementia.

Experimental Protocol

A fixed cohort of 3,777 individuals was identified in 1989 (T0) using stratified sampling on the basis of age, gender, and size of town or village. On four occasions, interviews were subsequently conducted by trained neuropsychologists during visits to each subject's home:

T1	in 1990	first neuropsych interview
T3	in 1992	second neuropsych interview
T5	in 1994	third neuropsych interview
T8	in 1997	final neuropsych interview

At each assessment point, participants were interviewed for approximately 90 minutes. Information was collected on demographics, social environment, and medical history (past and present). A set of neuropsychological tests was administered (items not specified in the team's report), and screening for dementia was performed using the criteria of the *DSM-IIIR (Diagnostic and Statistical Manual for Mental Disorders, Third Edition Revised)*. If dementia was suspected, subjects were referred to a neurologist in order to confirm or disconfirm the diagnosis.

<u>Case Definition</u>
Cases were identified on the basis of **new-onset dementia**, first recognized at any of the key index dates (T3, T5, or T8). Anyone who satisfied criteria for dementia at T0 (cohort selection) or T1 (first screening) was excluded, in order to ensure the monitoring of incidental (new) rather than prevalent (pre-existing) cases.

<u>Risk Factor Identification</u>
In order to identify potential risk factors for new-onset dementia, patients or caregivers were asked about the use of medications. In addition, each subject's medicine chest was visually inspected, and the brand names of all drugs were recorded. Of specific concern in this study was the identification of exposure to benzodiazepines. Anyone without clear information about benzodiazepine use *prior to the onset of dementia* was excluded from the study.

Benzodiazepine use was classified in one of two categories: "non-use" (zero use before dementia) or "ever use" (any exposure before dementia). The latter category was further subdivided on the basis of "former" or "current" use:

Epidemiology

> non-use = zero use prior to the onset of dementia
>
> ever use = any exposure to benzodiazepines prior to dementia
>
>> former = use ending 2 to 3 years before index date (T3, T5, T8)
>> current = use occurring at the time of index date (T3, T5, T8)

Statistical Analysis
Using a technique called unconditional logistic regression analysis, the researchers calculated the odds of developing new dementia based upon prior use of benzodiazepines. These odds were determined after controlling for the effects of multiple potential confounders, including age, gender, education, wine consumption, social environment (living alone), depressive symptoms, and any history of psychiatric disorders.

Limitations

Researchers did not distinguish the total chronicity or intensity of drug use (dose x duration). Also, the incidence of dementia may have been significantly underestimated by the ascertainment technique. Specifically, it was possible that the researchers missed some cases of dementia due to the relatively insensitive definition of dementia which appears in the *DSM-IIIR*.

Findings

After excluding 123 pre-existing cases of dementia at T0 and T1, there were 3654 subjects who were eligible for continuing analysis. Of these individuals, aged 65 or older, **4% developed dementia over the ensuing eight years**:

% of entire cohort developing dementia:	150/3654 = 4%	
	Cases n = 150	Controls n = 3519
mean age at last f/u	79	73.9
male	43%	65%
living alone	60%	41%
depressed	29%	21%

After controlling for potential confounders (e.g., age, gender, depression, education, social environment, alcohol), *the odds of developing dementia was twice as high for "former" and "ever" users of benzodiazepines relative to the dementia risk among "non-users":*

	Cases n = 150	Controls n = 3519	Adjusted Odds Ratios for Dementia
Benzodiazepine Use:			
Ever	**97 (64.7%)**	**1714 (48.7%)**	**1.7** [CI 1.2-2.4]
Former	**14 (12.9%)**	**203 (6.3%)**	**2.3** [CI 1.2-4.5]
Current	37 (34.3%)	1065 (33.2%)	1.0
None	53 (35.3%)	1805 (51.3%)	1.0
% of Ever Users Developing Dementia:		**5.4%**	= 97 ÷ (97+1714)
% of Former Users Developing Dementia:		**6.5%**	= 14 ÷ (14+203)
% of Current Users Developing Dementia:		3.4%	= 37 ÷ (37+1065)
% of Non-Users Developing Dementia:		2.9%	= 53 ÷ (53+1805)

When interpreting these findings, the research team was understandably perplexed by the association between dementia and former (but *not current*) benzodiazepine use. Not unreasonably, they speculated that this result may have reflected the pre-evaluation (T3, T5, T8) withdrawal of medication in precisely those individuals who had been unable to tolerate adverse drug-induced symptoms – a research artifact which some have referred to as the "sick quitter" or "depletion of susceptibles" effect.

Unfortunately, the methodology in this protocol did not permit the precise identification of the *temporal onset of dementia symptoms*. Thus, it was not possible to determine how often these deficits may have started in the setting of active (ongoing) drug use, and how often the recognition of these deficits may have provoked the termination of benzodiazepine therapy. However, the likelihood of the "sick quitter" phenomenon in this dataset is strongly supported by the subsequent findings of another research team, as described below.

Barker et al (2004a, 2004b) [55-56]

Reacting to the existence of numerous but conflicting studies on the cognitive consequences of chronic exposure to benzodiazepines, a team of Australian investigators embarked upon two expansive studies which used the technique of meta-analysis. [For readers who may be unfamiliar with this method, a *meta-analysis* is a research design which combines results from several different studies of the same outcome, in order to determine the overall strength of a given finding.][57]

Epidemiology

Experimental Protocols

Meta-Analysis #1 – Cognitive Effects of Recent or Continuous Chronic Drug Use
The researchers identified English language, peer-reviewed papers published between 1980 and 2000. Selection criteria for each study included the following elements: 1) the use of formal cognitive testing; 2) a period of benzodiazepine exposure for at least one year; 3) the inclusion of a control group or within-subjects design; and 4) the reporting of data which allowed for the calculation of a statistic known as effect size.

[For those who may be unfamiliar with epidemiology, a *within-subjects design* means that the same group of subjects has served as their own control – for example, by undergoing cognitive testing before and after exposure to a medication. An *effect size* or ES refers to the magnitude of a given relationship or difference between two groups. In meta-analyses, the calculation of the Effect Size permits comparisons of outcomes across a wide range of investigations, despite variability in the sample size of each study.] [58]

Key Features of the Barker Study – Meta-Analysis #1
In this first meta-analysis, 13 studies qualified for review. The results from each neuropsychological testing battery were classified according to 12 different cognitive domains (sensory processing, visuospatial ability, nonverbal memory, speed of processing, attention/concentration, general intelligence, working memory, psychomotor speed, problem solving, verbal memory, verbal reasoning, and motor control). Effect sizes for each domain were calculated for each component study, and then across the entire set of studies. The goal was to estimate the overall cognitive impact of **chronic, continuous exposure** to benzodiazepines. Nine of the 13 studies specified the time of testing relative to last medication dose: in five of the studies, testing was conducted between 1 and 18 days of the last dose; in four of the studies, testing occurred just prior to the usual, ongoing daily dose.

Findings

> **Meta-Analysis #1 – Chronic Drug Use (Active or Recently Withdrawn)**
>
> | | | |
> |---|---|---|
> | # of studies | 13 | |
> | total # of subjects | 384 | |
> | mean age | 47.6 years | (range: 21 to 75) |
> | mean duration of drug use | 9 yrs 11 mo | (range: 1 to 34 years) |
> | average daily dose | 17.2 mg per day | (diazepam equivalents) |
> | # of domains affected: | 12 (100%) | |
> | *Effect Size | **-1.30 to -0.40** | |
>
> * Effect Size (ES) in this study relied upon Cohen's *d*. As a rough estimate, an ES of 0.20 is small, 0.50 is moderate, and 0.80 or above is large. A negative sign refers to a negative association (here, it refers to cognitive decline).

Conclusion:

Long-term users of benzodiazepines were consistently more impaired than non-users across all cognitive domains. The magnitude of this impairment, based upon effect size, was moderate to large for every aspect of mental functioning (i.e., attention, verbal reasoning and memory, motor control, problem solving, working memory). This impairment occurred despite the consumption of clinically moderate doses.

Meta-Analysis #2 – Cognitive Effects of Former Drug Use

A similar protocol was employed, as described above. In the first of two related analyses (meta-analysis 2A), the goal of the research team was the identification and quantification of changes in cognitive functioning among chronic users of benzodiazepines who eventually stopped taking these drugs.

Based upon data from 10 different component studies, effects sizes were calculated for changes in the same 12 areas of mental functioning. In other words, this was a study which compared longitudinal, *intra-individual* **changes, before and after the cessation of long-term benzodiazepine use**.

In a second phase of this investigation (meta-analysis 2B), 9 different component studies were selected for a *cross-sectional, between-subjects* analysis. Participants were evaluated on the same 12 cognitive domains. The goal of this inquiry was to explore ***persistent deficiencies in former, chronic drug users*** by comparing their mental aptitudes with normative data (published test norms) or with the performance of unexposed (usually, non-anxious) healthy controls.

Meta-Analysis 2A – Improvements After Withdrawal of Therapy		
# of studies	10	
total # of subjects	297	
mean age	47.1 years	(range: 21 to 75)
mean duration of drug use	10 years	(range: 1 to 29 years)
average daily dose	15.3 mg per day	(diazepam equivalents)
median time off drug	**3 months**	**(range: 1 to 65 months)**
# of domains affected:	12 (100%)	
Effect Size	**0.06 to 0.70**	(mean weighted ES = 0.41)

Meta-Analysis #2B – Persistent Effects of Former Long-Term Use		
# of studies	9	
total # of subjects	284	
mean age	42.7 years	(range: 21 to 75)
mean duration of drug use	8 yrs 11 mo	(range: 1 to 29 years)
average daily dose	16.7 mg per day	(diazepam equivalents)
mean time off drug	**11.6 months**	**(range: 3 to 65 months)**
# of domains affected:	11 (92%)	
Effect Size	**-1.50 to 0.26**	(mean weighted ES = -0.48)

Conclusion:

The chronic use of benzodiazepines for a period of 1 to 29 years (average use: 9 years) was associated with persistent cognitive deficits. Although the withdrawal of medication resulted in relative improvements for former users, these changes were generally mild or moderate in size. Moreover, when compared to healthy controls or normative data (standards), most domains of mental aptitude in former, chronic users of benzodiazepines showed evidence of unrelenting dysfunction. Given the duration of time away from benzodiazepine therapy (mean time: 11.6 months), these findings were suggestive of lasting and possibly irreversible drug effects.

Mood Stabilizers – (Lithium and Antiepileptic Drugs)

Dunn et al (2005) [59]

In an effort to identify the long-term risks of lithium, researchers in the United Kingdom performed a nested case-control study of all new cases of dementia occurring within a ten-year period.

Experimental Protocol

Researchers reviewed a national database of approximately three million patients drawn from 300 separate primary care practices in the U.K. A cohort of patients, aged 60 or older, was identified based upon new diagnoses of dementia (Alzheimer's disease, vascular dementia, or dementia of "uncertain cause") occurring between 1992 and 2002. Controls were randomly selected from the same database based upon age and the absence of dementia. Treatment histories were reviewed in order to corroborate the use of lithium within four years prior to the detection of dementia. [For unclear reasons, this particular investigation *excluded dementia associated with Parkinson's disease.*]

Findings

In this case-control study, lithium did not protect against dementia. In fact, *after adjusting for age*, a statistical analysis revealed that the *odds of developing dementia were doubled by exposure to lithium*. This risk was positively associated with cumulative drug dose (# of lithium prescriptions within the four-year period before dementia):

	Dementia	Non-Demented Controls
# of subjects	9954	9374
lithium exposed	47 (0.47%)	40 (0.43%)
pre-dementia duration of lithium	4.0 to 9.1 years	3.7 to 8.5 years

Age-Adjusted Odds Ratio for Developing Dementia After Lithium	
1-13 prescriptions	1.4
14-24 prescriptions	1.5
25-39 prescriptions	1.8
≥ 40 prescriptions	3.1

> *Conclusion:*
>
> Despite some limitations in study design, this review of ten-year data from a primary care database revealed a positive association between the use of lithium and the **emergence of dementia in patients aged 60 and above**. The duration of therapy in this study ranged from 4 to 9 years.
>
> Exposure to lithium increased the odds of developing dementia by ***40 to 200%*** (odds ratios: 1.4 to 3.1).
>
> For patients aged 60 or older, the risk of new-onset dementia was also positively associated with the ***cumulative dose of lithium*** (i.e., total number of prescriptions).

Nunes et al 2007 [60]

Although the dates of their investigation were not provided, researchers associated with the University of Sao Paolo recently reported the results of a hospital-based, cross-sectional survey which examined the link between lithium use and Alzheimer's disease.

Experimental Protocol

The Brazilian team reviewed the hospital records of 184 patients who satisfied the following criteria: 1) age of 60 years or older; 2) diagnosis of bipolar disorder;
3) receiving continuous treatment within the past 6 months (or longer); and 4) stable mood symptoms (as evaluated by the Hamilton Depression Rating Scale and the Young Mania Rating Scale).

Excluded from the study were individuals with the following profile: 1) Electro-convulsive Therapy (ECT) within the past 6 months; 2) any acute physical illness;
3) organic brain syndrome (not defined); and 4) any other major psychiatric syndrome.

Ultimately, 118 bipolar patients agreed to undergo further interviews in order to assess current levels of cognitive functioning. All subjects were evaluated through the use of four different screening instruments. These consisted of the Cambridge Examination for Mental Disorders in the Elderly, the Informant Questionnaire on Cognitive Decline in the Elderly, Petersen's diagnostic criteria for Mild Cognitive Impairment (aka, MCI), and the *DSM-IV (Diagnostic and Statistical Manual of Mental Disorders, 4th Ed.)* criteria for dementia.

Study Design	
identification of bipolar patients	n = 184
removal of subjects for various reasons	
exclusion criteria	- 17
refused to participate	- 12
not located	- 27
no longer living	- 10
total remaining subjects	n = 118

Findings

Of the total sample (n=118), screening revealed the following levels of cognitive functioning: 59% with normal cognition (n = 70), 21% with mild cognitive impairment or MCI (n = 25), and 19% meeting *DSM-IV* criteria for dementia (n = 23). Next, researchers removed data from four individuals who were diagnosed with vascular dementia. This left 114 subjects for a statistical analysis of dementia risk based upon past, current, or no exposure to lithium:

	Current Lithium	**Past Lithium**	**No Lithium**
# of subjects	66	33	15
mean age	67.4	69.1	69.1
mean duration of drug therapy	6 yrs	4.5 years	n/a
normal cognition	50 (75.8%)	14 (42%)	6 (40%)
MCI	13 (20%)	9 (27%)	3 (20%)
Alzheimer's disease	**3 (4.5%)**	**10 (30%)**	**6 (40%)**

Although at first glance, these data appeared to imply that continuous lithium might have protected some patients against the onset of Alzheimer's disease (relative to bipolar drug regimens which excluded lithium), this conclusion was not supported by the research design.

First, the Sao Paolo research team provided no clarification about the temporal onset of Alzheimer's symptoms. As other epidemiologists have been careful to explain, one must be vigilant about the possibility of a "sick quitter" or "depletion of susceptibles" artifact when evaluating data from former users. (In other words, the "current users" category in this experiment may not have captured all of the demented patients whose symptoms began while under the influence of lithium.)

Second, by excluding individuals with acute physical illnesses, organic brain syndrome (never defined), and/or recent Electroconvulsive Therapy, the investigators may have excluded an important subset of patients in whom lithium had been the cause of cognitive dysfunction *along with* each of these specific disqualifying conditions.

Third, the researchers excluded 15% of their original hospital cohort due to death or unstable mood symptoms. It is unclear how many of these individuals may have been suffering from lithium-related MCI or dementia.

Fourth, the evaluation methods in this experiment consisted of symptom checklists, rather than more stringent psychometric test batteries or examinations by neurologists. This may have minimized the detection of cognitive dysfunction and deflated the strength of the findings.

Conclusion:

The Brazilian study was hindered by numerous deficiencies in methodology. Despite these limitations, several findings were remarkable.

The overall prevalence of dementia among elderly bipolar patients was ***three times higher*** than the background rate in the general population **(19% vs. 7%, respectively)**.

Past or continuing exposure to lithium was associated with readily detectable cognitive deficits in a large percentage of drug users **(25% to 57%).**

Even after relatively short drug exposures (mean duration: 4.5 to 6 years), **Alzheimer's disease had emerged in 5 to 30% of the lithium consumers**.

Kessing et al (2008) [61]

In a retrospective case-control study, investigators in Denmark identified a sample of all national residents who had purchased specific pharmaceuticals within a ten-year span. The goal of this project was to identify any potential association between exposures to drug therapies and the post-treatment onset of new dementia.

Experimental Protocol

A random sample consisting of 30% of all Danish inhabitants was selected from the national healthcare database. Researchers reviewed the records of 1,503,415 subjects aged 40 and older, identifying all individuals with presumptive exposures to lithium, anticonvulsants, or antidepressants. **Cases were defined as those individuals who had received drug treatment** *prior to the onset of dementia.* The study incorporated data from the years 1995 through 2005.

Findings

Four findings were remarkable. First, approximately *3 to 5% of the prescription drug users developed dementia within a ten-year period.* Given the relatively young age of this cohort (median age: 53 to 56 years), these findings reflected early-onset dementia. Second, when compared to individuals who remained unexposed, *lithium users experienced a 47% higher risk (relative risk = 1.47) and a 56% higher incidence of dementia (5% vs. 3.2%).* Third, the association between lithium and dementia was dose-dependent. As cumulative exposures increased, so did the relative risk of dementia. Fourth, an *elevated rate of dementia (3.9%) was also documented among consumers of anticonvulsant medications* relative to the background prevalence of dementia in the entire sample of drug purchasers (3.2%):

	Total Cohort	**Lithium Exposed**	**No Lithium**	**AED**
# of patients	1,503,415	16,238	1,487,17	102,644
median age	NA	52.5	52.7	55.8
# with dementia	**48,483 (3.2%)**	**820 (5%)**	**47,663 (3.2%)**	**3969 (3.9%)**
Alzheimer's disease	9,996 (1%)	134 (1%)	9,862 (1%)	NA
other dementia	38,487 (3%)	686 (4%)	37,801 (3%)	NA
death	290,075 (19%)	3156 (20%)	286,919 (19%)	NA
Relative Risk of Dementia		1.47	1.00	NA

NA = not available
AED = antiepileptic drug

Stimulant Drugs (Drugs Used for ADHD)

Garwood et al (2006) [62]

In the first and – to date - the only investigation of its kind, researchers affiliated with the University of California San Francisco explored the association between numerous environmental, dietary, and medicinal chemicals and Parkinson's disease.

Experimental Protocol

Using a case-control, cross-sectional research design, the investigators began their study by identifying any patient who had been evaluated in their institution's neurology clinics between 2001 and 2004 for the following conditions: 1) *early-onset* Parkinson's disease; 2) amyotrophic lateral sclerosis (aka, Lou Gehrig's disease – a neuromuscular disease that progresses to respiratory failure and death); and 3) peripheral neuropathy (a condition which involves dysfunction of the nerves in the limbs, face, or trunk). After screening these individuals to solicit voluntary and unpaid participation, telephone interviews were performed by trained research assistants. Each subject (case) was matched, whenever possible, to a spouse or caregiver (control) who agreed to respond to the same survey. Participants were asked about past exposures to pesticides, herbicides, fungicides, well water, solvents, caffeine, tobacco, alcohol, and "recreational stimulants" (recreational amphetamines, Ecstasy, cocaine). The survey also included questions about **past treatment with 10 different kinds of pharmaceuticals**: low potency and high potency neuroleptics, diet aids, metoclopramide, opiates (pain killers), tricyclic antidepressants, aspirin, other non-steroidal anti-inflammatory medications, **prescribed amphetamines** (such as Adderall, Dexedrine), and **methylphenidate** (Ritalin). Finally, subjects were asked about the duration and frequency of drug use. ***Prolonged exposures*** were defined as *weekly use for more than one year*, or at least *two uses per week for three or more months*.

Limitations

Cases were included only if their neurological diagnosis had originated between the ages of 40 and 64. The reason for this restriction was to maximize the study's capacity to detect pertinent risk factors. (For example, knowing that experimentation with certain street drugs was a relatively recent societal phenomenon, the researchers excluded older patients who had progressed through childhood prior to the onset of America's amphetamine epidemic.) Another limitation involved a failure to distinguish the prevalence of amphetamine use which had occurred for medical rather than recreational purposes. Finally, as a cross-sectional study which relied solely upon voluntary telephone interviews, information about the quality and intensity of past exposures was subject to recall bias and other possible distortions (e.g., minimization of past use due to fear of reprisal or stigmatization; memory changes associated with prior drug use).

Findings

Of more than 400 potential candidates with qualifying conditions, 306 patients and 174 controls completed the telephone survey. Among these subjects were 158 individuals with early-onset Parkinson's disease and 76 pair-matched controls (spouses or caregivers) *Eleven percent of these patients – but only 3% of controls – had experienced prolonged use of prescription and/or recreational amphetamines*. In most cases, these exposures had occurred in the distant past (on average: 27 years before diagnosis):

	Parkinson's disease n = 158	caregiver or spouse n = 76
mean age at diagnosis	52	n/a
male	60%	33%
prolonged amphetamine exposure	**11%**	**3%**

A statistical procedure was performed to adjust for the influence of possible disease modifiers. After controlling for age, gender, and the use of several neuroactive substances (caffeine, alcohol, and tobacco), the research team determined the odds of exposure to amphetamine for patients with each of the three neurological conditions, relative to the pair-matched controls. *For those individuals who developed Parkinson's disease, the odds of a prior exposure to amphetamine were 8-times higher than for their respective controls:*

*Odds Ratios for Amphetamine		
Parkinson's disease	**8.04**	**p = 0.013**
ALS (Lou Gehrig's)	2.75	p = 0.178
peripheral neuropathy	2.74	p = 0.171

* odds ratio – Odds ratios provide estimates of relative risk in case-control studies, such as this one. In this particular investigation, the ratio (quotient) reflected the odds that a patient *with* a given disease had been exposed to a specific risk factor, divided by the odds that a patient *without* disease (control) had been exposed to the same hazard.

Implications

Although the findings of this study were influenced by several important methodological limitations – among them, the failure of the research team to clarify the rate and intensity of stimulant *abuse* – the observed association between amphetamine and Parkinson's disease had several crucial implications.

First, it was significant that this study featured patients between the ages of 40 and 64. This was a study of ***early-onset* neurological disease**. To the extent that amphetamine was a cause or contributor of dysfunction in these subjects, this study implied that amphetamine was a risk factor for neurodegenerative changes *far in advance of old age*.

Second, it was significant that most exposures to amphetamine had occurred remotely, 25 years or more before the emergence of disease. This finding implied that amphetamine (like many other drugs) had been capable of provoking delayed and/or persistent changes in the structure and function of the brain.

Third, it was significant that this cohort was selected on the basis of treatments received between 2001 and 2004. In other words, this was a group of patients whose childhood preceded America's ADHD scourge by 20 to 40 years. Given the fact that 11% of *these* individuals had experienced some treatment or experimentation with amphetamine, one can only wonder what the future might hold with respect to the incidence and prevalence of Parkinson's disease once the Dexedrine/Adderall/Ritalin Generation enters middle age.

Fourth, the results of this study were compatible with basic science. With a degree of consistency that is remarkable for neurobiological research, studies involving lab animals, cell cultures, and human autopsies have all converged upon the link between amphetamines (and other stimulants), the induction of oxidative stress, and the degeneration of neurons within and beyond the dopamine pathways of the brain.

Summary:

Epidemiology refers to the study of disease within populations. This chapter has presented a sample of the international research evidence which demonstrates that psychiatric drugs are a major risk factor for death and neurodegenerative disease.

One of the most impressive features of the extant literature has been the consistency of several findings. First, regardless of geographic setting, calendar year, or research design, epidemiological studies have repeatedly documented a clinically significant association between psychiatric drugs, mortality, and dementia. Second, these findings have pertained to pharmaceutical treatments irrespective of drug class. Third, these findings have held even when drug therapies have been consumed by younger patients, at average or low doses, and for relatively short durations of time.

To review:

Cooper and Holmes — case-control study of dementia risk and mortality
U.K. registry of new dementia cases between 1993 and 1995
identified patients aged 60 and older in South London
13% of dementia cases had prior treatment with psych drugs
risk of dementia: 3X to 4X higher for drug users vs. controls

Kessing — case-control study of dementia risk and mortality
Danish study of all psych inpatients who progressed to dementia
mean age at beginning of study: 30 to 50 years
1/3 died within 20 years, another 3 to 9% developed dementia
neuroleptics and bipolar drugs > 14X higher risk of dementia

Brandt-Christensen — case-control study of new Parkinson's disease and mortality
population-based study of drug purchases: 1995 to 1999
patients aged 30 and older buying antidepressants or lithium
15% of antidepressant & 10% of lithium users dead in five years
Parkinson's disease risk was 2X higher among drug users

Chen — prevalence study of antidepressants and stroke
used large U.S. database encompassing 45 million insured
identified all depressed patients between 1998 and 2002
13% of antidepressant users had new stroke within 5 years
76% of strokes were atypical
72% of strokes occurred among patients < 65 years of age

Kessing	case-control study of dementia risk and mortality
sampled antidepressant purchasers between 1995 and 2005
median age: 53 to 58 (all subjects: 40 years of age or older)
4 to 6% of antidepressant group developed dementia in 10 years
(vs. 2% background rate in general population)
2X to 5X higher risk of dementia with antidepressant drugs
20% of those taking SSRIs or older drugs died within study period
Kales	cross-sectional study of mortality in dementia (outpatients)
10,600 veterans followed in 2002 and 2003
20 to 30% died within first 12 months of exposure to neuroleptics (vs. 15% of demented patients who remained neuroleptic-free)
Ballard	prospective case-control study of mortality in Alzheimer's disease - U.K. study of residential and nursing home patients
4-year survival rate for chronic users of neuroleptics: 25%
4-year survival rate for placebo: 50%
Paterniti	prospective case-control study of risks for cognitive decline
sample of 60- to 70-year-olds in Nantes, France
evaluated at baseline, 2 years, and 4 years (1991-1997)
2X higher risk of cognitive decline for chronic users
25% to 45% of chronic users experienced significant decline
Laganoui	prospective case-control study of dementia risk
sample of Bordeaux region inhabitants aged 65 and above
8-year duration (1990-1997) with multiple assessments
(T3: 1992   T5: 1994   T8: 1997)
5 to 6.5% of benzodiazepine users developed dementia
(vs. 2.9% of those who remained non-exposed)
2X higher risk of dementia for benzo users vs. non-exposed
Barker	meta-analyses of 9 to 13 studies published in 1980-2000
goal: examine cognitive effects of benzodiazepines
study #1: chronic continuous use of benzodiazepines
associated with moderate to large cognitive decline
study #2A: intra-individual changes after stopping chronic use
mild to moderate improvement in most domains
study #2B: persistent effects after former use
former chronic users remained moderately impaired
when compared to normative data or healthy controls;
these changes still present 1 year after drug cessation

Dunn	case-control study of new-onset dementia
identified subjects aged 60 and above between 1992 and 2002
cases = dementia subjects with lithium use in preceding 4 yrs
lithium users had 2X higher odds of developing dementia
average duration of lithium before dementia: 4 to 9 years
Nunes	cross-sectional study of new-onset dementia
identified bipolar patients aged 60 and older
overall rate of dementia in bipolar patients: 19%
(vs. 7% in general population)
mean exposure to lithium: 4.5 to 6 years
5 to 30% of lithium users developed Alzheimer's disease
25 to 60% of lithium users displayed cognitive dysfunction
Kessing	case-control study of risk factors and incidence of dementia
sampled data from 30% of Danish inhabitants in national
register to identify purchasers of specific psychotropic drugs
period of study: 1995 to 2005
(drugs of concern: lithium, antidepressants, antiepileptic drugs)
median age: 53 to 56
3 to 5% of psychiatric drug users developed **early dementia**
risk of dementia was 47% higher with lithium
Garwood	cross-sectional, case-control study of risk factors for
three neurological conditions: PD, ALS, neuropathy
identified patients with early-onset disease
all patients were between the ages of 40 and 64
period of study: 2001 to 2004
odds of amphetamine use 8X higher for Parkinson's vs. controls
11% of early Parkinson's disease patients had used
amphetamine (vs. 3% of pair-matched controls)
exposure to amphetamine had been remote (> 25 years on
average)

The epidemiological studies in this chapter were of two primary kinds. Some of these studies identified specific cohorts of individuals and followed them forward in time, in an effort to discern the prevalence and determinants of dementia. Other studies worked backwards from known cases of dementia, in an effort to identify chemical exposures which were over-represented in patients relative to non-demented controls.

Both methods of inquiry converged upon the finding that antidepressants, antipsychotics, anxiolytics, mood stabilizers, and stimulants shortened the lifespan of their recipients by 10 to 30 years, and/or contributed to higher risks and higher rates of dementia, relative to age- and gender-matched community controls.

The implications of these findings are clear and alarming. It should concern physicians, regulators, and patients that **life-threatening, brain damaging drugs** are precisely the interventions whose use is now:

* mandated by State Medical Boards

* promoted by professional organizations, the news media, and medical journals

* encouraged by academicians and Key Opinion Leaders

* coercively enforced by the edicts and policies of insurance companies, treatment facilities, and the courts.

At the very least, psychopharmaceuticals are *enhancing* a pre-existing outbreak of dementia and death. At the very worst, they are *actively and independently engendering* this problem.

The medical literature presents consistent findings from around the world that the drugs which are used to treat dementia praecox and other forms of *assumed* brain disease are – ironically – associated with precocious (early) dementia. Furthermore, the data which have been published thus far necessarily minimize the true extent of the carnage and disability attributable to these chemotherapies, since dementia itself has generally been ignored on death certificates, and since patients who die *before* reaching old age are not likely to be "seen" as demented.

Although epidemiological investigations cannot, and do not, constitute definitive proof that psychiatric drugs are the *cause* of death and dementia, they assume a critical role for at least two reasons. First, they demonstrate the reality of treatment effects as they occur in the real world. In other words, they show that the present "gold standards" of therapy neither prevent, nor reverse, the processes which result in neurodegeneration. Second, they provide an essential context for interpreting the results of basic science experiments and other clinical research. It is this *totality of evidence* which must be considered when assigning causation to drug effects. As the next chapters explore, it is this *totality of evidence* which powerfully validates the epidemiological observations of psychiatric drugs as agents of brain destruction.

Chapter Four

Antidepressant Drugs

The Western world has never lacked for a chemical explanation of depression. Beginning with the Greek physicians of the 5th century B.C., health was believed to depend upon the proper balance of vital substances known as the four humors: black bile, yellow bile, phlegm, and blood.[1-3] When Hippocrates tended to a patient with depression or melancholia, he attempted to correct a black bile excess using a regimen of proper diet, sleep, and exercise. If this failed, he subjected his patient to increasingly aggressive interventions: the use of poisons (e.g., hellbore), surgery, and burning (cauterizing).[4]

The humoral theory of medicine came to dominate Europe until the advent of modern science in the 18th and 19th centuries. As a formal precept of healing, humoralism was replaced by bacteriology and pathophysiology. As an informal precept, however, humoralism continued to inform the practice of psychiatry and psychopharmacology.

With the discovery and testing of two major classes of medication in the mid-1950s – the monoamine oxidase inhibitors, and the tricyclic antihistamines – new theories of depression were born. Based upon the effects of pharmaceuticals which boosted the availability of chemicals in the nerve synapse (the space between cells of the brain), researchers proposed that depression must involve a deficit in the synaptic levels of two substances: norepinephrine (a catecholamine) and serotonin (an indoleamine).[5-7] With the passage of time, other neurotransmitters were added to the list of depression-determining substances (e.g., dopamine, GABA, glutamate, glycine, acetylcholine).

***Although a pre-existing imbalance of monoamines – serotonin, dopamine, or norepinephrine – has never been clearly or consistently confirmed as a cause of depression*,**[8] and although there remains no biological test which can be used for diagnosis, an empirical, "pick and choose" philosophy is still emphasized in the textbooks and journals of psychiatry. Most recently, this trend has been updated and extended by the *neurogenesis theory* of depression and antidepressant activity.

The Neurogenesis Theory of Depression and Antidepressant Activity

The term **neurogenesis** refers to the birth of new neurons. Until the 1960s, the prevailing dogma in medicine held that neurogenesis was a time-limited event which did not, and could not, occur in the adult brain. Textbooks assured physicians that the body's complement of neurons was distributed *in utero* and remained fixed in number until old age. This belief was strongly reinforced by the observations of early researchers, such as Ramon y Cajal, who failed to detect any evidence of new neurons in the brain specimens which he viewed through microscopes of the early 20th century.

With innovations in scientific theory, instrumentation, and experimental techniques, various investigators – beginning with Altman and Das in 1965 – documented the existence of **new neurons** in the brains of adult mammals. Despite the fact that these findings were replicated at intervals by other research teams, it was not until the 1990s that scientists caught the attention of the medical world and the news media, proposing that neurogenesis might be harnessed for healing. Fundamental to this theory was the belief that humans – just like rodents – must be capable of regenerating neurons from primitive cells in the **subgranular zone of the hippocampus** and possibly the **subventricular** regions of the brain:

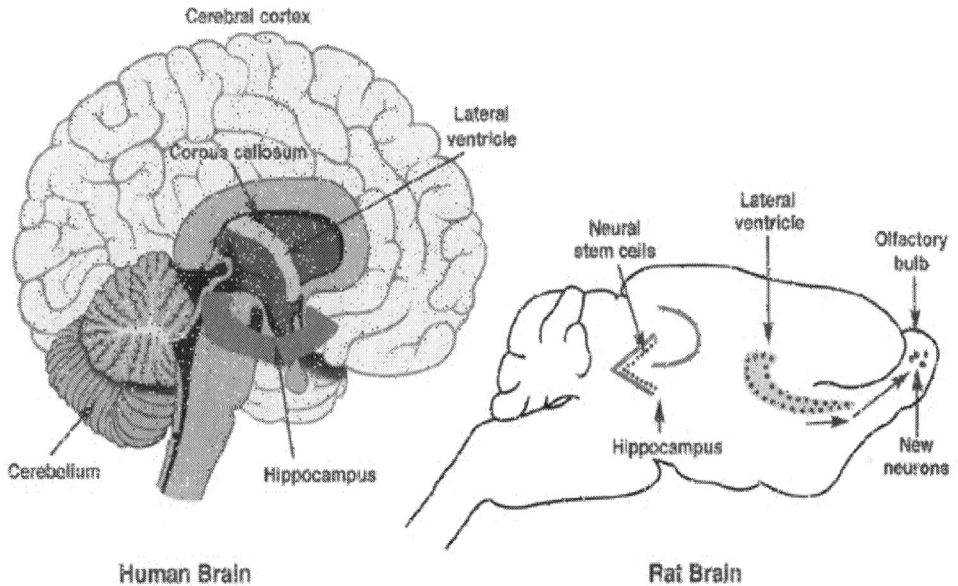

Source: National Institutes of Health, National Institute on Alcohol Abuse and Alcoholism, public domain.

These pictures show the location of the hippocampus and the lateral ventricles in the human brain (left) and in the brain of the rat (right).

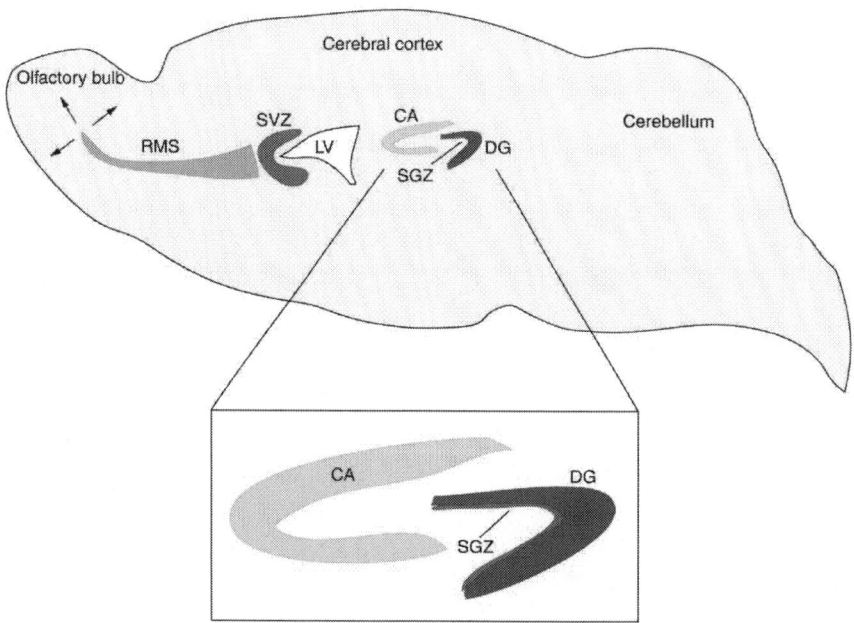

Source: Pozniak and Pleasure (2006), *Genome Biology* 7:207 (open access).

This image depicts the brain of an adult mouse. The **subventricular zone** or **SVZ** gives rise to new neurons in adult rodents. The other source of new neurons is the **subgranular zone** or SGZ, located within the hippocampus (see box above: thin stripe = SGZ, thick curve = dentate gyrus).

Knowing the importance of the hippocampus for the human functions of language, learning, and memory, and understanding the detrimental effects of injuries and diseases which strike this area, neuroscientists were understandably excited about the prospects of manipulating cell growth in this area. When subsequent investigations revealed that antidepressants appeared to stimulate neurogenesis in the hippocampus of the rodent brain, certain voices in psychiatry were quick to jump upon this finding as a new rationale for the cause and treatment of depression.

According to the *neurogenesis theory of depression*, dysphoric mood is caused by shrinkage of the hippocampus. The reasoning behind this proposition depends upon a series of arguments and other research evidence, which can be diagrammed in the following way:

Stress causes high cortisol and depression.
High cortisol causes hippocampal atrophy.
Therefore, hippopcampal atrophy causes depression.

Using the symbols of formal logic, however, a fundamental error in reasoning – known as **the fallacy of joint effects** – will hopefully be clear:

A causes B and C.
B causes D.
Therefore, D causes C.

Here is another example of the same fallacy or "mistake" in reasoning:

Rainstorms cause wet brakes and mud.
Wet brakes cause car accidents.
Therefore, car accidents cause mud.

Today, it seems that American physicians are either unable or unwilling to challenge the assumptions of the monoamine and structural theories of depression. Because the use of antidepressants continues to be based upon false premises and false promises, it is the purpose of this chapter to present just a small portion of the research evidence which contradicts these prevailing views.

Postmortem Studies of Animals

Study #1   Kalia et al   Philadelphia, Pennsylvania - Serotonin Fibers in Rats

Based upon past research evidence which had shown a link between serotonin depletion and structural changes in nerve terminals (boutons), a group of investigators affiliated with Thomas Jefferson University undertook a comparison of several medications which affect the release or reuptake of serotonin.[9] The working hypothesis in this experiment was that drugs which deplete serotonin would result in anatomical defects.

The experiment involved the use of male Sprague-Dawley rats. Following a four-day exposure to various chemicals, the animals were euthanized and their brains were examined for structural changes. Each rat received twice-a-day injections of a drug-free liquid or one of the chemicals listed below:

---

1) 5,7-dihydroxy-tryptamine (5,7-DHT): a potent neurotoxin used in animal research to intentionally produce brain lesions

2) **m**ethylene-**d**ioxy-**m**ethamphetamine (aka, MDMA): a euphoria-inducing stimulant known on the streets as "Ecstasy"

3) dexfenfluramine: a former appetite suppressant/weight-loss drug (removed from the U.S. market in 1997 because of its cardiotoxicity)

4) fluoxetine: a drug which blocks the reuptake of serotonin into neurons (a so-called "selective" serotonin reuptake inhibitor or SSRI; in reality, fluoxetine has many other effects upon the brain)

5) sertraline: like fluoxetine, another drug which blocks the reuptake of serotonin into nerve endings (another so-called "selective" serotonin reuptake inhibitor or SSRI)

6) sibutramine: a weight-loss drug which blocks the reuptake of serotonin and norepinephrine into nerve endings (serotonin norepinephrine reuptake inhibitor or SNRI)

Experimental Procedure

Lab animals were divided into three subgroups. Two groups of animals served as controls, allowing the investigators to replicate the effects of well known toxins. The third group participated in an experimental protocol which compared "high" and "low" doses of each prescription drug. **For the sake of emphasizing the findings which were most relevant for humans**, the following discussion will focus upon the low-dose subgroups.

Control Group #1   Ecstasy

Twelve animals received subcutaneous injections of MDMA (Ecstasy) over a period of eight hours (20 mg/kg every two hours x four). Half of these animals were euthanized and examined 18 hours after the last dose. The remaining animals were euthanized and examined 30 days later. The purpose of this control group was to verify the anatomic consequences of drug-induced serotonin depletion.

Control Group #2   5,7-DHT (Chemical Lesioning)

Six animals received single injections of 5,7-DHT (50 ug) into each lateral ventricle of the brain. The purpose of this control group was to verify the appearance of serotonin neurons in response to another known neurotoxin.

Experimental Protocol: (fluoxetine, sertraline, dexfenfluramine, sibutramine)

The remaining animals were divided into low- and high-dose groups, consisting of 6 medicated subjects and 6 unmedicated controls. **Daily intraperitoneal injections of the medications (or control fluid) were administered for a period of four days**. Most of the animals were then euthanized and examined 18 hours after the last drug dose. A few animals were euthanized and examined after a recovery period which lasted for 30 days.

Tissue Analysis

Following the surgical sacrifice of each animal, brain tissue was extracted and prepared for chemical and morphological (structural) analysis. First, tissue specimens from each animal were processed in order to measure the **total content of serotonin**. Second, slices of brain tissue were treated with an antibody labeling technique (immunocytochemistry) in order to **specifically identify the cells which contained serotonin**. These fibers were then visualized using light and dark field microscopy.

Background Information

Why were these methods and chemicals chosen?

A significant body of research by many other scientists had already connected the street drug, Ecstasy (MDMA), with structural changes involving the monoamine **nerve terminal**. The researchers in this experiment were curious to see if they could replicate this finding with Ecstasy or any of the other drugs.

In order to appreciate this focus of concern, it is important to consider the following features of brain anatomy. Recall, from chapter two, that one of the two major cell types is the **neuron**. Each neuron contains a tentacle-like extension (the axon) which culminates in **nerve terminals** (sometimes referred to as "knobs" or "boutons"):

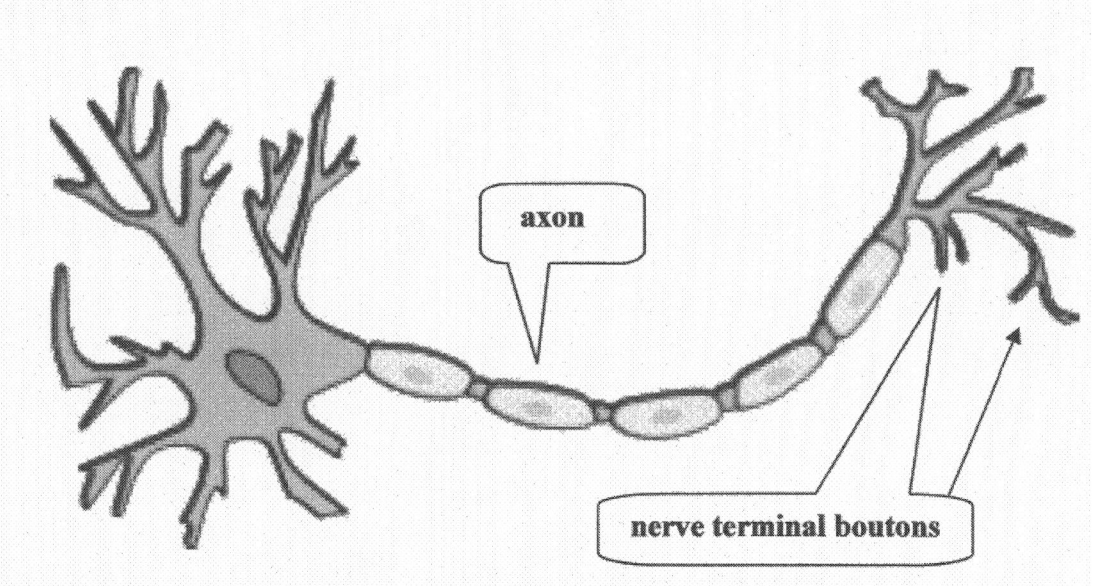

Source: Wikimedia commons, public domain (labeled by author).

The importance of the **nerve terminal** in considerations of drug toxicity arises from the fact that the each one is a critical cell zone which mediates one or more steps of neuronal signaling: neurotransmitter synthesis, release, reuptake, and metabolism. Within each nerve terminal, drugs and their metabolites also kick off the complex sequences of events which result in cell death.

What do the selected medications do to the brain?

Chemical communication between neurons and other cells involves the release of neurotransmitters – in this case, serotonin – into a gap known as the synapse or synaptic cleft. From this gap, serotonin either travels to special docking stations on the membranes of other cells, diffuses away to more distant regions, or recycles back into the originating neuron. When this recycling process occurs, serotonin travels back inside the nerve terminal through a special unit called the **Serotonin Reuptake Transporter** or **SERT** (aka, the serotonin reuptake pump).

All of the drugs which were used in the Philadelphia experiment (fluoxetine, sertraline, and sibutramine) block the Serotonin Reuptake Pump. Initially, this results in the accumulation of serotonin (5-HT) within the synapse (as shown below):

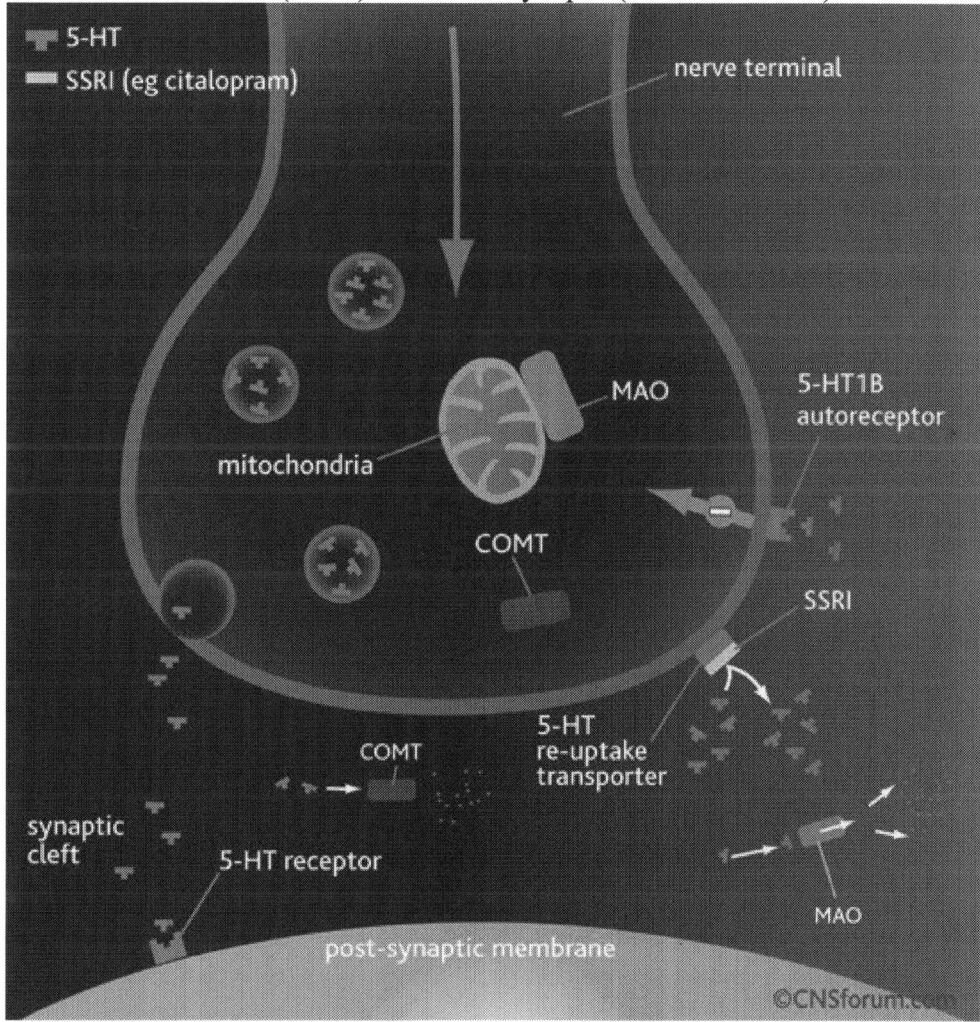

Source: CNS forum, The Lundbeck Institute - http://www.cnsforum.com/educationalresources/

Two of the drugs used in this experiment – dexfenfluramine and Ecstasy (MDMA) – not only block the reuptake of serotonin, but also force the release of *serotonin out of the nerve terminal through this pump.*[10]

What was special about the methods used in this investigation?

First, the Philadelphia study involved the microscopic analysis of ***axonal structures following exposure to fluoxetine and sertraline***. Although anatomic analyses had been performed previously with respect to *street drugs*, other scientists had never pursued this kind of inquiry using the pharmaceutical inhibitors of serotonin reuptake.

Second, this particular investigation involved the use of a ***special labeling technique*** (immunocytochemistry) and two different viewing methods (**light field** and **dark field** microscopy) in order to facilitate the detection of structural changes. Brain samples were treated chemically with antibodies to serotonin. Then, they were examined using a microscope. **Light field** imaging refers to the process of passing visible wavelengths of light through a series of lenses in order to focus a cone-shaped beam in the plane of the specimen. As the light passes through the specimen and changes its speed and direction, an image of the target is produced.[11] In **dark field** microscopy, a different system of illumination is applied. In the latter case, an inverted (hollow) cone of light is directed toward the specimen in such a way that light scatters *around* the target.[12-13] This makes the object appear light against a dark background. The potential advantage of combining both methods is the detection of structural features which are not discernable with standard techniques.

Third, the experiment involved a ***dosing protocol which was relevant for humans***. According to their published narrative, the research team envisioned (and apparently intended) an animal toxicity study by employing drug quantities that were 10 to 100 times greater than human doses on a dose-per-body-weight (mg per kg) basis. However, their strategy failed to consider the metabolic differences between rodents and humans which justified the use of higher and more frequent drug dosing.

While it is often difficult to extrapolate the findings of animal research to the human situation, a commonly accepted practice involves the determination of *dose equivalence* between species. This is achieved by delivering a drug dose to animals which yields a bloodstream concentration equal to (or less than) the therapeutic drug level in humans.

The **low-dose fluoxetine group** in the Philadelphia experiment received intraperitoneal injections (**11.4 mg/kg**) twice a day for just four days. On the basis of *dose-related drug levels and elimination half-lives*, this was biologically comparable to the standard oral dose range in humans (20 to 80 mg per day).[14-18]

Antidepressant Pharmacokinetics: rat vs. human

fluoxetine doses and blood levels

	dose	route	frequency	*blood level	time of blood draw
rat:	10 mg/kg	IP	one time	372 ng/mL	2 hrs after dose
	10 mg/kg	p.o.	one time	186 ng/mL	3 hrs after dose
	12.5 mg/kg	p.o.	b.i.d. x 3 wks	2786 ng/mL	4 hrs after last dose
human:	20-60 mg	p.o.	once a day	261-2321 ng/mL	after 6 wks of use

** fluoxetine and norfluoxetine elimination half-lives

		fluoxetine	norfluoxetine
rat:	IP	4 hrs	10 to 12 hrs
	intravenous	4 to 6 hrs	15 to 17 hrs
	p.o.	7 to 13 hrs	14 to 16 hrs
human:	p.o.	4 to 6 days	4 to 16 days

* blood level = fluoxetine + norfluoxetine (the latter, a major metabolite of fluoxetine)

** elimination half-life: time required to reduce the drug concentration by 50%

b.i.d. = twice a day
IP = intraperitoneal (injection into the intra-abdominal cavity)
p.o. = per orum (by mouth)

The **low-dose sertraline group** in this experiment received intraperitoneal injections (28.6 mg/kg) twice a day for a period of four days. On the basis of dose-related drug levels and elimination half-lives, the regimen given to these animals should have produced blood levels comparable to those which have been measured in human patients (e.g., 300 mg per day): [19-22]

Antidepressant Pharmacokinetics: rat vs. human

sertraline doses and blood levels

	dose	route	frequency	*blood level sertraline	primary metabolite
rat:	25 mg/kg	IP	one time	310 ng/mL	110 ng/mL
	10 mg/kg	p.o.	daily x 3 mo	100-270 ng/mL	NA
human:	200 mg	p.o.	daily	190 ng/mL	140 ng/mL
	300 mg	p.o.	daily	99-309 ng/mL	135-423 ng/mL

** sertraline and desmethyl-sertraline elimination half-lives

		sertraline	primary metabolite
rat:	intravenous	6.5 hrs	10.5 hrs
human:	p.o.	20-37 hrs	65 hrs

* blood level = maximum concentration (C_{max})

** elimination half-life: time required to reduce remaining blood level by 50%

IP = intraperitoneal (injection into the intra-abdominal cavity)
NA= not available
p.o. = by mouth
primary metabolite in rats and man = N-desmethyl-sertraline

[Note: N-desmethyl-sertraline is a mildly active metabolite in humans.]

Findings

All of the drugs in this experiment reduced the serotonin content of the prefrontal lobe. *Even low doses reduced the serotonin content of this brain region, based upon comparisons of medicated animals and the drug-free controls*:

Prefrontal Cortex *Serotonin			
	# of animals	low dose	high dose
fluoxetine	6	284.0	264.8
control	6	323.5	316.8
sertraline	6	306.3	213.0
control	6	319.8	311.3
5,7-DHT nerve toxin	6	only one dose given:	112.57
control	6		335.12
fluoxetine low dose vs. control		**12% decline**	
fluoxetine high dose vs. control		**16% decline**	
sertraline low dose vs. control		**4% decline**	
sertraline high dose vs. control		**32% decline**	
5,7-DHT nerve toxin vs. control		66% decline	

* The published report did not provide units for these measurements. Presumably, these values referred to nanograms of serotonin per milligram of brain tissue (ng/mg).

In addition to the reduction in brain levels of serotonin, **anatomic deformities** were observed in response to the antidepressants and other agents. These changes occurred in a dose-dependent manner (the higher the dose, the more numerous the defects). Furthermore, these changes were detected in multiple regions of the brain: the frontal cortex, the occipital cortex, the hippocampus, and several regions of the brainstem (superior and inferior colliculi).

Low and high doses of fluoxetine and sertraline resulted in four major kinds of **anomalies**:

> - axonal thickening (thicker fibers)
> - swollen profiles (swollen nerve terminals and varicosities)
> - corkscrew malformations (abrupt hairpin turns or "kinks")
> - axonal truncation (shorter axons)

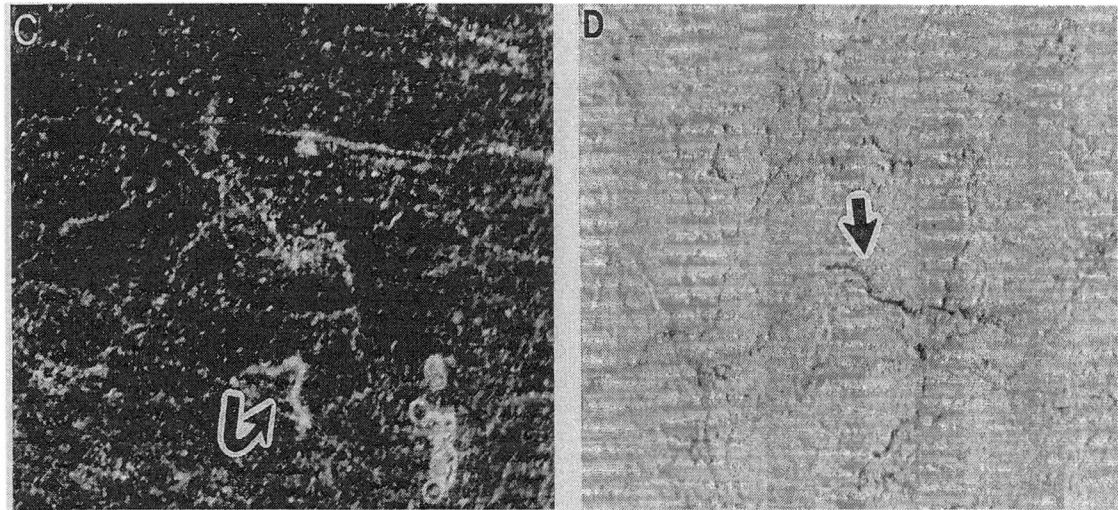

Reprinted from *Brain Research* with permission from Elsevier.

Structural Abnormalities in the Prefrontal Cortex Following Low-Dose Fluoxetine

Dark field image (left) shows the abnormal kinking or corkscrew appearance of a serotonin neuron. Light field image (right) shows abnormal truncation (shortening) of a serotonin axon.

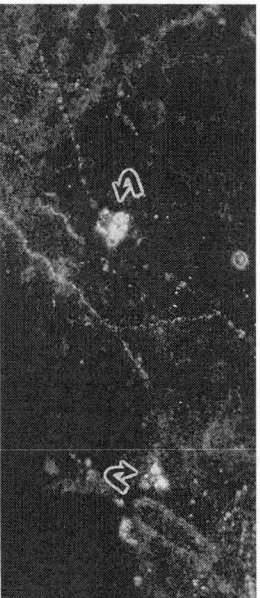

Reprinted from *Brain Research* with permission from Elsevier

Abnormal nerve terminals in the Prefrontal Cortex After Low-Dose Fluoxetine

This dark field image shows abnormal swelling of neuronal boutons (nerve terminals).

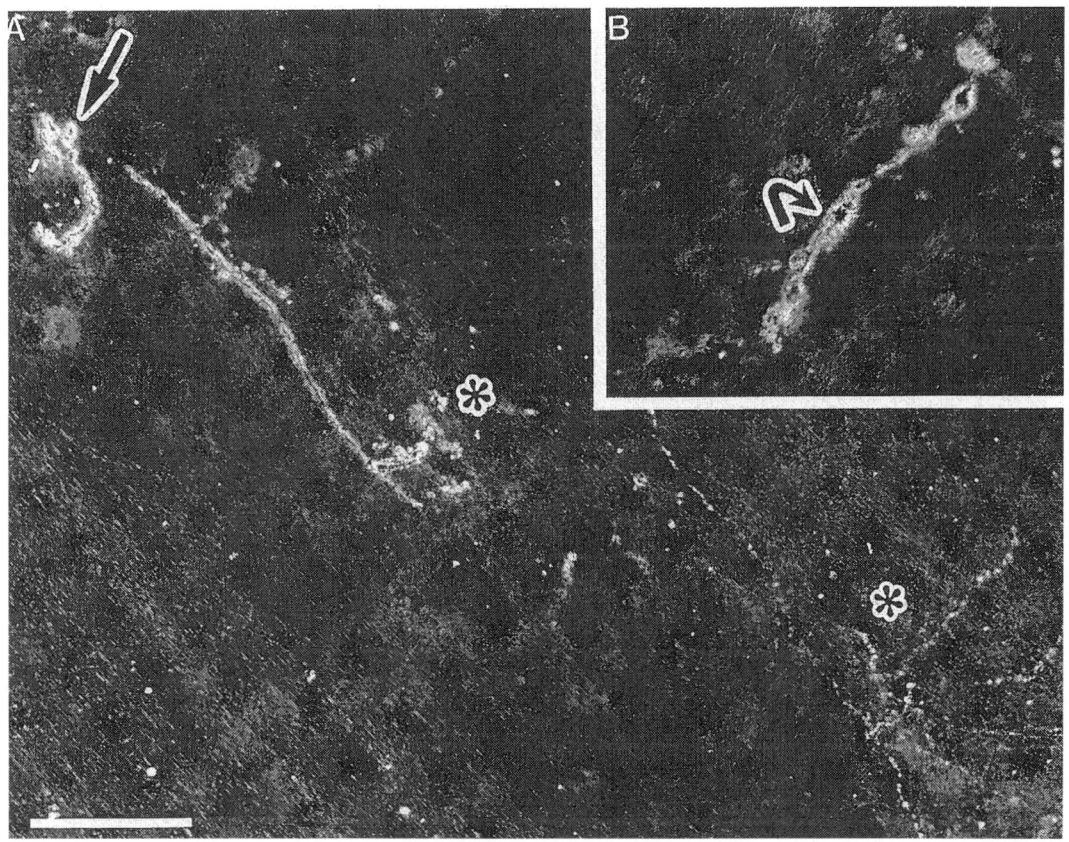

Reprinted from *Brain Research* with permission from Elsevier

Structural Abnormalities in the Prefrontal Cortex After Low-Dose Sertraline

Left: Dark field image reveals axonal truncation and kinking (arrow, left) as well as numerous regions of abnormal swelling (asterisks).

Inset: Dark field image reveals abnormal swelling along the length of an axon.

The significance of these findings hinges upon their clinical relevance for humans. In this regard, it is important to appreciate the similarity between the defects observed in this experiment and the abnormalities which have only recently been confirmed in human subjects who had suffered from neurodegenerative disease.

In 2008, researchers affiliated with New York University reported the results of their postmortem analysis – presumably, the first in history – which compared *the integrity of the serotonin network* in three different neurological disorders: Parkinson's disease, frontal lobe dementia, and Lewy body dementia.[23] Using the technique of immunocytochemistry to identify serotonin axons in multiple regions of the human brain, the scientists documented diffuse serotonin abnormalities in all three forms of dementia, including:

> reduced fiber density
> abnormal clustering of fibers
> swollen and twisted varicosities (protrusions from the sides of axons)
> splayed fibers with irregular shape (bloated and twisting axons)

Interestingly, at least half (and as many as 75%) of the patients in the New York study had received treatment with antidepressants. While this fact certainly does not prove that the antidepressants were the cause of the documented defects, it surely suggests that the antidepressants which these patients had consumed were incapable of reversing deterioration of the serotonin pathways in the brain.

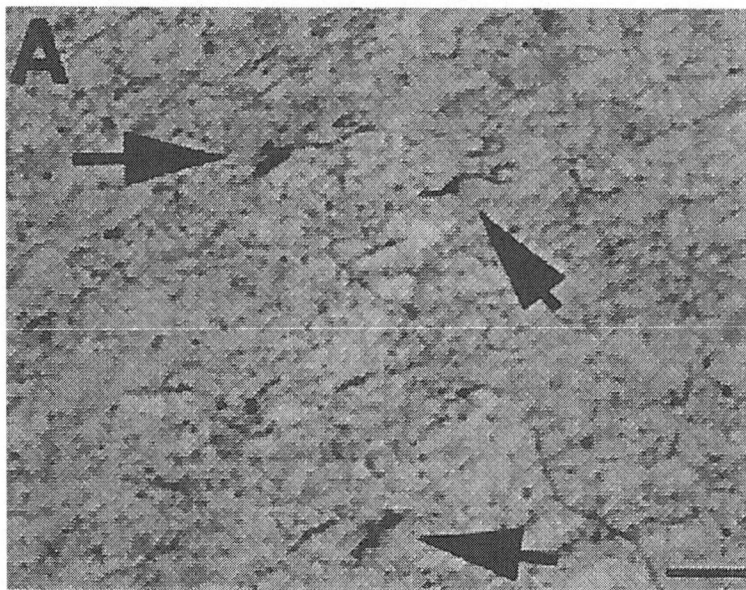

Reprinted from *Brain Research* with permission of Elsevier.

Abnormal Serotonin Axons in Parkinson's Disease (Upper Midbrain)

This image depicts abnormal axonal swelling (arrows), axonal truncation, and kinking.

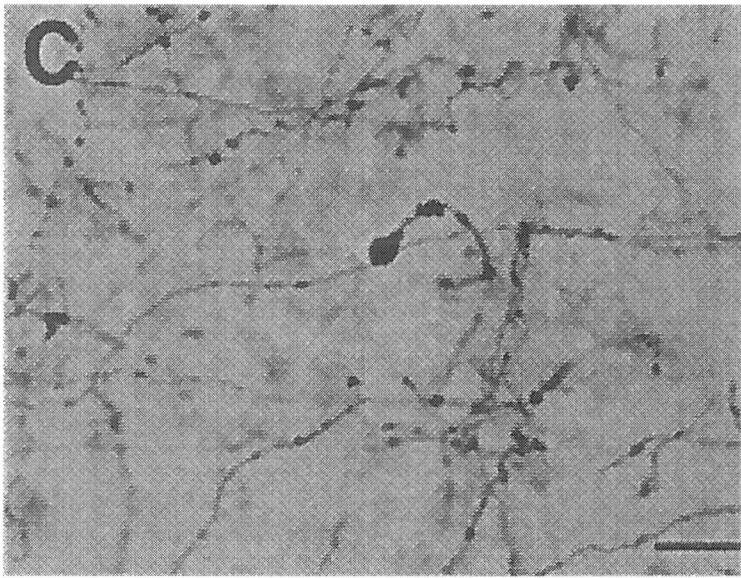

Reprinted from *Brain Research* with permission of Elsevier

Abnormal Serotonin Fibers in Parkinson's Disease (Hippocampus)

This image shows a neuron (center) with an abnormally enlarged nerve terminal and enlarged axonal varicosities.

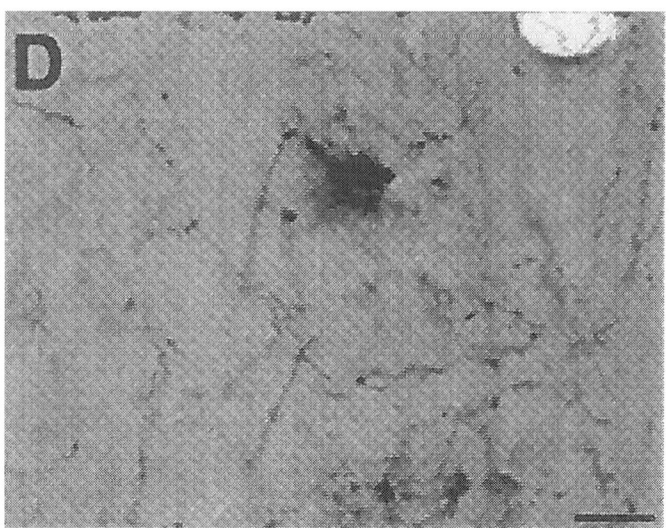

Reprinted from *Brain Research* with permission of Elsevier.

Abnormal Axons in Parkinson's Disease (Entorhinal Cortex)

This image shows axonal "splaying" or bloating of a serotonin nerve terminal (center). Other abnormalities include a marked depletion of serotonin fibers, as demonstrated by the pale areas in the background. Like the hippocampus to which it connects, the brain region shown here (entorhinal cortex) plays a key role in spatial and verbal memory.

In Sum:

To date, the investigation by Kalia et al is the only animal study which has analyzed the appearance of serotonin nerve fibers following exposure to pharmaceutical inhibitors of serotonin reuptake.

Unexpectedly, these researchers found that fluoxetine and sertraline were toxic to serotonin neurons in multiple regions of the brain.

Based upon dose-equivalent levels of medication in the peripheral bloodstream, the antidepressant regimens used in this experiment were comparable to those consumed by humans.

Although the four-day exposures to fluoxetine and sertraline did not fully deplete serotonin, both drugs contributed to reductions of serotonin content (12% and 4%, respectively) within the prefrontal cortex of the brain.

Fluoxetine and sertraline produced the same quality and same distribution of structural defects caused by two other drugs (Ecstasy and dexfenfluramine) whose toxicities preclude sale on the U.S. market.

Damage to the structural integrity of the serotonin network cannot be regarded as a trivial phenomenon. Most notably, the anatomical changes which occurred in rats in response to fluoxetine and sertraline were similar to the defects observed in humans in association with several neurodegenerative conditions.

Study #2   Czeh et al   Gottingen, Germany   -   Astrocytes in Tree Shrews

In 2006, researchers in Gottingen, Germany reported the results of an experiment which explored the anatomic consequences of stress and medication (fluoxetine) in the brain of the tree shrew.[24]

Source: Tree shrew, public domain, http://farm3.static.flickr.com

Based upon a controversial theory which dates back to 1945, some biologists have considered this squirrel-like creature to be an evolutionary relative of *homo sapiens*. For those who embrace this view of a common lineage (placing the tree shrew in the order *Primates* rather than the order *Scadentia*), this small animal has become a valued alternative to monkeys in lab research.[25-26] This value arises from the fact that the predictive validity of animal experiments, relative to humans, depends upon similarities in genes, anatomy, and physiology.

Experimental Procedure

The German experiment involved the examination of twenty-two adult male tree shrews (aged 9 months). Animals were divided into four subgroups in order to compare the effects of environmental and chemical stimuli. Half of the animals were exposed to a psychosocial stress procedure (with and without drug treatment). The remaining animals served as "no stress" controls (again, with and without drug treatment). A schematic of the study design is shown below:

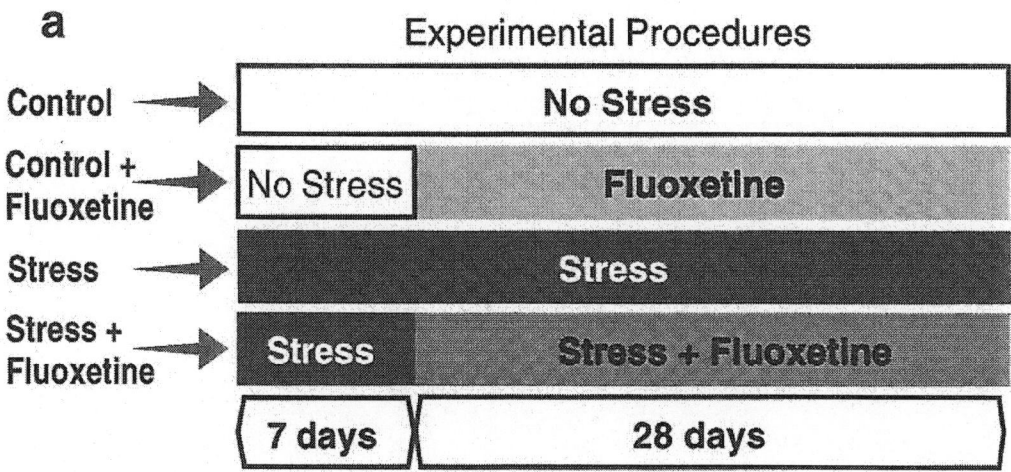

Reprinted with permission of Macmillan Publishers Ltd. *Neuropsychopharmacology*, B. Czeh, M. Simon, B. Schmelting, C. Hiemke, and E. Fuchs, "Astroglial Plasticity in the Hippocampus is Affected by Chronic Psychosocial Stress and Concomitant Fluoxetine Treatment," volume 31, p. 1617, copyright 2006.

Animals assigned to the medication arms of the study received oral doses of fluoxetine (15 mg/kg) on a daily basis, for a total treatment period of 28 days. This dose was carefully selected for its capacity to produce drug metabolite blood levels in the tree shrew (norfluoxetine: 81-634 ng/mL) which were comparable to those observed in humans (norfluoxetine: 73 to 1,055 ng/mL).[27]

Following the completion of the experimental protocol, the animals were euthanized. Brains were excised and prepared for several kinds of analyses. In the analysis reported here, the research team focused upon the identification, quantification, and measurement of astrocytes as visualized within samples of the hippocampus. [For the technically inclined, the specific techniques involved immunocytochemistry using a mouse monoclonal antibody against glial fibrillary acidic protein or GFAP, followed by stereological assessment with light microscopy.]

Background Information

To review, the brain consists of two major populations of cells: neurons and glia. The **astrocyte** is one of four subtypes of glia. Its many functions include the regulation of metabolism, cerebral blood flow, and electrical and chemical transmission.

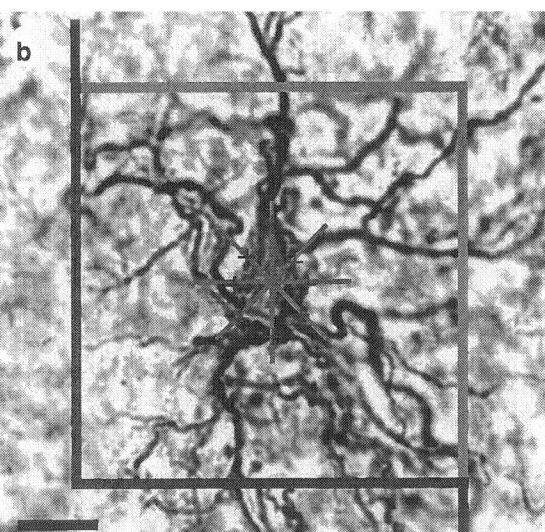

Reprinted with permission of Macmillan Publishers Ltd. *Neuropsychopharmacology*, B. Czeh, M. Simon, B. Schmelting, C. Hiemke, and E. Fuchs, "Astroglial Plasticity in the Hippocampus is Affected by Chronic Psychosocial Stress and Concomitant Fluoxetine Treatment," volume 31, p. 1617, copyright 2006.

This image shows an astrocyte obtained from the hippocampus of a tree shrew in the German experiment. Note the soma, or cell body, in the middle (triangular shape) and the numerous extensions which protrude like the points of a star.

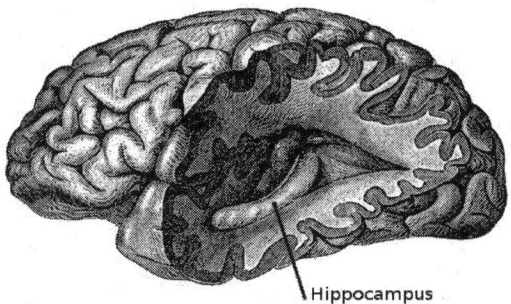

Source: Wikimedia Commons, public domain. File:Gray739-emphasizing-hippocampus.png

In humans, the **hippocampus** is a subcortical structure located deep within the temporal lobe. Through its rich connections to the limbic (emotional) and cortical (higher cognition) systems of the brain, the hippocampus plays a major role in learning, memory, and mood.

Findings

Following the four-week period of fluoxetine exposure, the tree shrews manifested several anatomic abnormalities. (The focus of the publication summarized here was the identification of changes in only one cell type – the astrocyte.)

First, there was a ***decrease in the volume of the hippocampus***. Remarkably, researchers identified hippocampal atrophy in fluoxetine-treated animals exposed to "stress" and "no stress" conditions:

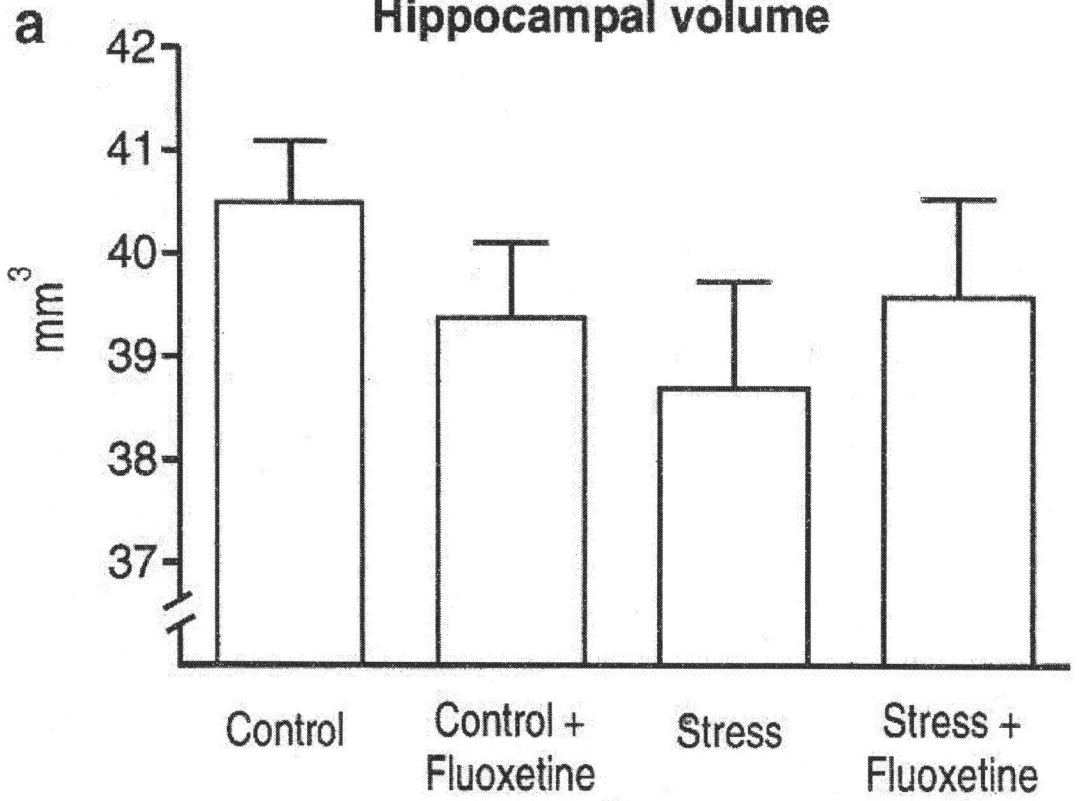

Reprinted with permission of Macmillan Publishers Ltd. *Neuropsychopharmacology*, B. Czeh, M. Simon, B. Schmelting, C. Hiemke, and E. Fuchs, "Astroglial Plasticity in the Hippocampus is Affected by Chronic Psychosocial Stress and Concomitant Fluoxetine Treatment," volume 31, p. 1622, copyright 2006.

Second, the investigators confirmed a statistically significant *decrease in the somal volume of each astrocyte* (reduced volume of the cell body). This was seen in all of the subgroups with the exception of the no-stress, non-medicated controls. Notably, the animals assigned to the fluoxetine/stress-free condition developed astrocytic shrinkage which was almost identical to the effects of stress alone. Furthermore, *the combined effects of fluoxetine and stress resulted in the most severe reduction in somal size*:

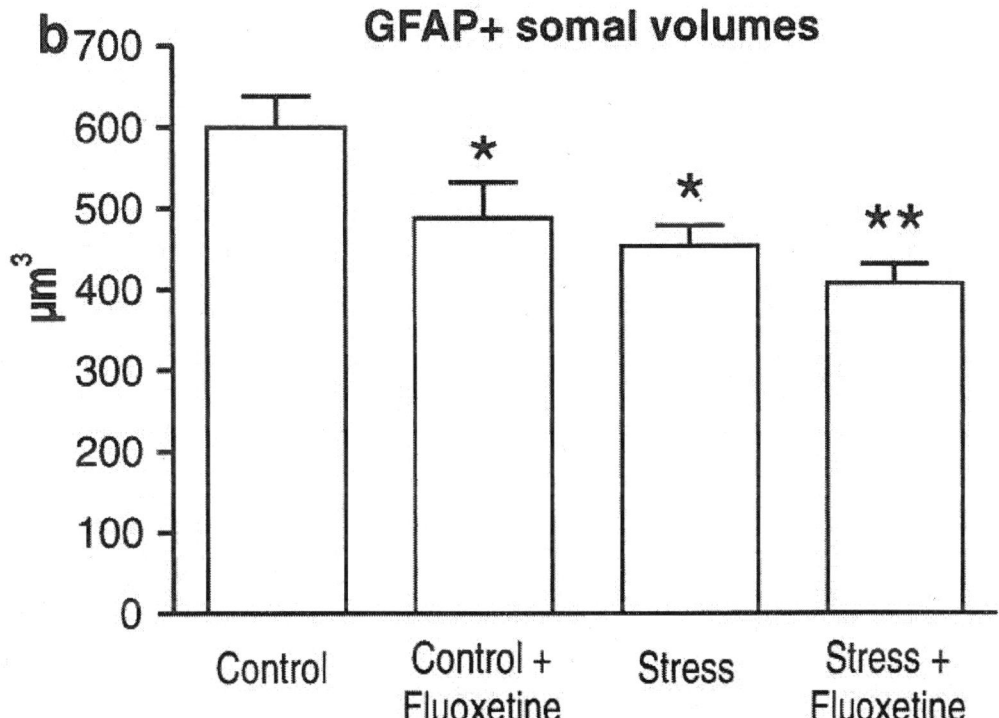

Reprinted with permission of Macmillan Publishers Ltd. *Neuropsychopharmacology*, B. Czeh, M. Simon, B. Schmelting, C. Hiemke, and E. Fuchs, "Astroglial Plasticity in the Hippocampus is Affected by Chronic Psychosocial Stress and Concomitant Fluoxetine Treatment," volume 31, p. 1622, copyright 2006.

The report of the tree shrew experiment included a number of limitations. For example, it would have been important to know if the German research team had found other structural abnormalities consistent with degeneration, such as beaded, curly, or retracted glial processes. Similarly, it would have been important to know if the astrocytes which they observed were *mature* or *immature* cells.

Nevertheless, it is important to appreciate the broader implications of these discoveries in terms of their relevance to dementia. Although Czeh et al did not mention it, astrocytic degeneration has been detected in a recent postmortem investigation of patients with frontal and temporal lobe dementias. Moreover, in that study, changes in the size and number of astrocytes corresponded to the progression and severity of disease.[28]

In Sum:

In a four-week study involving tree shrews, exposure to the antidepressant, fluoxetine, resulted in the shrinkage of the hippocampus (reduced volume).

Fluoxetine was also associated with astrocytic atrophy in this brain region.

Given the fact that astrocytes are ordinarily quite resilient to the effects of chemicals and injury –usually responding with proliferation and an *increase* in cell size – the results of this experiment implied that fluoxetine was a glial toxin (at least in the hippocampus).

The clinical significance of these findings arises, in part, from the fact that astrocytic degeneration has recently been documented in human subjects suffering from frontal and temporal lobe dementias. In those individuals, the intensity of astrocytic pathology corresponded to the progression and severity of dementia symptoms.

Background Information

Neurogenesis and Neurotoxicity in the Temporal Lobe

In the past five years, standard textbooks of psychopharmacology have been revised in order to highlight the **neurogenesis theory** of depression and antidepressant activity.[29-30] These popular resources feature illustrations similar to the one below:

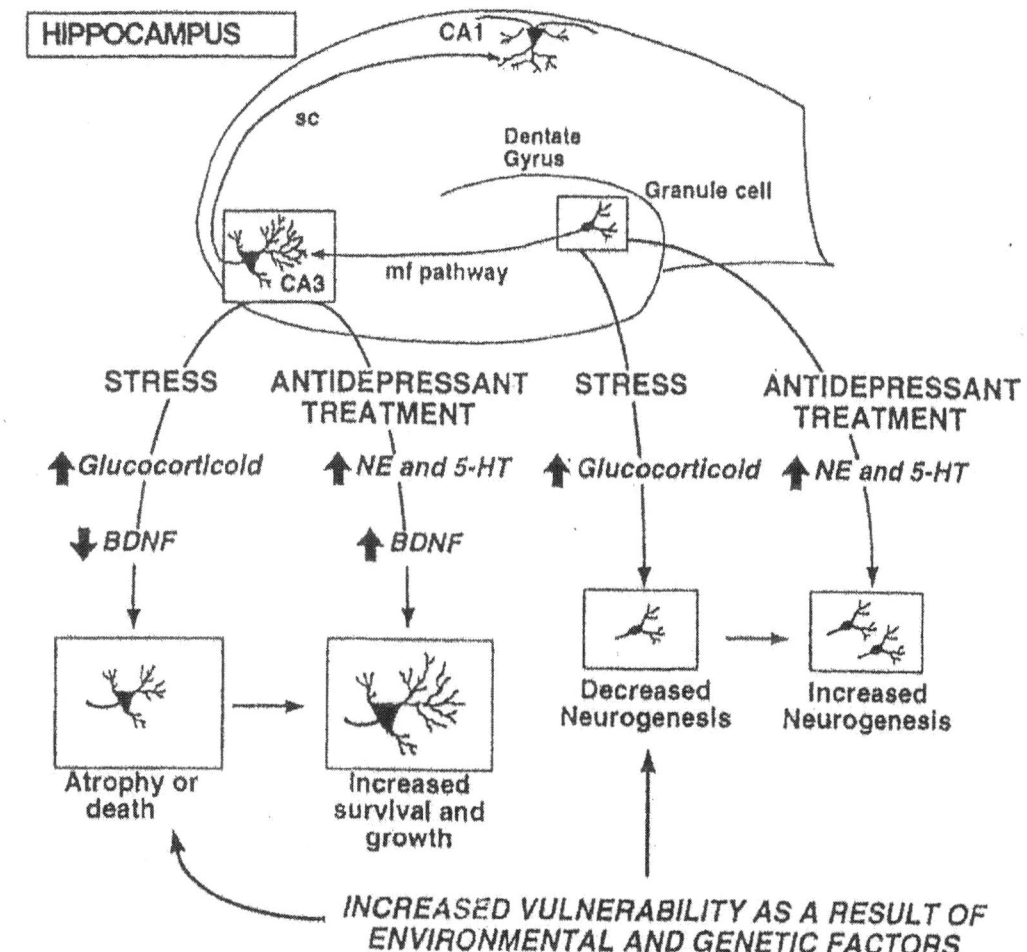

Reprinted from *Biological Psychiatry*, vol. 46, R.S. Duman, J. Malberg, and J. Thome, "Neural plasticity to stress and antidepressant treatment, p. 1182, copyright (1999), with permission from Elsevier.

This diagram shows the hippocampus (top) and the mechanisms through which stress and medications are alleged to affect it. According to Duman et al (referenced above), stress increases glucocorticoids (such as cortisol) and suppresses growth factors (BDNF). This results in the shrinkage or death of neurons. Meanwhile, antidepressant treatment is depicted as neuroregenerative. *Unfortunately, this perspective about antidepressant drugs is not entirely true.*

Challenges to the Neurogenesis Theory of Depression and Antidepressant Efficacy

In addition to the fallacy that a possible *effect* of stress is the *cause* of depression, a substantial body of research evidence contradicts the premise that enhanced *neurogenesis* in the hippocampus is either necessary or sufficient for mental and cognitive health. Equally important, however, is the credibility of the research method (i.e., the **BrdU labeling technique**) which has been repeatedly *misused and misinterpreted* by many neurogenesis researchers. To understand why the BrdU labeling technique casts a shadow across the validity of the neurogenesis theory of depression and antidepressant efficacy, it is essential to understand the biology of cell division.

Cell Division and the BrdU Labeling Technique

In order for scientists to confirm the presence of neurogenesis, they must employ reliable methods for detecting new neurons. Before any new neuron can emerge in the adult brain, a neuronal precursor – either a stem cell or neural progenitor cell – must progress through a sequence of activities known as the **Cell Cycle**. These activities involve the replication of a cell's genetic blueprint or DNA (S phase), and the eventual division of the cell via the process of mitosis (M phase).

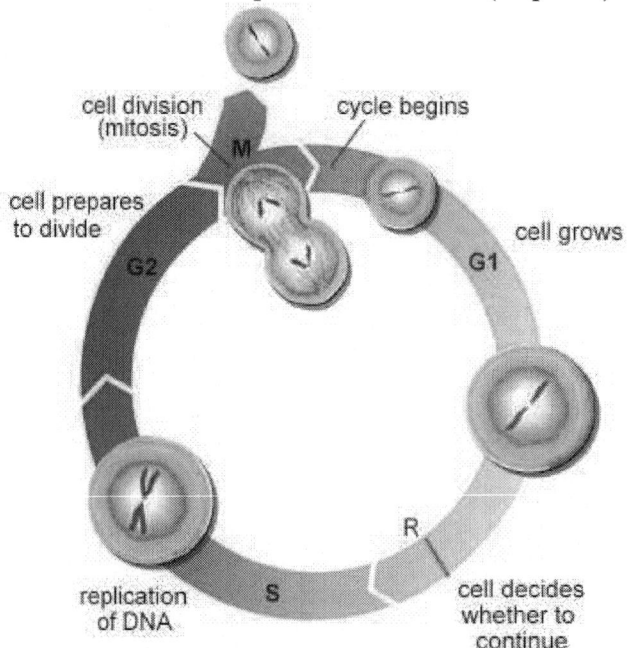

Source: Wit Sombatworapat, reproduced with permission.

The complex lab technique known as **bromo-deoxyuridine** (BrdU) labeling makes use of the Cell Cycle in the following way. BrdU is a nucleotide – a DNA building block – which can be administered to animals or humans by injection or ingestion. When it is taken up by the brain, **BrdU becomes incorporated into new strands of DNA within any cell that is replicating genes in S-phase.** In other words, BrdU becomes a marker for **DNA synthesis**.

Following the death of experimental animals (or humans), scientists are able to use immunological methods to identify cells which have acquired the BrdU label.
When researchers stain brain tissue and detect BrdU in cell nuclei, this means they have found cells which have recently synthesized DNA. It does not necessarily mean that a stem cell has progressed through M phase (mitosis), nor does it necessarily mean that a stem cell has matured to become a new neuron. [To prove either of those cell fates, other techniques must be applied.]

Unfortunately, the BrdU labeling technique has become one of the most misunderstood research methods in the history of psychiatry.[31-34] Possibly with no exception, the rat studies which have been used to advance the theory of antidepressant neurogenesis have used the BrdU method. ***Many authors have erroneously implied that BrdU uptake signifies neurogenesis, rather than DNA synthesis***. None have addressed the essential features of BrdU and cell biology which complicate the interpretation of these experiments:

1) BrdU can induce cell growth.

Under certain conditions, BrdU has been shown to stimulate cell division and proliferation. This means that experiments which investigate neurogenesis via BrdU labeling run the risk of wrongly attributing cell changes to other phenomena – such as pharmaceuticals –when, in fact, it is the BrdU *itself* which has triggered new rounds of DNA replication.

2) BrdU can function as a cytotoxin (cell poison).

As a halogenated pyrimidine, BrdU alters the stability of the DNA chains into which it incorporates. This can result in mutations, strand breaks, alterations in the cell cycle, and cell death. In fact, it is precisely for this reason that BrdU has been used as an antiviral and anti-cancer agent in humans. In animal experiments, high doses (60 to 600 mg/kg) and low doses (10-20 mg/kg) of BrdU have been found to kill neurons during embryonic and neonatal development.

3) BrdU labeling cannot distinguish between DNA synthesis which occurs as a part of *neurogenesis* (cell division and proliferation) and DNA synthesis which occurs as a part of *cell death*.

One of the defining features of neuronal cells is their extreme sensitivity to exogenous chemicals and other sources of injury (such as blunt trauma, seizure activity, ischemia, hypoxia, hypoglycemia, oxidative stress). During several processes which either accompany or cause cell death (described below), neurons may acquire extra copies of DNA. In these cases, there is a risk that the BrdU labeling technique may be misconstrued as a sign of neurogenesis, rather than as a marker of DNA synthesis or transfer *associated with cell death*.

Abortive cell cycle re-entry (fast death)

Injury to a neuron may cause acute changes which *simultaneously* activate the synthesis of DNA (S-phase) in the nucleus – a phenomenon known as **abortive cell cycle re-entry** – and the breakdown of the cell via apoptosis (programmed cell death). BrdU which enters a dying cell during this process will label a neuron that is actually on its way to extinction.

Gene duplication (slow death)

In some cases, a lethally damaged neuron may duplicate its DNA (S-phase) *far in advance of the final stages of dying*. This is another example of **abortive cell cycle re-entry** but in this case, the processes of gene duplication and mortality are not simultaneous. In other words, **DNA can be replicated in a dying neuron during a process of slow death which unfolds over a period of weeks, months, or even a full year**. Here, the danger of BrdU labeling arises from the misidentification of mortally wounded neurons as healthy, functional new cells.

BrdU transfer

Stem cell transplantation studies have demonstrated that BrdU (introduced into progenitor cells that were then killed by freezing) can be passed along from dying or dead cells into the genome of a living adult neuron. This phenomenon has serious implications for neurogenesis research. Given the fact that apoptosis (programmed cell death) is the most common fate of brain cells following injury, it is possible that some of the BrdU labeling which occurs in antidepressant experiments reflects the passage of the BrdU nucleotide from dying cells to neural stem cells or progenitors. When this occurs, the detection of BrdU may be misinterpreted as a marker of cell vitality, rather than as a marker of cell proliferation *and* cell death.

With these caveats in mind, it is possible to proceed with a cautious and careful inspection of several studies which not only invalidate the theory of antidepressant neurogenesis, but which also demonstrate that **antidepressants act as neurotoxins within the hippocampus of the temporal lobe**.

Study #3   Sairanen et al   Finnish Study   -   Antidepressant Toxicity in Mice

In 2005, researchers in Finland reported the effects of two different antidepressants upon the fate of hippocampal cells in mice.[35] The study employed male and female members of several strains, including two different kinds of genetically modified animals. For the sake of simplicity, this summary will be limited to the observations which were made in the non-genetically modified ("wild type") mice.

Experimental Procedure

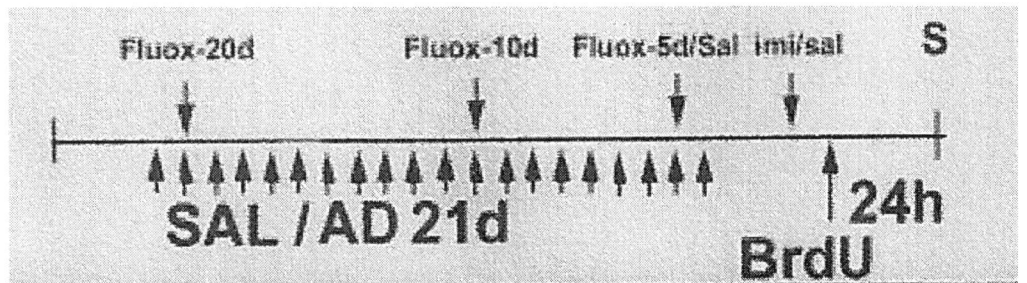

Source: Reprinted with permission, *Journal of Neuroscience*, M. Sairenen et al, vol 25, Brain Derived Neurotrophic Factor and Antidepressant Drugs Have Different But Coordinated effects on Neuronal Turnover, Proliferation, and Survival in the Adult Dentate Gyrus," p. 1090, copyright (2005).

The experiment proceeded in two phases, only one of which will be reviewed here. [Phase II was not interpretable due to the research team's assumptions about BrdU.] In Phase I, animals were exposed to three weeks of **daily intra-abdominal injections** of saline, imipramine (an older, tricyclic antidepressant which blocks the reuptake of serotonin and norepinephrine: **20 mg/kg IP injections each day**), or fluoxetine (a so-called "selective" inhibitor of serotonin reuptake: **10 mg/kg IP injections each day**). Animals in the fluoxetine group were exposed to the drug for 5, 10, or 20 days, in order to permit a temporal analysis of drug effects at several intervals of exposure.

Twenty-four hours after the last drug treatment, the animals were injected with successive doses of **BrdU** (50 mg/kg IP injections every 2 hours x 4). After another twenty-four hour period of recovery, the animals were sacrificed. Brains were excised, and random slices of the hippocampus (three sections per animal) were processed using several staining techniques:

1) **BrdU staining** with an anti-BrdU antibody

2) **TUNEL labeling** (to identify dying cells)

3) **double-labeling with BrdU and TUNEL** (presumably, capturing those cells which had incorporated BrdU due to abortive cell cycle re-entry)

Cell counts of the granule cell layer in the dentate gyrus were performed using bright field microscopy.

Findings

When the researchers examined brain tissue *immediately after* the three-week exposure to antidepressants, they observed an increase in TUNEL staining (which they correctly associated with cell death); an increase in BrdU staining (which they incorrectly associated with neurogenesis); and an increase in double-labeled cells (which they did not interpret or explain, but which other scientists would interpret as a marker of abortive cell cycle re-entry and death).[36]

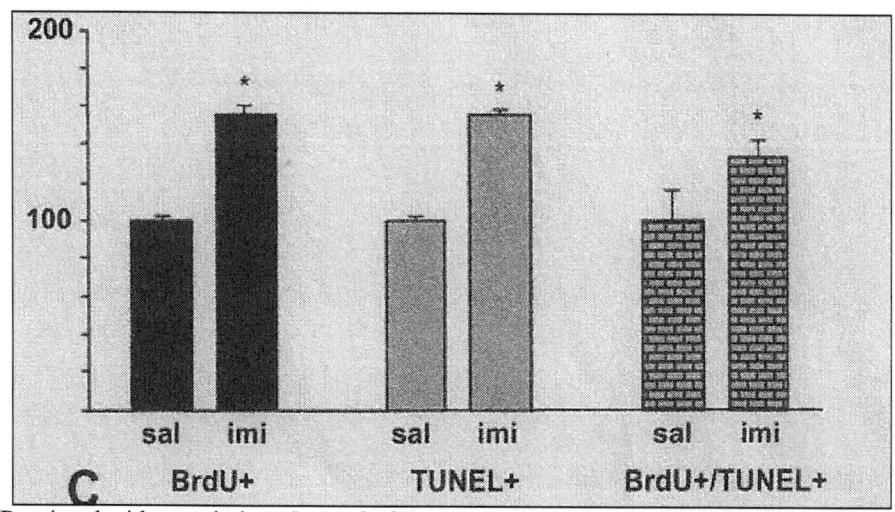

Source: Reprinted with permission, *Journal of Neuroscience*, M. Sairenen et al, vol 25, Brain Derived Neurotrophic Factor and Antidepressant Drugs Have Different But Coordinated effects on Neuronal Turnover, Proliferation, and Survival in the Adult Dentate Gyrus," p. 1090, copyright (2005).

Results of imipramine treatment: 20 mg/kg IP daily injections x 3 weeks

y-axis shows cell changes relative to SALINE   (saline controls = 100%)

Left: Relative to saline: imipramine resulted in a 60% increase in DNA synthesis
Center: Relative to saline: imipramine resulted in ~ 60% more cell death (TUNEL+)
Right: Relative to saline: imipramine resulted in ~ 40% more double-labeled cells

The Finnish investigators interpreted these results as evidence of *"enhanced cell turnover."* In other words, they implied that imipramine had induced a ***balanced process*** of neuronal death and neuronal birth in the hippocampus. Unfortunately, their published report provided no raw data for the actual numbers of cells which stained positive for BrdU+ or TUNEL+. **[Without a comparison of actual *cell counts*, their conclusion about "cell loss = cell birth" cannot be sustained.]**

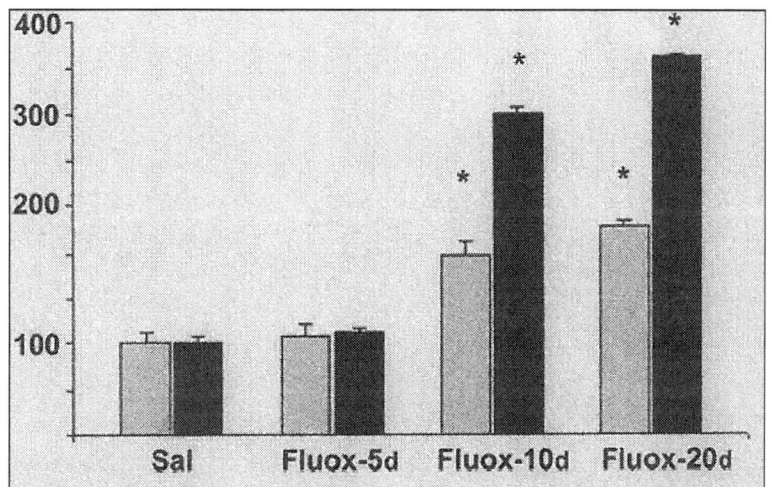

Source: Reprinted with permission, *Journal of Neuroscience*, M. Sairenen et al, vol 25, Brain Derived Neurotrophic Factor and Antidepressant Drugs Have Different But Coordinated effects on Neuronal Turnover, Proliferation, and Survival in the Adult Dentate Gyrus," p. 1090, copyright (2005).

Results of fluoxetine treatment: 10 mg/kg IP daily injections (5, 10, or 20 days)

light bars = TUNEL labeled cells
black bars = BrdU labeled cells

Similar results were observed in response to fluoxetine. Following 10 and 20 days of exposure to the SSRI, mice displayed significant increases in BrdU *and* TUNEL staining. ***Like imipramine, fluoxetine was found to enhance DNA synthesis and cell death in the hippocampus***. Regrettably, the researchers again failed to report raw data (cell counts). Also, they either failed to perform or failed to report the results of *double-labeling* (BrdU + TUNEL) in the animals exposed to fluoxetine. [The y-axis above, as published by Sairenen et al, demonstrates an expansion of the 100 to 200% interval.
It is unclear if this reflects a simple mislabeling of the y-axis (200 should really be 250, 300 should be 350, 400 should be 450). Regardless, this anomaly precludes any reliable comparison to the magnitude of the changes associated with imipramine, as shown before.]

> In Sum:
>
> The Sairanen et al study appears to be the only published study of living animals, to date, which has simultaneously performed an analysis of **DNA synthesis *and* cell death**.
>
> BrdU labeling increased under the influence of serotonergic antidepressants. Although this may have been a marker of neurogenesis, it may also have reflected BrdU uptake by damaged, mature neurons (and glia) in response to chemical injury.
>
> To the extent that TUNEL labeling is an accepted marker of apoptosis and late-stage necrosis, *this experiment confirmed that two different antidepressants were a cause of cell death within the hippocampus.*

Study #4   Castren   Helsinki, Finland - Unpublished Findings of Cell Death

Interestingly, a 2004 publication from the same Finnish research team *covertly* acknowledged the toxicity of antidepressants.[37] Although their narrative minimized the significance of previous, **unpublished** results, a table and illustration demonstrated the lethality of monoamine reuptake inhibitors upon neurons in the hippocampus.

Table 1

The effects of antidepressant treatment on the differentiation of newborn hippocampal neurons (see Figure 1).

Treatment	1[a]	2	3	4	5	6
Stress	↓ [43]	↓ [43]	↓ [43]	↑ [43]	↓ [43]	↑ [43]
Antidepressants	↑ [53]	?	?	↑[b]	↑ [2]	↑[b]
ECT	↑ [53]	↑ [50]	?	?	↑ [2]	?

[a]Numbers 1–6 correspond to the numbers in Figure 1. [b]Sairanen and Castrén, unpublished. ↑, increased; ↓, decreased; ?, no data.

Reprinted from *Current Opinion in Pharmacology*, volume 4, E. Castren, "Neurotrophic effects of antidepressant drugs," p. 60, copyright (2004), with permission from Elsevier.

According to the data in Table 1 (above), *antidepressants had been found to replicate two of the hippocampal effects of stress.* Castren demonstrated these changes with a diagram of the pertinent brain regions (show below).

Antidepressants

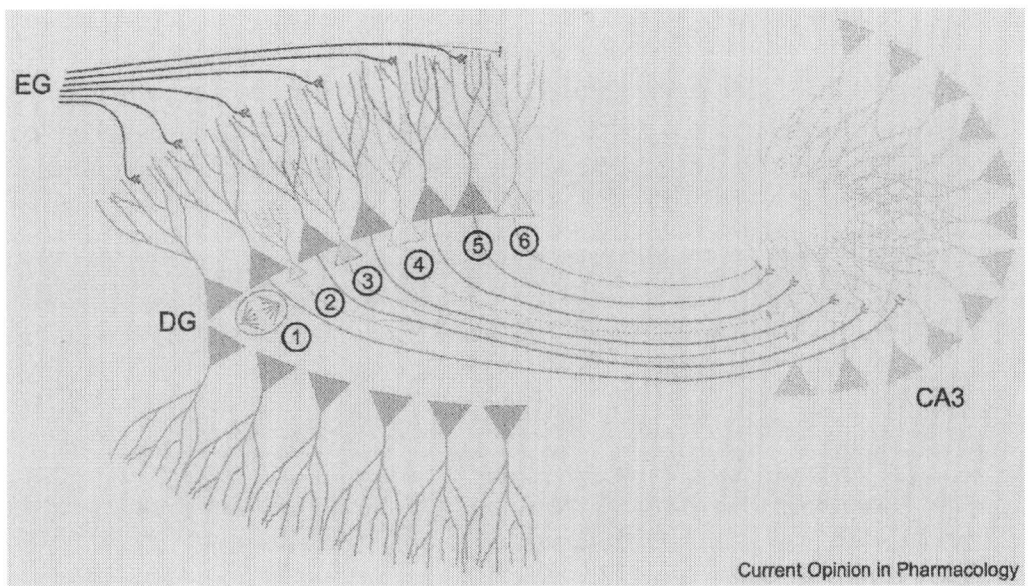

Reprinted from *Current Opinion in Pharmacology*, volume 4, E. Castren, "Neurotrophic effects of antidepressant drugs," p. 60, copyright (2004), with permission from Elsevier.

EG	= Entorhinal Cortex	(presumably, EG is a typo – should read EC)
DG	= Dentate Gyrus	(a region of the hippocampus)
CA3	= Cornu Ammonis 3	(a region of the hippocampus)

Castren's schematic of a rat *hippocampus* was intended to portray the effects of antidepressants. However, only a very careful reading of Castren's paper would permit the reader to appreciate the fact that *unpublished* animal studies had revealed that antidepressants harm this region of the brain.

First, antidepressants had been found to eliminate new cells which failed to form active synapses between the dentate gyrus (DG) and the CA3 region of the hippocampus. This toxic effect was identified in Castren's illustration by a dying neuron (located at #4).

Second, antidepressants had also been found to eliminate mature neurons which had already formed functional circuits (located at #6). [These mature pathways, which conduct input from the entorhinal cortex to the CA3 region of the hippocampus, play important roles in visual and verbal memory.]

Study #5   Cowen, et al   -   No Neurogenesis in the Hippocampus

In 2008, researchers in New Jersey reported the results of an investigation in which they failed to detect neurogenesis in response to fluoxetine.[38] Hypothesizing that there might be age-related differences which mediate the anatomic effects of serotonin reuptake inhibition, this team exposed three different groups of male Sprague-Dawley rats to a chronic drug regimen.

Experimental Procedure

Sixteen rats in each age group were randomly assigned to receive saline (control) or fluoxetine (daily intraperitoneal injections of a 5 mg/kg dose):

adolescent rats	1 month old	8 controls, 8 fluoxetine
young adult rats	2.5 months old	8 controls, 8 fluoxetine
aged rats	12 months old	8 controls, 8 fluoxetine

The experiment began with a single injection of BrdU (200 mg/kg) in an effort to label "new" cells in the hippocampus. Treatments were initiated 24 hours later. Following an exposure period which lasted for 25 days, the animals were euthanized. Brains were removed, processed with immunocytochemical staining techniques, and inspected by light microscopy. This particular experiment involved the analysis of specific sections of the hippocampus, using antibodies against BrdU (a marker of DNA synthesis) and Ki67 (the latter, a marker of cell proliferation). A separate analysis of coronal slices was also performed in order to estimate changes in the volume of the hippocampus (tissue area x thickness).

Background Information

The goal of neurogenesis research is the detection of new neurons. Previously, the limitations of the BrdU labeling were briefly explored. One technique which is used by scientists to improve upon the BrdU method involves exposing biological samples (in this case, brain tissue) to antibodies which are directed against specific cell components. The goal is to apply an antibody which will latch onto the cell and provide a chemical signal: "pay attention, here is a brand new cell." **Ki67** is a protein which is present in the nucleus of cells throughout many phases of the cell cycle (G1, S, G2, and M). Using sophisticated lab procedures, scientists can track increases in Ki67 as a rough indicator of cell division. Although other methods must still be used to confirm and quantify the appearance of new neurons, Ki67 is considered to be a more reliable marker of neurogenesis than BrdU.

Returning to the New Jersey fluoxetine experiment, the goal of the research team was to identify the exact location of new neurons within the hippocampus (shown below).

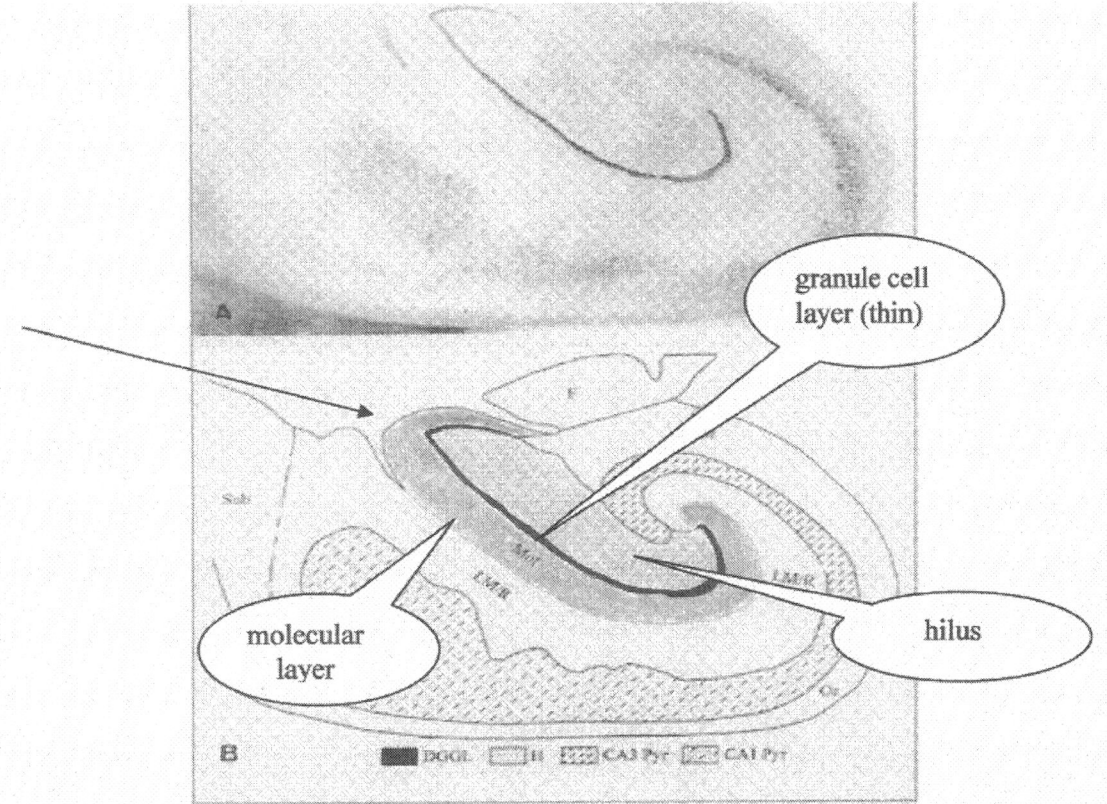

Reprinted and modified with permission of *Arquivos de Neuro-Psiquiatria*, volume 65:4B, M.J. Sa, C. Ruela, and M.D. Madeira, "Dendritic Right/Left Asymmetries in the Neurons of the Human Hippocampal Formation," p. 1106, (2007).

Top (Slide A): microscopic view of the hippocampus

Bottom (Slide B): hand-drawn legend of the hippocampus (key for Slide A)
arrow points to dentate gyrus

dentate gyrus = molecular layer + granule cell layer + hilus

The dentate gyrus is one of several regions in the hippocampus. Based upon differences in cellular, chemical, and electrical properties, scientists subdivide the dentate gyrus into discrete components: a molecular layer, a granule cell layer, and the hilus. A fourth subzone – the subgranular zone or SGZ – lies at the intersection of the granule cell layer and the hilus.

Findings

Cowen's team detected no differences in hippocampal volume between the recipients of fluoxetine and saline, regardless of age. (However, this was only a microscopic estimate of volume based upon an examination of select slices – not the entire hippocampus.) Although the New Jersey researchers erroneously interpreted BrdU and Ki67 to be markers of neuronal proliferation (rather than DNA-synthesis and cell cycle activation, respectively), ***they were unable to confirm a "neurogenesis effect" of fluoxetine in any age group.*** In fact, their unique protocol revealed ***a growth-suppressive effect of fluoxetine*** in several regions of the hippocampus, regardless of age:

	# of Ki67 "proliferating" cells		
	dentate gyrus	subgranular zone	hilus
adolescent			
saline	13401	9611	3791
fluoxetine	13102	9320	3783
flx vs. saline	**-2.2%**	**-3.0%**	**-0.2%**
young adult			
saline	8676	6357	2319
fluoxetine	7976	5714	2262
flx vs. saline	**-8.1%**	**-10.1%**	**-2.5%**
aged			
saline	2273	1044	1229
fluoxetine	2196	1020	1176
flx vs. saline	**-3.4%**	**-2.3%**	**-4.3%**

	# of BrdU Labeled Cells		
	fluoxetine vs. saline controls		
	dentate gyrus	granule cell layer	hilus
adolescent	**-12.5%**	**-7.1%**	**-25.7%**
young adult	**-0.5%**	**+1.4%**	**-5.1%**
aged	0%	**-3.8%**	+3.3%

Neuroimaging Studies of Humans

Studies of the Hippocampus

Surprisingly, the research basis of the **neurogenesis theory of depression** in humans has been built upon a small number of neuroimaging studies which have been repeatedly *mischaracterized* and *inadequately critiqued*. For example, in a 2001 commentary which appeared in the prestigious journal, *Proceedings of the New York Academy of Sciences*, a Stanford University professor was allowed to make the following claim:

> "...major depression is associated with atrophy within the central nervous system Such atrophy is centered in a brain region called the hippocampus and the magnitude of the hippocampal volume loss (nearly 20% in some reports) helps explain some well documented cognitive deficits that accompany major depression..." [39]

Referring to only three investigations in the medical literature, the same author opined:

> "...These were careful and well-controlled studies, in that the atrophy was demonstrable after controlling for total cerebral volume and could be dissociated from variables such as history of antidepressant treatment..." [40]

Interestingly, a careful examination of these and other studies reveals no attempt to account for the influence of medication.

Study #1   Sheline et al 1996   Small Hippocampi in Medicated Patients

Investigators used magnetic resonance imaging to compare the brains of ten depressed females (mean age: 68, age range: 51 to 86) with a control group matched for age, education, and height. Smaller hippocampal volumes were detected in the depressed patients.[41] However, 90% of the patients had experienced multiple episodes of depression, and **80% of them were taking antidepressants at the time of the brain scans. Despite the past and current use of medication, the researchers did not address the role of drugs in mediating the observed effects.**

Study #2   Sheline et al 1999   Small Hippocampi in Medicated Patients

In 1999, members of the same research team published a second report which extended the size and age range of their previous study.[42] This time, the brain features of 24 female patients (mean age: 54, age range: 23 to 86) were compared with an equal number of age-, gender-, and height-matched controls. All of the patients had histories of recurrent depression and psychiatric treatment. *Two-thirds of these patients were receiving antidepressants at the time of neuroimaging.* **Despite the fact that the patients displayed smaller gray matter volumes (10% smaller hippocampi, 13% smaller amydgalae), the investigators again neglected to address the role of pharmaceuticals in mediating these results.**

Study #3   Bremner et al 2000   Small Hippocampi in Medicated Patients

In this study, researchers used magnetic resonance imaging (MRI) to compare 16 outpatients who were in remission from non-psychotic depression (mean age: 43) and 16 non-psychiatric controls (mean age: 45). Participants were matched for gender, handedness, and whole brain size.[43] Hippocampal volumes were 12 to 19% smaller in the formerly depressed patients. However, *all of the subjects were taking antidepressants (either paroxetine, fluoxetine, or desipramine) at the time of their brain scans*. Despite the fact that all of the patients had extensive treatment histories (which included an average of three inpatient hospitalizations), and despite the fact that all of them were actively using medication, *the researchers did not address the role of antidepressants in mediating the observed effects*.

Study #4   Campbell et al 2004   Meta-Analysis of Hippocampal Imaging Studies

Motivated by a desire to account for discrepancies in the neuroimaging literature, Campbell et al undertook a comprehensive meta-analysis of all depression/hippocampus studies indexed in Medline, and published between 1960 and 2002.[44]

Notwithstanding the careful selection of fifteen reports for a final analysis, interpretability was undermined by numerous confounders in the research methodologies:

> 88% of the studies had included patients who were taking antidepressants at the time of the brain scans

> 67% had failed to correct for height or head size

> 40% had included patients with histories of alcohol dependence (a well known cause of cerebral atrophy).

Of the 12 studies which had focused solely upon comparisons of the hippocampus, 50% had found no significant difference between the brain size of depressed patients and non-depressed controls. A defining feature of these "no difference" studies was the inclusion of younger or first-episode patients with limited exposures to antidepressant drug therapy.

Although no neuroimaging studies have *intentionally* explored the role of antidepressants as inducers of temporal lobe pathology, two publications have provided preliminary evidence for just such an effect.

Neuroimaging Studies of First-Episode vs. Recurrent Depression

Study #5   MacQueen et al   1st episode vs. recurrently depressed

In 2003, MacQueen et al published the results of their cross-sectional study which compared the hippocampal volumes of 17 recurrently depressed patients (mean age: 36 years, average # of past episodes: 6) and 20 individuals who were experiencing a first episode of unipolar depression.[45] These subjects were matched to non-psychiatric controls on the basis of age, gender, and IQ.

It is significant that all of the patients with recurrent depression had extensive histories of treatment with pharmacotherapy, including an average of three drug trials per patient, and at least one exposure to a serotonergic agent. In contrast, the first-episode patients were drug naïve at the time of enrollment. (An undisclosed number were placed on medication shortly before scanning).

Based upon the results of magnetic resonance imaging, researchers detected no differences between the size of the hippocampus in first-episode patients and the "normal" controls. ***In comparison, the hippocampus of the formerly and extensively medicated subjects was 12% smaller than their respective controls***:

	Comparisons of Hippocampal Volume (mm^3)			
	first episode n = 20	controls n = 20	multiple episode n = 17	controls n = 17
mean age:	28.4	28.4	35.9	36.2
right HC	2,793	2,784	**2,392**	**2,692**
left HC	2,738	2,761	**2,381**	**2,703**
HC = hippocampus				

Study #6   **Frodl et al**   **1st episode vs. recurrently depressed**

In what appears to be the only longitudinal investigation of hippocampal volume among the non-psychotically depressed, Frodl et al examined the course of 30 inpatients between 2000 and 2003 (mean age: 48, age range: 18 to 65).[46] The study consisted of 11 "single-episode" and 19 "multiple-episode" patients, *most of whom had started antidepressant drug therapy prior to hospitalization*. A control group consisted of individuals who were matched according to height, weight, education, and approximate age. Patients received MRI scans within 2 weeks of hospitalization and again one year after discharge. Controls participated in serial neuroimaging across the same time span.

Several findings were pertinent to the subject of brain atrophy.

First, in contrast to the MacQueen study, there were no differences between the first-episode and the recurrently depressed patients. This finding may have been influenced by the fact that all but two of the eleven "single-episode" patients had been taking medication prior to scanning.

Second, a subgroup analysis of longitudinal changes was suggestive of a detrimental drug effect. Among those patients who progressed to remission (18 of the initial 30), 22% had stopped antidepressant drug therapy by the time of follow-up. For this subgroup, hippocampal volumes appeared to "normalize" over time. Among the patients who failed to achieve remission (12 of the original 30), 92% remained on drug therapy. **Despite the continuing use of medication, their hippocampal volumes remained significantly smaller than controls.**

	remission n = 18	no remission n = 12
% continuing drug therapy	78%	92%
"normal hippocampus" (group average)	yes	no

Brain Scans Showing White Matter Abnormalities

Antidepressant effects upon the brain extend beyond the hippocampus. The potential hazards of antidepressant drugs are of particular concern for patients at risk for cerebrovascular disease and neurodegeneration.

In 2008, researchers affiliated with four major American research centers published the results of a longitudinal imaging study which featured almost 2000 older patients.[47] The project involved "before and after" brain scans (MRI) of patients, aged 65 and older, who had agreed to participate in the Cardiovascular Health Study (CHS). This was a long-term investigation of individuals at risk for heart attack and stroke.

The goal of this research was to compare the progression of **white matter hyperintensities** in the periventricular and subcortical regions of the brain.

Background Information

By way of clarification, white matter hyperintensities refer to "bright spots" which appear in MRI scans of the brain. Although their precise cause remains an item of dispute among neurologists and radiologists, *many pathologists have confirmed a link between white matter hyperintensities and a variety of structural and functional abnormalities.* These include damage to cerebral blood vessels (e.g., clogging or hardening of the arteries, resulting in reduced or occluded blood flow); breakdown of the blood-brain barrier; disintegration of myelin (demyelination); and – in some studies – frank atrophy of gray matter in the brain (e.g., neuronal cell bodies, dendrites, astrocytes).

Experimental Procedure

The study involved the comparison of 1,826 patients who agreed to undergo serial brain scans. Scan #1 was performed between 1991 and 1994. A second scan was obtained approximately five years later (1997 through 1999).

Neuroimaging results were rated by trained readers who were blinded to clinical data. These readers graded the severity of white matter abnormalities on a scale from 0 to 9, with nine representing the most severe pathology. White matter worsening was defined as an increase of one or more grades over time.

Comparisons in brain anatomy were made between two subgroups: 163 individuals who initiated antidepressant medication between the time of the two head scans, and 1,663 individuals who remained antidepressant drug free. A statistical analysis using logistic regression was performed to confirm the strength of association between demographic features, brain changes, and exposure to medication.

Findings

A worsening of white matter was experienced by 36% of the patients who received antidepressant therapy, versus 27% of the patients who remained antidepressant drug free. These findings were not attributable to differences in age, gender, race, lifestyle (smoking), or the baseline prevalence of diabetes, blood pressure anomalies (hypo- or hypertension), or vascular disease:

	worsening white matter n = 509	no worsening n = 1317
age	74.2	74
female	58.7%	58.8%
Caucasian	84.1%	83.4%
smoker	10.0%	7.9%
orthostatic hypotension	13.3%	14.8%
hypertension	42.2%	41.7%
diabetes	11.4%	11.3%
heart attack	7.1%	8.0%
TIA	2.2%	2.0%
stroke	3.1%	2.9%
White Matter Grade:		
2	27.1%	23.2%
3	16.7%	12.7%
4 to 9 (severe)	16.1%	10.9%

Using logistic regression to test the strength of the relationship between medication and the worsening of structural brain defects, the **researchers confirmed a positive association between white matter pathology and exposure to any of the serotonergic antidepressants (old and new)**:

	odds ratio for white matter worsening	% of users who worsened
SSRI only	1.30	NA
any serotonergic agent (SA)	1.36	33%
tricyclic antidepressant (TCA)	1.77	35%
SA *and* TCA	NA	60%

SA = selective serotonin reuptake inhibitor or trazodone or nefazodone
SSRI = fluoxetine, sertraline, paroxetine, fluovoxamine, citalopram, or escitalopram
TCA = tricyclic antidepressant (e.g., amitriptyline, imipramine, clomipramine)
NA = data not provided

Postmortem Studies of Humans

Given the attention which has been lavished upon the hippocampus and its alleged role in the cause of depression, one might expect to find a large number of postmortem investigations on this subject. However, as far as this writer has been able to confirm, the medical literature has, thus far, featured just one study of this kind.[48]

In 2001, Lucassen et al published the results of an autopsy analysis which compared three groups of subjects:

> - a group of mood-disorder patients who had been diagnosed with recurrent depression (n=12) and manic depression (n=3)

> - a group of non-psychiatric patients (n=9) who had received high doses of glucocorticoids (steroids) for medical conditions

> - a group of age- and gender-matched controls (n=15) who lacked histories of psychiatric or neurological disorders, and who lacked histories of treatment with glucocorticoids

Notably, *67% of the psychiatric patients had received treatment with antidepressants*. A full 93% (14/15) had received therapy with an antidepressant *and/or* neuroleptic (antipsychotic) drug:

Drug Treatments of the Depressed Patients			
antidepressant and/or neuroleptic		14 of 15	(93%)
any antidepressant		10 of 15	(67%)
monoamine oxidase inhibitor	3		
SSRI	4		
tricyclic antidepressant	3		
other antidepressant	2		
lithium		3 of 15	(20%)
benzodiazepine		10 of 15	(67%)
neuroleptic		5 of 15	(33%)

Experimental Procedure

Comparisons of brain anatomy and chemistry were made after processing hippocampal tissue using a variety of intricate lab techniques. These included **In Situ End Labeling (ISEL)** for the purpose of identifying fragmented DNA associated with cell death (e.g., apoptosis and necrosis), and a variety of additional staining procedures to identify neurons and glia.

A semi-quantitative analysis was performed according to the **intensity of ISEL labeling**: 0 to 2 cells (negligible death), 3 to 8 cells (mild death), 8 to 15 cells (moderate death), 15+ (severe death). Changes in cell morphology were interpreted in an effort to classify the mode of death (definite apoptosis, definite necrosis, and/or other signs of brain degeneration.)

Background Information

One of the most active areas of research in biology is the study of how cells die. Three major processes have come to be emphasized: apoptosis, necrosis, and autophagy.[49-53]

Apoptosis (from the Greek for "dropping or falling off," as leaves from a tree) refers to a process which can be compared to a well controlled demolition. Frequently described as programmed cell death, apoptosis involves a highly regulated sequence of processes in the mitochondria which result in the dissolution of cells without generating an inflammatory response.

Necrosis refers to a process of dramatic injury, swelling, and disintegration. If one conceptualizes apoptosis as a controlled demolition (implosion), necrosis could be compared to an act of war (violent explosion). Unlike apoptosis, necrosis *does* result in inflammation due to the leakage of cell components into the surrounding environment.

Autophagy (from the Greek: "eating the self") represents a third kind of cell catabolism (breakdown) which involves the fusion of cell parts with lysosomes (digestive sacs). Some scientists believe that autophagy represents a special form of programmed cell suicide. Others believe that it represents a survival mechanism which provides nutrients and energy during times of stress.

Although scientists once viewed apoptosis as the only form of cell death that involved precise timing and regulation, recent research has implied that all forms of cell disintegration may, in fact, be "programmed" to some degree. Where they differ is in their dependence upon energy (ATP): apoptosis requires energy, while autophagy and necrosis do not. Another recent discovery has been the temporal overlap between these events. For example, in response to injury, it is possible for the processes of apoptosis, necrosis, and autophagy to occur concurrently (i.e., they can co-exist).

Potential Confounders

The hippocampal tissue of each subject was dissected and fixed in formaldehyde, then stored for variable intervals (24 to 3,867 days) prior to formal analysis. **To detect apoptosis or late-stage necrosis using ISEL, however, fixation periods should be less than 21 days.** Hence, the storage times in this study may well have *hindered the detection of programmed cell death*. Similarly, postmortem delays (the interval between death and tissue fixation) greater than 48 hours would have compromised the ability of the researchers to capture cells which were still experiencing apoptosis.

In other words, the postmortem processing delays and long fixation times in this experiment most likely contributed to low estimates of hippocampal damage – particularly, among the depressed patients:

	depressed n=15	steroid n=9	control n=15
% with lengthy postmortem delay ≥ 48 hours	40%	0%	1%
duration of fixation (days):			
range	28-406	28-3867	28-849
mean	75.8	496	121.5
# of subjects with fixation < 21 days	1	4	3

Findings

	% of Each Patient Subgroup Displaying Hippocampal Cell Death	
	ISEL + apoptosis	ISEL+ necrosis
depressed	11/15 (73%)	4/15 (27%)
steroid-treated	3/9 (33%)	2/9 (22%)
controls	2/15 (13%)	0 (0%)

Hippocampal cell death was far more common among the patients who had received treatment with psychiatric drugs: 73% showed definite signs of apoptotic cell death, and 27% showed signs of necrosis.

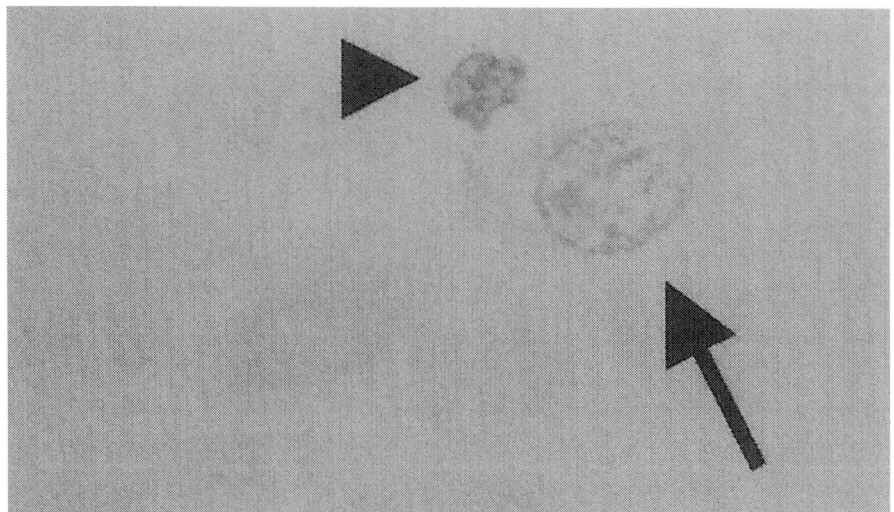

Reprinted from *American Journal of Pathology*, 2001, vol 158: 453-468 with permission from the American Society for Investigative Pathology.

This image depicts dying neurons in the hippocampus of a patient who had received antidepressant and/or neuroleptic drug therapy. The top image shows a cell which is undergoing **apoptosis**. The bottom image shows a swollen cell which is disintegrating via the process of **necrosis**.

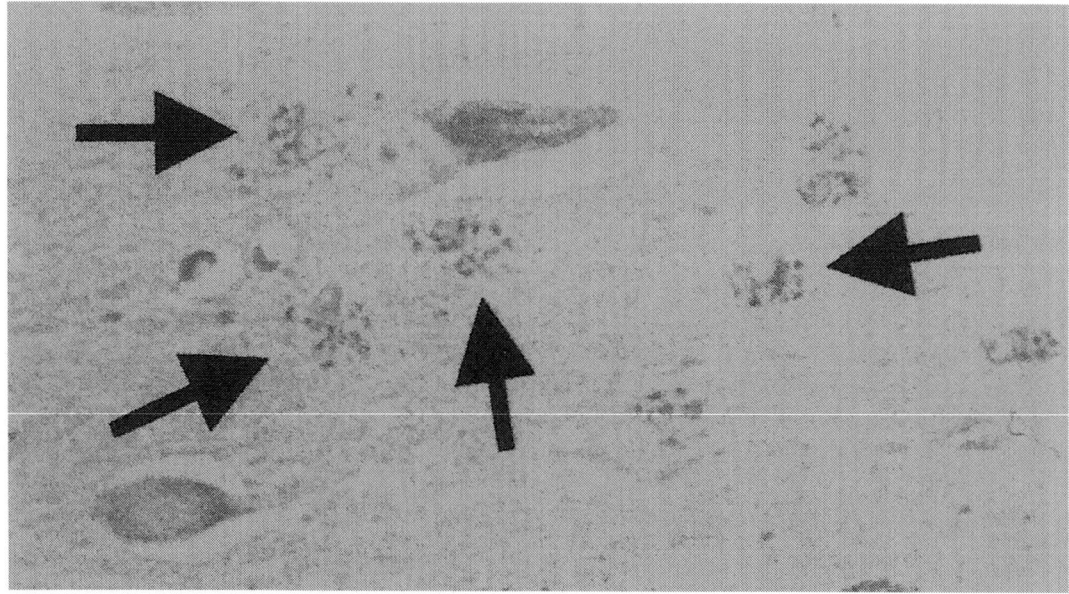

Reprinted from *American Journal of Pathology*, 2001, vol 158: 453-468 with permission from the American Society for Investigative Pathology.

This image reveals **chromatolysis** (dissolution of DNA) within the hippocampal neurons of another psychiatric patient. Here, the nuclear membrane has already disintegrated and the genetic material (chromatin) within the nucleus has started to break down. This is another example of cell death via the process of **necrosis**.

In addition to comparing the overall prevalence of hippocampal cell death among the three groups of subjects, the investigators were also interested in identifying the *distribution* and *intensity* of these changes. First, they compared the *prevalence* of **apoptosis** in several regions of the hippocampus:

# of Subjects with Clear Evidence of Apoptosis (by Hippocampal Region)						
	DG	CA1/2	CA3	CA4	any region	
depressed	10/15	2/15	1/15	5/15	**11/15**	**(73%)**
steroid	2/9	1/15	0/15	2/15	3/9	(33%)
control	0/15	1/15	0/15	0/15	2/15	(13%)
DG = dentate gyrus CA = cornu ammonis (regions 1, 2, 3, 4)						

The researchers also evaluated the *severity* of cell damage on the basis of the intensity of ISEL stains:

# of subjects with mild ISEL staining				
(mild = 3 to 8 cells with DNA fragmentation)				
	DG	CA1/2	CA3	CA4
depressed	7	6	7	6
steroid	5	1	3	4
control	6	2	2	1
# of subjects with moderate to severe ISEL staining				
(8 or more cells with DNA fragmentation)				
	DG	CA1/2	CA3	CA4
depressed	3	2	0	5
steroid	0	0	0	1
control	0	1	0	4

For the recipients of antidepressants and/or neuroleptic drugs, the use of medications was associated with more frequent, more severe, and more diffuse cell death in the hippocampus. This damage was especially prominent in the dentate gyrus and CA4.

In Sum:

The Lucassen study appears to be the only postmortem analysis, to date, which has focused upon the hippocampus of depressed patients.

Potential confounders included lengthy postmortem delays and protracted storage intervals. However, due to the unequal distribution of these features among the three groups of subjects, this study more than likely *underestimated* hippocampal cell death among those who had been medicated with psychiatric drugs.

Despite the use of antidepressants (and/or other psychopharmaceuticals), depressed patients were more likely to experience hippocampal cell death than the recipients of high-dose steroids and steroid-free controls.

Despite the use of antidepressants (and/or other psychiatric drugs), the hippocampal cell death in the brains of depressed subjects was more diffuse and more intense than that which occurred among the users of high-dose steroids and steroid-free controls.

Chapter Five

Antipsychotic Drugs

The modern history of psychopharmaceuticals dates back to the early 1950s, when derivatives of the synthetic dye and rocket fuel industries were found to have medicinal properties. Among these early compounds, an antihistamine known as chlorpromazine (Thorazine) found favor as a replacement for invasive treatments in the care of the institutionalized mentally ill.

Omitted from most textbooks of psychiatry is the fact that the initial popularity of chlorpromazine arose from its capacity to cause the signs and symptoms of serious brain damage. On the one hand, the drug was welcomed as an effective substitute for surgical lobotomy. On the other hand, the drug was embraced as a chemical which replicated the symptoms of *encephalitis lethargica* (sleeping sickness).[1-2] The latter condition – a severely disabling, often lethal illness characterized by high fever, delirium, and unconsciousness – had ravaged Europe and North America during early 20th century outbreaks of pandemic flu. Unfortunately for many survivors of that affliction, either the virus or secondary infections (such as *Streptococcal* bacteriae), or the body's *response* to pathogens, incited changes in the brain which engendered long-lasting disturbances. These included the features of **Parkinson's disease** (tremor, abnormal gait, and difficulty initiating movement), psychosis, dementia, and a variety of ***dyskinesias*** (abnormal, involuntary movements of the face, limbs, or trunk).[3]

Believing that it would be clinically useful to calm the denizens of the world's asylums via psychological and physical immobilization, many of the first psychiatrists who experimented with chlorpromazine came to value the emergence of any and all symptoms of *encephalitis lethargica*. ***In time, psychiatrists grew to believe that the infliction of Parkinsonian features was, itself, part and parcel of the drug's therapeutic effects***. (According to psychiatrist and pharmacohistorian, Dr. David Healy, this notion may have been based upon the fact that post-encephalitic hallucinations or delusions receded in some patients at the precise moment that movement abnormalities and other problems emerged.)[4]

It took more than fifteen years for astute clinicians to uproot the Parkinson's dogma, and twenty more to alert their peers to the unnecessary harmfulness of other drug-induced effects, identical to the unanticipated consequences of encephalitis lethargica. Among these delayed-onset ("tardive") phenomena were the problems of ***akathisia*** – an inner and outer restlessness characterized by the inability to sit still – various dyskinesias, psychosis, mania, and **dementia**.

How The New Drugs Got Their Name

Due to the limitations of neuroscience and medical technologies in the 1950s, the pharmacological and physiological actions of chlorpromazine and other compounds were not verified by the techniques of contemporary molecular biology. Rather, the drugs were classified empirically, according to their behavioral and symptomatic effects. Thus, in 1955, Drs. Jean Delay and Pierre Deniker identified the new drugs as ***neuroleptics*** (from the Latin, for "brain seizers") based upon five essential features: [5]

- the creation of psychic indifference and reduced spontaneous activity

- the suppression of excitation, agitation, mania, or aggression (sedation)

- the gradual reduction of psychosis

- the production of extrapyramidal and vegetative syndromes

- predominant subcortical effects (accounting for the previous states)

Over time, the term *neuroleptic* was eventually sanitized in favor of the idea that the dopamine antagonists were somehow specifically and powerfully ***anti-psychotic***. With the arrival of new potions in the 1960s (clozapine) and 1990s (risperidone, olanzapine, quetiapine, ziprasidone, etc.), the psychiatric and pharmaceutical industries collaborated in the creation of new mythologies. First, the term *"atypical"* was applied to the newer chemicals, based upon misconceptions of enhanced safety and efficacy. Second, the term ***neuroleptic*** became a pejorative appellation which was reserved for the older drugs, in deference to their belatedly acknowledged toxicities.

With the passing of time, all of the dopamine antagonists have been shown to satisfy the criteria used by Delay and Deniker when they invented the word neuroleptic. For this reason, the terms neuroleptic and antipsychotic shall henceforth be used in this book interchangeably, as ***all of these chemicals are inherently destabilizing and dementing***.

Postmortem Studies of Humans

Studies in the Era of Parkinson's Disease

At the start of the neuroleptic era in the 1950s, psychiatrists believed that it was beneficial to administer drugs in doses which damaged the deep structures of the brain. As a result of these initial misadventures, early autopsy studies of medicated patients were understandably preoccupied with an analysis of the neural pathways associated with Parkinson's disease. One such study is particularly important in the history of pharmacology and toxicology because of its methodology and findings.

The Jellinger Study

In 1977, the Austrian neuropathologist, Dr. Kurt Jellinger, published a comprehensive review of studies which had explored the anatomic changes associated with neuroleptics.[6] Although other investigators had previously divulged findings from research involving animals and humans, the distinguishing feature of Jellinger's work was the inclusion of *new autopsy results* from three different experimental groups:

1. psychiatric patients who had been exposed to neuroleptic drugs

2. psychiatric patients with the *same diagnoses* who had *avoided* neuroleptic drugs

3. non-psychiatric patients with no history of neuroleptic drug therapy.

Jellinger's analysis began with a review of the diagnoses and treatment histories of twenty-eight patients, all of whom had received long-term treatment with dopamine blocking agents. Half of these patients had developed movement abnormalities which included the symptoms of Parkinson's disease (pill rolling tremor, abnormal gait, difficulty initiating movements, blunted emotionality) and/or dyskinesias (abnormal involuntary movements of the facial muscles, the limbs, and/or the trunk). The other fourteen patients had remained motorically "normal." The features of these subgroups are summarized below:

The Jellinger Study		
	patients with movement abnormalities n = 14	patients without movement abnormalities n = 14
mean age at death:	54 yrs	57.8 yrs
mean duration of drug treatment:	5 years	4.4 years
diagnoses:	9 schizophrenia 2 hebephrenia 1 vascular dementia 1 manic/dep., alcoholism 1 involutional psychosis	11 schizophrenia 2 depression 1 mixed psychosis
drugs consumed:	chlorpromazine trifluoperazine reserpine thioridazine tranquilizers tricyclic antidepressants chlorprothixene	chlorpromazine trifluoperazine reserpine thioridazine tranquilizers tricyclic antidepressants chlorprothixene
cause of death:	8 pneumonia 2 overdose 1 heart attack 1 kidney disease 1 pulmonary edema 1 heart failure	6 pneumonia 2 tuberculosis 1 heart attack 1 GI cancer 1 hypertension 1 shock 1 hemorrhagic cystitis 1 pulmonary embolism
damage to basal ganglia:	**9 patients (64%)**	**5 patients (37%)**

Antipsychotics

When Jellinger compared the brains of these 28 patients at autopsy, he was careful to distinguish between changes which he considered to be nonspecific signs of ageing, versus changes arising from each specific cause of death. Only after eliminating all of these features did he move on to consider the implications of any residual pathology appearing in the brain tissue of these patients.

Although damage was noted in several regions of the *basal ganglia* – the collective term for the brain zones which are generally associated with the involuntary control of movement – ***approximately half (46%) of the patients displayed prominent injury to the* caudate nucleus**.

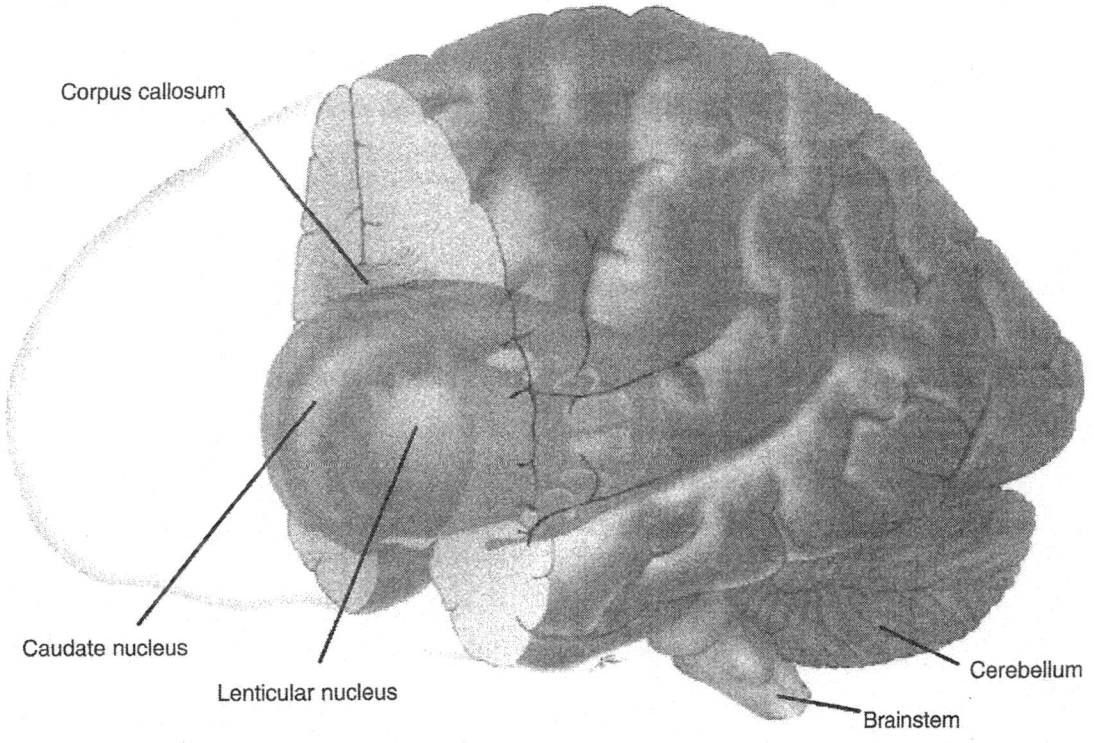

Source: W.J. Hendelman. *Atlas of Functional Neuroanatomy* (Washington, D.C.: CRC Press, 2000), p. 55. Reprinted with permission of Taylor & Francis Group LLC books.

This image shows the location of the **caudate nucleus** deep within the brain.
As a component of the *basal ganglia*, the caudate plays a key role in the modulation of movement, cognition, and mood.

Interestingly, it is possible that Jellinger did not appreciate the significance of his findings in 1977. Although the caudate was not historically privileged in discussions about the cause of Parkinson's disease and dyskinesias, neuroscience has since revealed its role in the control of movement *and other functions*. For physicians and patients in the new millennium, the Jellinger study is arguably one of the most important investigations in pharmacological history. This is because of the fact that Jellinger included two control groups of patients who had *avoided* treatment with neuroleptics (a feat that would be almost impossible to achieve in the current climate of psychiatric care).

Most importantly, the *prevalence* of brain pathology in the drug-exposed was more than 10 to 20 times higher than that which Jellinger had witnessed in previous examinations of patients:

> "As these lesions in the caudate nuclei were only observed in 4%
> of an age matched control group of psychotics without long-term
> neuroleptic treatment and in less than 2% of a large neuropathologic
> routine group of patients, they were tentatively considered significant." [7]

Prevalence of Brain Pathology in the Jellinger Study	
	% with pathology
patients exposed to antipsychotic drugs	**46%**
psychotic patients who *avoided* antipsychotic drugs	4%
routine sample of "neuropathology" patients	2%

The following photos (reproduced from the Jellinger chapter) demonstrate a few of the abnormalities which were visible in the brains of chronically neuroleptized patients.

Antipsychotics

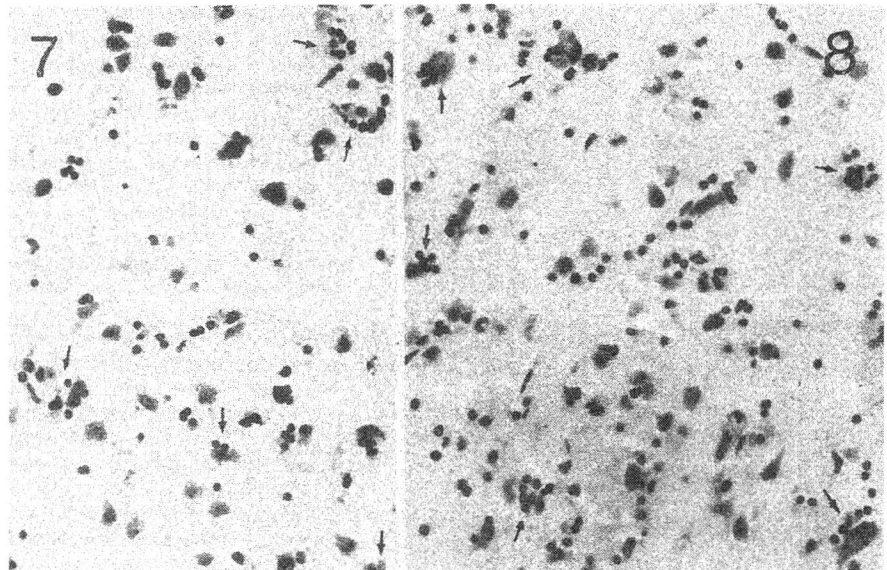

Source: K. Jellinger, "Neuropathologic Findings after Neuroleptic Long-Term Therapy," in L. Roizin, H. Shiraki, and N. Grcevic, Ed. *Neurotoxicology*. (New York: Raven Press, 1977), 34, with permission.

These images depict the ***proliferation of glial cells*** (scar tissue) surrounding neurons in the caudate of a patient who had consumed neuroleptics. Note the rims or "satellites" of glia (black dots) in these slides. This phenomenon, known as ***glial satellitosis***, reflects the brain's attempt to heal in response to chemical injury.

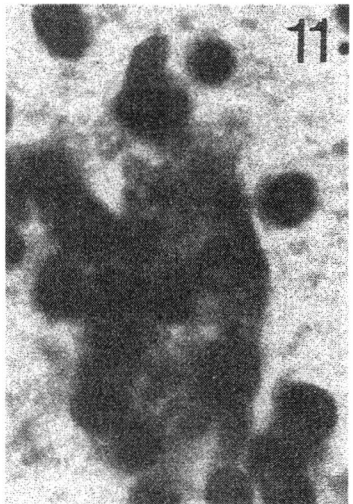

Source: K. Jellinger, "Neuropathologic Findings after Neuroleptic Long-Term Therapy," in L. Roizin, H. Shiraki, and N. Grcevic, Ed. *Neurotoxicology*. (New York: Raven Press, 1977), 34, with permission.

Close-Up View of Glial Satellitosis

In this photo, glia have surrounded a large neuron within the caudate. This cell was obtained from a patient who died at the age of 65 following chronic exposure to neuroleptic drug therapy.

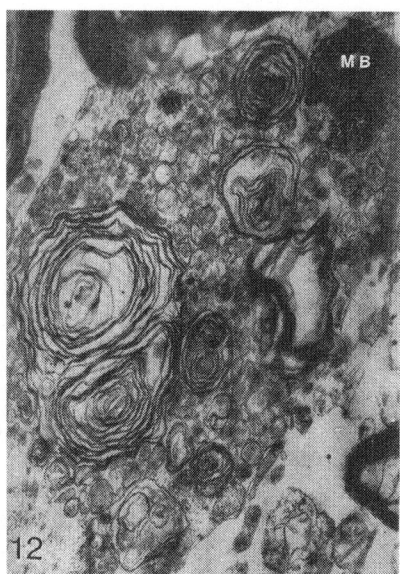

Source: K. Jellinger, "Neuropathologic Findings after Neuroleptic Long-Term Therapy," in L. Roizin, H. Shiraki, and N. Grcevic, Ed. *Neurotoxicology*. (New York: Raven Press, 1977), 35, with permission.

This image shows the **axon terminal** of a neuron obtained from a chronically neuroleptized patient. Several prominent figures known as lamellar bodies or myelin whorls can be seen. This is an example of Drug-Induced Phospholipidosis or DIPL – another marker of chemical harm (see chapter two).

Other Changes Observed by Jellinger ....

Although Jellinger did not publish photos of additional kinds of brain damage from these 28 patients, the text of his report explained other pathologies. For example, he reported "occasional neuronophagia" in the basal ganglia of the medicated patients. *Neuronophagia* refers to the digestion or "eating" of neurons by other cells, such as microglia. Other findings included **astrocytic proliferation** (another response to chemical injury), **swollen neurons** displaying signs of **chromatolysis** (necrotic disintegration), and **cerebral phlebitis** (inflammatory changes within the walls of cerebral veins – also found in patients with systemic lupus).

Validating Jellinger . . .

It is important to appreciate the existence of several insightful publications which *preceded* the 1977 Jellinger paper by more than a dozen years. However, as these earlier investigations had failed to include findings from unmedicated psychiatric controls, it had not been possible for those scientists to definitively confirm the relation between dopamine antagonists (neuroleptics) and degenerative brain changes. *With* the publication of the Jellinger study, however, it has become possible for discerning researchers to retrospectively appreciate the full implications of the historical record.

Antipsychotics

For example, as early as 1959, Roizin et al published a preliminary summary of autopsy results from six patients who had received treatment with antipsychotics.[8] Lacking a human control group, Roizin's team resorted to studies of rats and monkeys in order to characterize the nature of drug-induced, rather than illness-related, brain pathology. Given the importance of replication in scientific research as a method for proving or disproving a theory (in this case, a theory about drug-induced harm), it is significant that Roizin made several observations which were identical to the findings of Jellinger:

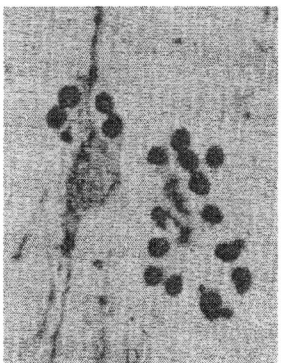

Source: L. Roizin, C. True, and M. Knight, "Structural effects of tranquilizers," *Research Publications – Association for Research in Nervous and Mental Disease* 37: (1959), 306.

Above: This image (Roizin study) was taken from the cerebral cortex of a patient who had received chlorpromazine (Thorazine). The photo depicts glial satellitosis (left) and neuronophagia (right). The latter phenomenon is demonstrated by a ring of glial cells which surround the pale center of a dying neuron.

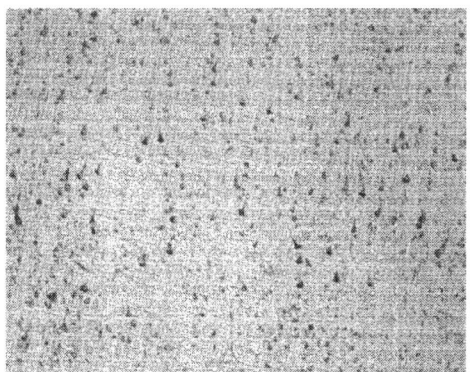

Source: L. Roizin, C. True, and M. Knight, "Structural effects of tranquilizers," *Research Publications – Association for Research in Nervous and Mental Disease* 37: (1959), 305.

Above: This tissue specimen of the **cerebral cortex** was obtained from the brain of a 49-year-old female who had been diagnosed with the condition of paranoid schizophrenia. Following three weeks of treatment with chlorpromazine (Thorazine, 68 mg per day), the suppression of infection-fighting white blood cells resulted in death from pneumonia. The black spots in this picture are the cell bodies of neurons. The *pale regions represent abnormal zones of neuronal depletion (cell death)*.

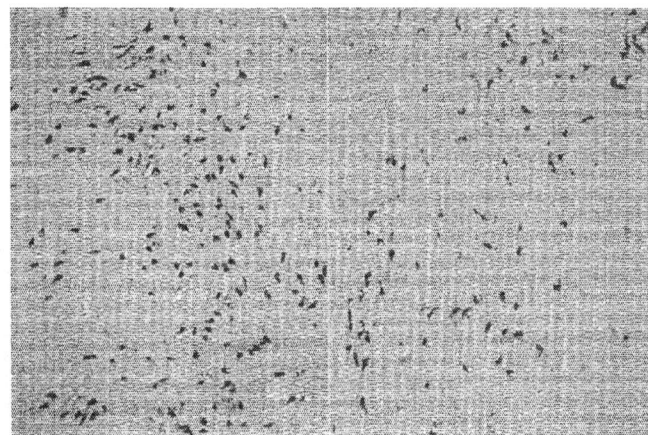

Source: L. Roizin, C. True, and M. Knight, "Structural effects of tranquilizers," *Research Publications – Association for Research in Nervous and Mental Disease* 37: (1959), 309.

Similar evidence of cell death was detected within the **midbrain** of the same patient (shown above). This image shows neuronal loss within the **substantia nigra**. Pathology of this type is one of the main causes of **Parkinson's disease**.

From Parkinson's to Alzheimer's Disease . . .

A puzzling omission from textbooks of psychiatry and neurology is the tie between neuroleptics and neurodegenerative conditions *beyond* Parkinson's disease. These include the currently fashionable "proteinopathies" of **Alzheimer's disease** and **Lewy body dementia**, as well as the increasingly neglected condition of **vascular dementia**. A word about each of these phenomena is in order.

Until 1977, the diagnosis of Alzheimer's disease was reserved for patients *under* the age of 65. Subsequently, changes in diagnostic technologies, disease definition (nosology), screening methods, research funding, global demographics, and the marketing and production of pharmaceuticals contributed to an epidemic of "new" dementias. Critical observers have appropriately challenged the legitimacy of branding illnesses according to age of onset, the temporal sequence of symptoms, and minute details of neuropathology. The nagging truth of the matter is that all dementias – by definition – share overlapping clinical features, and none of them can be distinguished on the basis of *wholly unique* or *exclusive* biochemistry (via the detection of ubiquitin, alpha-synuclein, tau, or beta-amyloid).

Background Information

Before returning to the discussion of neuroleptics and their role in the new dementias, it may be helpful to review some basic biology. First, it is important to recognize the **neuron** as one of two major cell types in the nervous system.

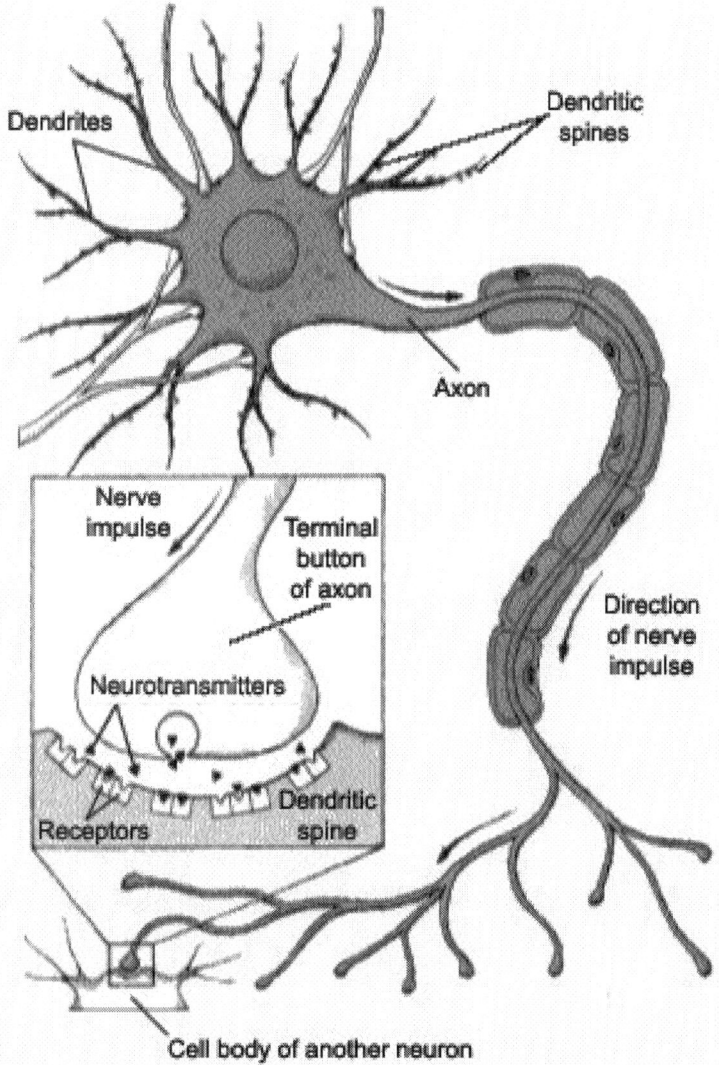

Source: http://thebrain.mcgill.ca/flash/i/i_01/i_01_cl/i_01_cl_ana/i_01_cl_ana.html#1

When patients are diagnosed with the condition of Alzheimer's disease – a diagnosis which can presently be confirmed only by autopsy – a pathologist must identify specific abnormalities *within or around the neurons*. In addition to neuronal death and tissue atrophy, the structural anatomy of an Alzheimer's brain is also characterized by a number of specific tissue features: [9-10]

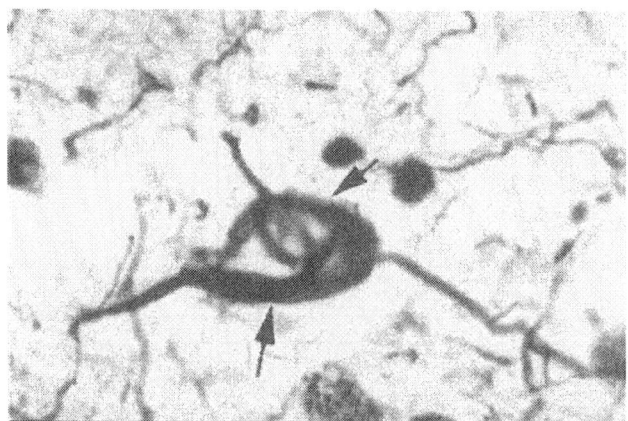

Reproduced with permission of Dr. William I. Rosenblum, VCU Department of Pathology.

neurofibrillary tangles – abnormal clusters inside of neurons, usually comprised of tau protein

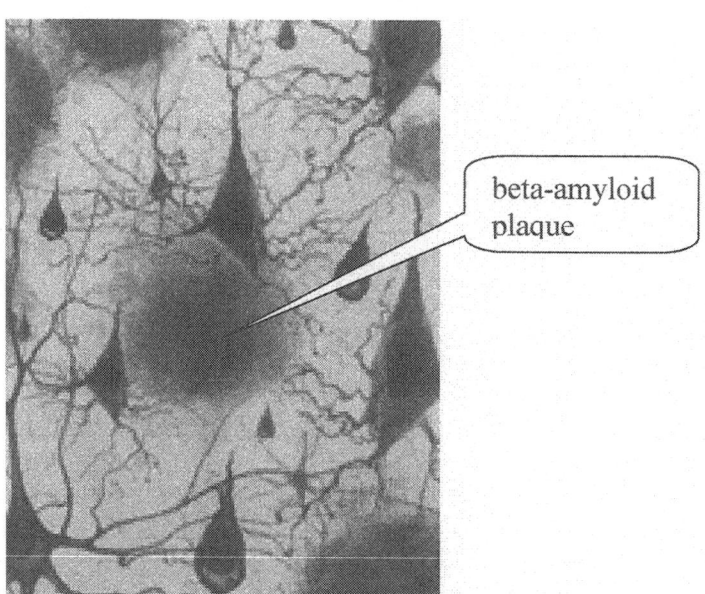

Source: National Institutes of Health. The Hallmarks of AD.

senile plaques – abnormal accumulations of amyloid (and other proteins) which form outside of neurons. So-called "classical" or "diffuse" plaques are frequently found with normal ageing. They are not generally accompanied by glial reactions or neuritic pathology (tangles). In contrast, **beta-amyloid plaques** refer to dense deposits of protein which assume a specific conformation known as a beta-pleated sheet. Thus, the Alzheimer's abnormality is sometimes referred to as a *compact plaque* for the sake of distinguishing pathological versus age-appropriate changes.

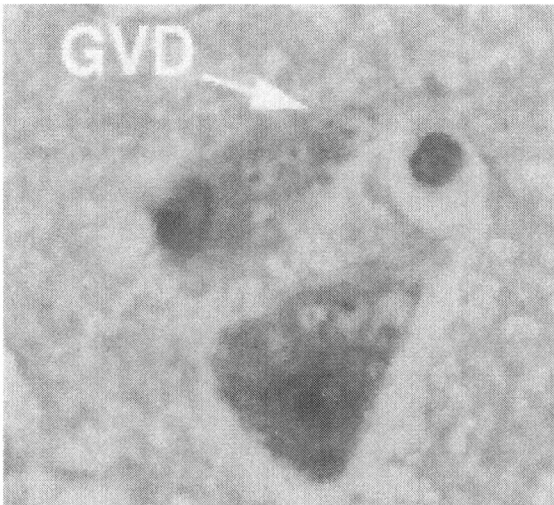

Reprinted with permission of Charles Y. Shao, MD, PhD, Department of Pathology, SUNY Downstate Medical Center, Brooklyn NY.

granulovacuolar degeneration (GVD) - a process which refers to the intraneuronal presence of abnormal, membrane-bound cavities (vacuoles) which contain dense cores of protein. Some researchers have found that these cores contain enzymes which modify tau protein and participate in the formation of neurofibrillary tangles. Others have proposed that these vacuoles hold materials (such as ubiquitin or aluminum) which the neuron has digested and stored via the process known as autophagy [see page 136].[11-12]

For the past three decades, the prevailing doctrine in American medicine has stressed the amyloid-cascade hypothesis of Alzheimer's disease. According to this theory, it is the accumulation of beta-amyloid and other proteins which disrupts neurons and finally destroys the brain. However, neuroscientists are still not sure if the changes which have been briefly described above (tangles, plaques, GVD) are the *cause* rather than the *effect* of brain dysfunction, and a growing minority of clinicians have reasonably questioned the amyloid dogma.[13]

In light of these controversies and uncertainties, the rest of this chapter shall proceed according to three assumptions: 1) that tangles and plaques may truly be a direct cause of disability; 2) that *tangles and plaques are even more likely to be the **effects** of a preceding toxic process which is, itself, the primary cause of cognitive dysfunction*; and 3) that vascular abnormalities – such as clots, inflammation, or bleeds – are due far more attention than they currently receive, based upon their role in the induction or progression of *all* dementias.[14]

Study #1   Prohovnik et al (1993)   -   New York State Office of Mental Health

In 1993, Prohovnik et al published the findings from more than 800 patients, aged 50 or older, who had been consecutively autopsied between January 1978 and December 1987.[15]   Although subjects had been drawn from a network of mental institutions throughout New York State, all autopsies were performed at the New York Psychiatric Institute. The goal of the investigation was to identify the prevalence of Alzheimer's pathology among three groups of patients: 544 patients with the diagnosis of schizophrenia, 258 with the diagnosis of dementia, and 47 patients with disorders of mood.

Two methods were used to identify Alzheimer's changes.   First, the researchers examined existing records of autopsy findings and antemortem conditions. Second, the investigators performed a verification experiment to check the validity of the previous autopsies. After preparing their own tissue specimens from a sample of cases, the team conducted their own analysis of brain structure (Bielschowsky staining method).

Using conservative criteria to define the presence or absence of Alzheimer's disease, ***the researchers identified full-blown Alzheimer's disease in approximately 30% of the schizophrenic subgroup***. However, by *applying a *less restrictive definition to define pathology,* ***58% of the same patients were adversely affected***:

% of patients with dementia pathology		
	schizophrenia n = 544	affective disorder n = 47
full Alzheimer's pathology	**28%**	15%
partial Alzheimer's pathology	**58%**	13%
vascular dementia pathology	25%	34%

Although Prohovnik's paper did not divulge treatment histories, the fact that these subjects had been chronically institutionalized during the neuroleptic era was highly suggestive of an association between antipsychotic drug therapy and the emergence or progression of tangles and plaques. Remarkably, when compared to postmortem findings from the general population of the same time period, the ***prevalence of Alzheimer's pathology among the schizophrenia subgroup was approximately 3 to 5 times higher than expected*** (30 to 60% of the patients, versus 13% of the general population aged 65 and above).

---

* Note: The conservative definition required "moderate tangle or plaque pathology affecting the neurites in cortical regions other than the inferior temporal lobe;" the definition of *possible* cases referred to cortical neuritic pathology that was *limited* to the inferior gyrus of the temporal lobe.

Study #2   Purohit et al (1998)   -   Pilgrim Psychiatric Center, NY

In 1998, investigators in the state of New York reported results from a consecutive autopsy analysis of 100 patients with the diagnosis of chronic schizophrenia (age range: 52 to 101).[16] Brain specimens of these patients were compared to 47 "non-schizophrenic" controls (age range: 53 to 106) and to 50 general hospital patients who lacked any history of dementia or psychiatric condition (age range: 52 to 99).

Employing a conservative standard for the detection of Alzheimer's pathology (Khachaturian criteria), the researchers discovered both a **_higher prevalence_** and **_greater intensity of plaque and tangle pathology_** among the schizophrenic subgroup when compared to the age-matched controls.

Although 9% of the schizophrenic patients qualified for Alzheimer's disease according to a strict definition of neuropathology, **_47% exhibited senile plaque formation in the neocortex, and 16% displayed severe tangle pathology_**. Moreover, 23% of the schizophrenic patients showed pathological evidence of ischemic cerebrovascular disease.

	Pilgrim Psychiatric Center Autopsies		
	schizophrenia n = 100	other psych n = 47	non-psychiatric n = 50
mean age:	78.5	76.9	76.5
% with senile plaques	**47.0**	43.0	40.0
plaque density per cm^2	**364.8**	275.2	192.6
average NFT score	3.20	3.10	2.34
% with severe NFT score	**16.0**	10.0	8.0
NFT = neurofibrillary tangle			

Neuroleptics and Lewy Body Dementia

For more than 25 years, an alert minority of clinicians have appreciated the special hazards of neuroleptics for individuals afflicted by Lewy body dementia. Unfortunately, it is only at rare intervals that their observations have appeared in the pages of the mainstream literature.[17-19] Using the term **acute neuroleptic sensitivity reaction**, several authors have reported accounts of severe side effects and abrupt lethality in these patients. Given the fact that Lewy body dementia is both underdiagnosed and misdiagnosed, it is quite likely that many elderly patients have been the unwitting recipients of medications which were contraindicated for their condition. It is also likely that the recent accounts of drug failure and mortality among dementia patients – even for those exposed to the "new and improved" atypical treatments – reflect this underlying vulnerability. Why the neuroleptics pose such dangers for this subgroup will now be briefly explored.

Background Information

The term **Lewy body dementia** refers to a specific neurological condition which is characterized by movement abnormalities identical to Parkinson's disease; by fluctuating deficits in cognitive abilities; and by changes in mood and perception. In fact, one of the distinctive hallmarks of this illness is the early onset, followed by intermittent recurrence, of hallucinations and delusions. Like Alzheimer's disease, Lewy body dementia is another protein-related dementia which can be *definitively* diagnosed only on the basis of pathology. Named for the physician who first observed it, the so-called **Lewy body** refers to an abnormal deposit of protein materials (usually, ubiquitin and alpha-synuclein) within neurons throughout the brain.

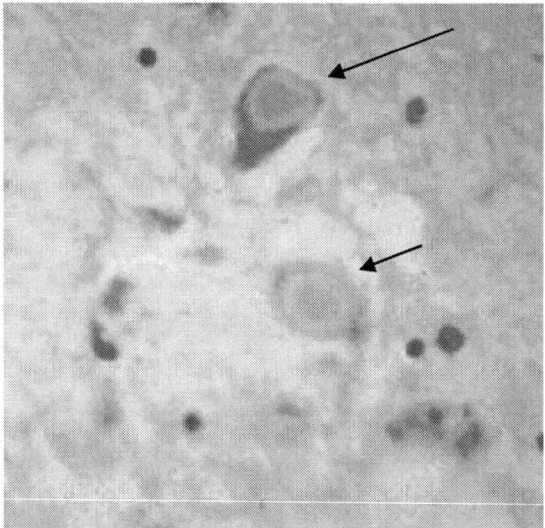

Reproduced with permission of Dr. William I. Rosenblum, VCU Department of Pathology.

This image shows two neurons which contain Lewy bodies. These cells were obtained from the substantia nigra of the midbrain. Lewy bodies can accumulate within the soma (as shown here) or within the axon and dendrites of neurons. They are commonly found within the neocortex, the basal ganglia, and the hippocampus.

Study #3   Ballard et al (2005)   -   Newcastle, UK

In 2005, researchers in Newcastle, England published the results of autopsy findings from 40 individuals who had been treated and closely monitored for Lewy body dementia.[20] Curious to understand more about the problem of **neuroleptic sensitivity** in patients with this condition, the scientists sought to confirm the effects of dopamine antagonists upon the progression of symptoms and brain pathology.

One quick point must be made about the Newcastle patients. Although the defining feature of Lewy body dementia is the presence of Lewy bodies, approximately 10 to 25% of patients demonstrate the same structural abnormalities which are seen in Alzheimer's disease. Some authors refer to these cases as the **Lewy body variant of Alzheimer's** disease, and it is apparently this form of dementia which was prevalent in this U.K. study.

When the investigators compared the density and distribution of Lewy bodies, tangles, and plaques according to treatment history (17 patients had received neuroleptics for a period of at least 6 months, continuing drug therapy through the time of death), they found that *exposure to neuroleptics was associated with a higher density of cortical plaques (30% more) and tangles (65-367% more).* Additional analyses revealed that tangle pathology was significantly and positively associated with the duration of drug treatment. Interestingly, with respect to frontal lobe pathology, risperidone (a newer neuroleptic) was associated with more tangles than the older drugs (1.1 vs. 0.9 per mm^2).

Newcastle Study of Lewy Body Dementia		
	neuroleptics n = 17	no neuroleptics n = 23
median duration of drug therapy	> 1 year	n/a
age at time of death	78.5	78
yrs of dementia	3.2	2.8
% experiencing psychosis (at first assessment of dementia)	65%	71%
tangle density per mm^2 **frontal cortex** **entorhinal cortex**	 0.42 5.6	 0.09 3.4
plaque density (frontal)	9.0	6.9

While these findings were dramatic, it is important to appreciate the fact some physicians have overtly denied a link between neuroleptics and Alzheimer's disease.[21-24] In the past, contradictions in the medical literature have arisen from differences in the neuropathology criteria which have been used to classify "definite" versus "probable" cases. Even at this moment, there are at least three major classification schemes in use by pathologists worldwide. Inconsistencies in the research literature have also reflected the misclassification of psychological versus organic syndromes. Also, without any universal standard for tissue fixation and staining techniques, some research results have been influenced by lab methods which precluded the detection of Alzheimer's pathology.[25-26] Ultimately, when one considers a wide range of historical and emerging research, it is difficult to ignore the breadth and depth of the evidence which has demonstrated neurodegenerative changes following treatment with neuroleptics.

Other Studies of Alzheimer's Pathology After Neuroleptics [27-30]

1986 Brown
autopsy analysis of 77 patients
diagnosed with schizophrenia
    (mean age: 67):   30% with Alzheimer's pathology or vascular dementia

1988 Buhl and Bojsen-Moller
consecutive autopsies of 100 patients
diagnosed with schizophrenia
    schizophrenia, mean age 80   >>>>>>>>   35% with Alzheimer's pathology
    non-psych controls, mean age 78   >>>>>>>>   0% with Alzheimer's pathology

1989 Soustek
225 patients with chronic schizophrenia   41% with Alzheimer's pathology
dying between 1975 and 1985   6X higher than general population

1994 Wisniewski
reviewed histories of 102 patients with schizophrenia
reviewed neurofibrillary tangles (NFTs) and other changes
in temporal lobe:
    41 died prior to neuroleptic era (1932-1952)   >>>>>>   36% with NFTs
    61 died after receiving neuroleptics (1954-1990)   >>>>>>   74% with NFTs

2002 Bozikas
autopsies of 18 institutionalized patients   patients had 400% higher tangle
with schizophrenia vs.14 age-matched controls   density in entorhinal cortex (layer II)
early-onset and late-onset schizophrenia   ↑ senile plaque density (overall)

Note: To put these numbers in context, Alzheimer's disease currently affects approximately 10% of all Americans aged 65 and older.

Biomarkers in Living Humans

If it is true that neuroleptics *cause* the structural abnormalities associated with Alzheimer's disease (or any other form of dementia), then one would hope to find evidence for this process *prior* to death. Ideally, medical professionals would be able to isolate a biomarker which could be used to diagnose and monitor the disease. As the name implies, a biomarker refers to a substance which can be sampled from a subject (from urine, blood, sputum, cerebrospinal fluid) in order to confirm the presence or absence of a given condition.

For more than a century, researchers in neuroscience have been desperate to identify an antemortem biomarker for Alzheimer's disease. Although this work is still ongoing, preliminary results have suggested a positive association between brain pathology and elevated levels of several proteins which can be measured in the spinal fluid. These include:

> **tissue transglutaminase (tTG)** – an enzyme which cross-links proteins, including components of tangles and plaques; some scientists have speculated that tTG is a cause or marker of programmed cell death [31-34]
>
> **tau** – a component of intra-neuronal fibrillary tangles
>
> **14-3-3** – a scaffolding protein which affects the structure of neurons

and

> **apolipoprotein D** – a component of compact amyloid plaques.

It is significant that several research teams have confirmed an association between neuroleptic use and high levels of each of these proteins.

Tissue Transglutaminase

In an effort to investigate the validity of tissue transglutaminase (tTG) as a marker for Alzheimer's disease, a team of Austrian researchers compared the levels of this protein and its association with various illnesses and treatments.[35]

Reporting their findings in 2005, the investigators described the tTG content of spinal fluid specimens obtained from 84 patients, all of whom had been *hospitalized for neurological conditions*. In the course of their hospitalization, twenty-nine of these individuals received neuroleptics for behavioral or pain control (not for psychosis). Subsequently, the researchers were able to compare tTG levels according to their patients' diagnoses (Alzheimer's disease vs. non-Alzheimer's) and according to drug therapy (neuroleptics vs. no neuroleptics).

It is notable that none of these patients had been diagnosed with an acute infectious disease, such as encephalitis or meningitis. This was a critical exclusion, in order to rule out an elevation in tTG which might have originated from white blood cells in the spinal fluid. While the precise role of tTG in dementia remains uncertain, three findings from the Austrian study were disquieting.

Findings

First, the ***consumption of neuroleptics was associated with a 3- to 6-fold elevation in the concentration of tTG*** relative to those who remained neuroleptic drug free:

	Alzheimer's patients n = 33		non-Alzheimer's patients n = 51	
	NL n = 8	no NL n = 25	NL n = 21	no NL n = 30
mean age:	71	74	59	58
mean tTG level (ng/dL):	**14.21**	5.45	**5.72**	1.04

NL = neuroleptics

Drugs included first-generation (flupentixol, haloperidol, melperone, prothipendyl) and second-generation (olanzapine, risperidone, zotepine) dopamine antagonists.

Non-Alzheimer's conditions included vascular dementia (n=18), somatoform disorders (n=10), low back pain, headache, neuropathy, Meniere's disease, trigeminal neuralgia, and spasmodic torticollis.

Second, for the patients who were younger and primarily *without* dementia, the use of neuroleptics was associated with a tTG level that was *indistinguishable* from unmedicated (non-neuroleptized) Alzheimer's disease.

Third, on the basis of associated tTG levels, the allegedly "atypical" second-generation drugs (olanzapine, risperidone, zotepine) failed to differ from the first-generation chemicals in terms of their neurotoxicity:

tTG concentrations (mean level, ng/dL)		
first-generation antipsychotics		mean levels (ng/dL)
total	n = 15	7.81
flupentixol	n = 7	7.86
haloperidol	n = 1	7.3
melperone	n = 4	14.95
prothipendyl	n = 3	3.16
second-generation antipsychotics		mean levels (ng/dL)
total	n = 14	8.11
olanzapine	n = 8	8.5
risperidone	n = 2	5.25
zotepine	n = 4	8.78

14-3-3 and Tau Proteins

In a 2005 publication, researchers in Verona, Italy reported the results of their investigation which measured the concentrations of *two dementia-related proteins* in the spinal fluid of 60 subjects.[36] Participants were divided into five subgroups retrospectively, on the basis of treatment history and disease:

> normal, unmedicated controls (n=8)

> patients undergoing evaluation for the symptoms of dementia,
> all receiving treatment with neuroleptics (n=12)
> [The drugs were chlorpromazine, haloperidol, and olanzapine.]

> patients with definite Alzheimer's disease but no neuroleptics (n=5)

> patients with vascular dementia but no neuroleptics (n=4)

> patients with Creutzfeldt-Jakob disease but no neuroleptics (n=44)

The significance of this study hinged upon the protein pattern which was detected in the patients exposed to neuroleptics. (For the technically curious, tau protein levels were quantified by ELISA; various isoforms of 14-3-3 protein were characterized by gel electrophoresis and immunoblotting.)

Findings

Patients who consumed neuroleptics exhibited biological changes that were consistent with three neurodegenerative conditions: elevated tau protein (as in Alzheimer's disease and vascular dementia) *and* detectable levels of 14-3-3-zeta (as in sporadic CJD):

Biomarkers of Neurodegeneration		
	tau protein mean (pg/mL)	14-3-3ζ protein
normal controls	196	absent
dementia, receiving NL	**244**	***present**
vascular dementia (no NL)	327	absent
Alzheimer's disease (no NL)	278	absent
Creutzfeldt-Jakob disease (no NL) (wide range of values according to genotypes)	607-10,172	**present**

NL = neuroleptics (chlorpromazine, haloperidol, olanzapine)
* 14-3-3ζ was unphosphorylated in CJD but phosphorylated in patients exposed to NLs

Although this study involved a limited number of subjects and pharmaceuticals, its results were highly concerning for *a causal link* between three popular dopamine antagonists and two suspected signals of brain pathology.

Apolipoprotein D (apoD)

As is true for many biomarkers which involve the brain, there is a large and confusing trail of evidence which assigns both toxic and heroic properties to a substance known as apolipoprotein D or **apoD**.

On the one hand, apoD has been found in glial cells and neurons in *response to injury*. The fact that apoD is a protein which participates in the transport of cholesterol, lipids, and certain heme molecules has implied, to some, that apoD plays a role in re-myelination and brain repair.[37-38] On the other hand, not all elevations in apoD have been linked to healing. In 1997, Ong et al found that apoD elevations *preceded,* and then peaked with, neuronal death in the hippocampus of animals injected with a potent excitotoxin (kainic acid).[39] In 1999, Franz et al found that apoD was upregulated in the glia and neurons of rats following traumatic brain injury.[40] In the latter experiment, the cortical increase of apoD was *concurrent* with neuronal death. The significance of these experiments lies in the explicit rejection of the premise that apoD always or predictably prevents or reverses cell death.

It is important to note that several groups of scientists around the world have confirmed an association between elevated levels of apoD in human patients with Alzheimer's disease: [41-43]

> **in 1998:** Canadian researchers implicated apoD as a marker of neuropathology. In their study, they documented a 60% increase in apoD in the hippocampus, and a 350% increase of apoD in the spinal fluid of Alzheimer's disease patients, relative to controls;

> **in 2001:** Swiss scientists replicated the finding of increased apoD in temporal lobe specimens obtained from five patients with Alzheimer's disease;

> **in 2003:** German investigators documented a positive association between apoD levels in the hippocampus, and the overall intensity of Alzheimer's pathology (i.e., Braak staging of tangles and plaques).

More recently, neuroscientists affiliated with the University of Pittsburgh have confirmed that ***apoD is a constituent of the senile plaques that are associated with Alzheimer's disease.***[44] In their complex analyses of postmortem brain tissue, these researchers discovered that all of the patients with Alzheimer's disease manifested plaques that contained apoD and/or beta-amyloid. All of the apoD plaques included beta-amyloid, and 63% of the beta-amyloid plaques contained apoD as a core component:

	Alzheimer's disease n = 36	controls n = 12
mean age:	78	63
% of apoD plaques with beta-amyloid	100%	n/a
% of beta-amyloid plaques with apoD	63%	n/a
apoD protein level (Western blot, relative optical density)	413.87 (n = 15)	100 (n = 5)

By using special immunological staining techniques, the Pittsburgh team was able to examine plaques via double-labeling. When they did so, *they observed apoD in the center of many compact, beta-amyloid plaques*. In other words, they discovered evidence that apoD – at least in these cases – was associated with Alzheimer's pathology:

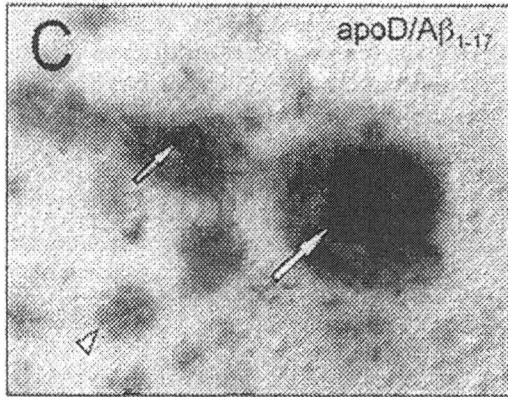

Reprinted from *Neurobiology of Disease*, volume 20, issue 2, P.P. Desai et al, pages 574-582, copyright (2005), with permission from Elsevier.

This photo depicts two large beta-amyloid plaques from a patient with Alzheimer's disease. At the core of these **compact plaques** is the protein known as apoD. (The arrows point to compact plaques with apoD.) In the lower left corner, one can see a small **diffuse plaque** (arrowhead) which is devoid of apoD. Diffuse plaques (which consist of amorphously arranged beta-amyloid protein) are commonly seen in normal ageing.

It is against this background that recent findings in psychiatry and psychopharmacology research become worrisome. First, it is significant that *several animal experiments have confirmed the capacity of neuroleptics – including risperidone, clozapine, and olanzapine – to induce the synthesis of apoD*. In research paradigms involving the short-term administration (14 to 45 days) of these drugs, clinically relevant dosing regimens in mice and rats have resulted in higher mRNA and protein levels of apoD. These changes have been detected in cortical and subcortical structures of the brain.[45-47] Moreover, the **results of animal experiments have been consistent with human treatment effects**.[48] For example, in a postmortem analysis of more than two dozen subjects, an international team of researchers compared the levels of apoD in several brain regions of individuals diagnosed with schizophrenia or manic depression (bipolar disorder). All but four of the patients in this study had received treatment with first-generation neuroleptics:

	schizophrenia n = 20	bipolar disorder n = 8	*controls n = 19
# taking 1st generation neuroleptics	17	6	0
# taking clozapine	1	0	0
# taking no neuroleptics	2	2	19

* 19 individuals were age- and gender-matched for the schizophrenia subgroup; 8 of these 19 individuals were used again as controls for the bipolar subgroup

When brain tissue was analyzed for the content of apoD, *both patient subgroups displayed higher levels of this protein throughout the brain, relative to non-psychiatric controls*:

apoD levels, ug/mg protein			
brain region	schizophrenia subgroup	controls	
dorsolateral prefrontal cortex	**0.244**	0.127	**p = 0.0002**
occipital cortex	0.201	0.196	
caudate	**0.132**	0.078	**p = 0.04**
hippocampus	0.069	0.059	
cerebellum	0.088	0.086	
brain region	bipolar subgroup	controls	
dorsolateral prefrontal cortex	**0.233**	0.115	**p = 0.02**
occipital cortex	0.205	0.220	
caudate	**0.112**	0.059	**p = 0.04**

Interestingly, the investigators failed to identify any differences between the effects of continuous versus interrupted drug therapy. They also found no significant differences between the apoD effects of the so-called "typical" neuroleptics and clozapine.

Unique Georgia Study [49]

To date, very few studies have examined **apoD as a biomarker for illness in living humans**. In one such investigation involving active duty Army personnel, researchers in the state of Georgia were able to evaluate twenty-two patients during their hospitalizations for a first episode of psychosis.

Important elements of this study included the enrollment of young individuals who lacked any history of head injury or seizure disorder, recent substance abuse, and past or present exposure to neuroleptics.

Although the study was limited by the assessment of apoD levels in the peripheral bloodstream (rather than apoD obtained from cerebrospinal fluid), the researchers were able to distinguish significant differences in apoD among three subgroups: subjects with first-episode psychosis, subjects with chronically medicated psychosis, and non-psychiatric controls.

	first-episode psychosis n = 22	chronic psychosis n = 18	controls n = 14
mean age	22.4	44.16	24
age at onset of symptoms	22.4	22.51	n/a
years of illness	4.5 days	23.55 years	n/a
taking clozapine	--	11	n/a
taking haloperidol	--	2	n/a
taking olanzapine	--	2	n/a
taking risperidone	--	1	n/a
taking other neuroleptics	--	2	n/a
plasma apoD (ug/mL)	**43.05**	**52.27**	**33.11**

While it is important to appreciate the fact that *plasma levels of apoD* have not always been found to distinguish psychiatric subjects from healthy controls, the chemical methods of analysis have varied widely. (For example, a 2001 study performed by Thomas et al employed a 1:750 dilution of plasma samples during the apoD measurement protocol, while the present investigation employed a 1:3000 dilution of plasma samples.)

Despite some contradictions in the medical literature, two findings from the Georgia study deserve emphasis here. First, the researchers detected an elevation of apoD which was concurrent with the emergence of psychosis. Second, the elevation of apoD was confirmed again in chronically neuroleptized patients.

Although the Georgia team of researchers attributed beneficial or "curative" properties to the apolipoprotein, based upon their discovery that apoD was further enhanced by psychopharmaceuticals, the opposite conclusion was arguably more probable.

Just as apoD elevation is a marker of brain pathology when it occurs in the context of Alzheimer's disease, inherited storage disorders (such as Niemann-Pick type C), meningoencephalitis, stroke, and demyelinating neuropathies (such as Guillain-Barre syndrome), so, too, is it a likely marker of brain pathology when it is provoked or enhanced by neuroleptic drugs.[50-52]

Neuroimaging Studies of Humans

A long record of radiological evidence corroborates the neurodegenerative effects of neuroleptics. For example, in 1986, a team which included Dr. Nora Volkow (now, the director of the National Institute of Drug Abuse) was able to locate four patients with chronic diagnoses of schizophrenia and *no history of treatment with neuroleptics* (median age: 28, median duration of symptoms: 8 years).[53] The research protocol used positron emission tomography, or PET, which provides an indirect measure of brain activity based upon changes in metabolism. When the PET scans of the patients were compared to twelve healthy controls, the results showed *higher activity in the frontal lobes of those diagnosed with schizophrenia.*

Surprised by these discoveries, the investigators offered the following interpretation:

> "We did not find metabolic hypofrontality in this group of patients with "naïve" [sic: neuroleptic naïve] chronic schizophrenia. This finding is similar to the one reported by Sheppard and by Widen in a group of young patients with schizophrenia…***the results suggest that there may be an association between exposure to neuroleptics and the metabolic hypofrontality pattern.***"[54]   [emphasis added]

In the aftermath of the events which ushered in the so-called **Decade of the Brain** – the moniker given by America's Congress and President G.H.W. Bush to the 1990s – it became increasingly difficult for chronic patients to *avoid* neuroleptics. As clinicians lost the freedom to deliver non-pharmaceutical therapies, so, too, did they lose the chance to discern the natural progression and *resolution* of psychotic phenomena. This development subsequently undermined the ability of contemporary practitioners to appreciate the reality of iatrogenic brain defects.

As a partial solution to this dilemma, some research teams began to focus upon younger patients entering treatment for a *first episode of psychosis*. Here was a population which provided the possibility of tracking changes in brain structure through the repeated performance of neuroimaging exams over time. Although this type of research activity was rarely featured in the medical literature until the late 20th century, there has recently been an explosion of brain scan studies of this kind. While the sheer volume of this work prohibits a comprehensive analysis, the consistency of findings in this area merits a brief review.

Denmark (1998)

In 1998, Madsen et al reported the findings of their study which used x-ray technology (**C**omputerized axial **T**omography, or C.T.) to identify progressive changes in the structure of the brain. Scans were obtained on thirty-one previously unmedicated psychotic patients and nine healthy controls at baseline, and again after a period of five years.[55-56] During this time interval, the patients received neuroleptic therapy consisting of first-generation antipsychotics and/or clozapine (the latter, construed by some to be an "atypical" second-generation drug). Findings were remarkable for the progression of *frontal lobe atrophy* in all patients, relative to controls. Most notably, **the researchers detected a dose-dependent association with brain shrinkage, estimating the risk of frontal degeneration to be 6% for every 10 cumulative grams of chlorpromazine (or other neuroleptic in terms of equivalent dose)**.

University of Pennsylvania (1998)

In a 1998 publication from the University of Pennsylvania, Gur et al revealed the results of a study which used magnetic resonance imaging (MRI) in the analysis of 40 patients with psychosis. These individuals were followed prospectively for 2 ½ years.[57] At the beginning of the study, half of the individuals had already received treatment with neuroleptics and half were neuroleptic-naive. Subsequent to enrollment, all of the participants received treatment with one or more dopamine-blocking drugs: chlorpromazine, fluphenazine, haloperidol, loxapine, mesoridazine, molindone, perphenazine, pimozide, thioridazine, thiothixene, trifluoperazine, clozapine, and risperidone. *At the end of thirty months, the patients displayed significant loss of brain volume (4 to 9%) in the frontal and temporal lobes*. For both patient groups, this change was associated with unimpressive changes in target symptoms (inability to experience pleasure, restricted affect, limited speech) and *significant deteriorations in cognitive functions* (verbal memory, spatial memory, abstraction).

Utrecht, Netherlands (2002)

In 2002, Cahn et al published the initial results of their longitudinal neuroimaging study involving 34 patients diagnosed with schizophrenia (10 treated previously with neuroleptics, 24 neuroleptic naive) and 36 healthy controls matched for age, gender, parental education, and handedness.[58] All subjects received MRI brain scans at baseline and again after one year. During the between-scan time period, the patients received treatment with first-generation neuroleptics (n = 5, specific identities not divulged); second-generation drugs (n = 14: clozapine, olanzapine, quetiapine, risperidone, sertindole, or sulpiride) or both (n = 14: drugs were given in successive trials). *Following one year of pharmaceutical therapy, patients demonstrated an 8% increase in lateral ventricle volume, a 1% reduction in total brain volume, and a 3% reduction in whole brain gray matter*. These changes were *significant statistically* ($p < 0.05$), *clinically* (brain changes were related to poor outcome in terms of psychotic symptoms, physical health, social intimacy, and independence), and *pharmacologically* (gray matter changes corresponded to cumulative neuroleptic dose).

University of Iowa (2003)

In 2003, Ho et al published the first results from a University of Iowa study which enrolled patients between 1991 and 2001.[59] This investigation was designed to monitor changes in the brain anatomy of 73 psychotic patients (diagnosed with relatively new-onset schizophrenia) and 23 healthy controls. The research protocol called for a series of MRI examinations to be performed at various intervals (2, 5, 9, 12 years), but the first period extended slightly beyond the planned duration (three years rather than two). Even at an early stage of development, however, several findings were striking.

First, *following three years of exposure to neuroleptics, patients demonstrated significant reductions in the volume of their frontal lobes (0.2% decrease per year). This loss was associated with an increase in the severity of negative symptoms,* including loss of pleasure, limited speech, and blunting of affect.

Second, *40% of these patients failed to achieve remission of psychosis* (defined as eight or more consecutive weeks with nothing more than mild symptoms).

Third, *these outcomes occurred despite the closely monitored use of neuroleptics (drugs were consumed throughout 84% of the follow-up interval) and despite the preferential administration of the newest therapies* (second-generation neuroleptics were used throughout 62% of the follow-up interval).

Reflecting upon these disappointing results, the research team conceded:

> "*...the medications currently used cannot modify an injurious process occurring in the brain, which is the underlying basis of symptoms...We found that progressive volumetric brain changes were occurring despite ongoing antipsychotic drug treatment.*"[60]   [emphasis added]

U.S.A./Canada/Netherlands/U.K. (2005)

In 2005, Lieberman et al reported the results of a multi-center international investigation (including the Utrecht study mentioned above).[61] Conducted between 1997 and 2001, the study enrolled 161 first-episode psychotic patients and 58 healthy controls with the goal of evaluating longitudinal changes in brain volume. Most patients (67-77%) had received prior neuroleptic drug therapy for a period of four months or more. Psychiatric participants were randomly assigned to receive treatment with olanzapine (5 to 20 mg per day) or haloperidol (2 to 20 mg per day). In retrospect, the study suffered from many experimental difficulties, including the allowance of many additional drugs (e.g., minor tranquilizers, antidepressants, and mood stabilizers) and an extraordinarily high rate of attrition (*only 29% of the olanzapine and 13% of the haloperidol patients returned for the 2-year brain scan*). Nevertheless, because of its consistency with the rapidly expanding literature, this study warrants brief review.

Following one year of exposure to drug therapy, recipients of olanzapine displayed shrinkage of the *frontal, parietal, and occipital lobes* ($p < 0.05$). At the end of two years (24 patients re-scanned), *atrophy persisted in the occipital lobe.* By comparison, *haloperidol recipients experienced early and relentless reductions in gray matter volume throughout the brain* ($p < 0.007$):

Average Change in Tissue Volume (cubic centimeter) by Week 52			
	olanzapine	haloperidol	controls
frontal gray	- 3.16	- 7.56	+ 0.54
parietal gray	- 0.86	- 1.71	+ 0.70
occipital gray	- 1.49	- 1.50	+ 0.99
whole brain gray	- 3.70	- 11.69	+ 4.12

Most importantly, these and other changes – such as increases in whole brain fluid and increases in the volume of the lateral ventricles – were associated with an unfavorable response to treatment.

Other 1st Episode Psychosis Studies: Before and After Neuroleptic Drugs [62-73]

Publication	Scan / Interval	Brain Changes After Neuroleptics
1997 DeLisi et al 50 patients vs. 20 controls	MRI ≥ 4 yrs	↓ cortex/cerebellum/corpus callosum
2001 Thompson et al 12 teens (schiz.), 12 controls	MRI 4.6 yrs	severe gray matter loss in frontal, parietal, temporal lobes
2001 Mathalon et al 24 ♂ schizophrenia 20 ♂ controls	MRI 4 yrs	↓ frontal and temporal gray ↑ left lateral ventricle
2002 Bustillo et al 10 schiz. vs. 10 controls	MRS 1 yr	↓ frontal N-acetyl-aspartate (NAA)
2003 Kasai et al 13 1st episode schizophrenia 15 non-psychiatric controls	MRI 1.5 yrs	↓ left temporal lobe gray matter
2005 Whitworth et al 21 first-episode psychosis 20 controls	MRI 2-4 yrs	↓ amygdala and hippocampus ↑ ventricles
2006 Whitford et al 25 1st episode schizophrenia 26 controls	MRI 2-3 yrs	gray matter atrophy in frontal, parietal, temporal lobes
2007 van Haren et al 96 patients (schizophrenia) 113 healthy controls	MRI 5 yrs	↓ left frontal and left temporal gray ↓ right caudate, ↓ right thalamus
2007 Theberge et al 16 1st episode schizophrenia	MRI 2.5 yrs	widespread gray matter loss
2007 Nakamura et al 29 1st episode schizophrenia	MRI 1.5 yrs	↓ frontal and temporal gray ↑ left ventricle, ↑ sulcal CSF
2008 Reig et al 21 psychotic teens, 34 controls	MRI 2 yrs	↓ frontal gray matter (males)
2008 Koo et al 39 1st episode schizophrenia 40 healthy controls	MRI 1.5 yrs	↓ cingulate gyrus gray

Postmortem Studies of Animals

Without the technological developments of computer processing, genetics, molecular biology, immunocytochemistry, and advanced microscopy to guide them, psychiatrists in the 1950s characterized the first neuroleptics solely by their empirical effects.

It is instructive to review some of the early commentaries which reflected the initial philosophy of pharmaceutical care. For instance, in 1964, physicians in Australia recorded some of these early treatment ideas:

> "We are quite certain that *Parkinsonism is reversible and therefore consider it important to push the drug until psychiatric improvement has been maintained..." [74]

and, similarly:

> "The most important effect on the central nervous systems of both Largactil [chlorpromazine] and Serpasil [reserpine] is the production of Parkinsonism. This is entirely of the classic variety. Practically any patient who receives either Largactil or Serpasil can exhibit Parkinsonism if the dose is sufficiently high, and if the medication is continued sufficiently long." [75]

---

* Parkinsonism – The term *Parkinsonism* or even *pseudo-Parkinsonism* is frequently used in the medical literature to refer to **drug-induced symptoms of Parkinson's disease**. In contrast, the term *Parkinson's disease* refers explicitly to features which are caused by a specific anatomic abnormality: the loss of dopamine neurons from the substantia nigra in the midbrain. There are good reasons to avoid the use of "Parkinsonism" when speaking about consumers of neuroleptics. First, the origins of Parkinson's disease are no longer as simple or sharply localized as pathologists once considered them to be (i.e., damage to the caudate, the locus coeruleus, and the cholinergic neurons of the forebrain may be just as important as damage to the substantia nigra). Second, it is now clear that neuroleptics (and other drugs) do far more than disrupt neuronal functions. By destroying the actual *structures* of the basal ganglia, these chemicals truly replicate both the symptoms *and* neurodegenerative pathologies of Parkinson's disease.

Eventually, as neuroscientists gradually uncovered the chemical and physiological determinants of Parkinson's disease, it became clear that the neuroleptics were a cause of the brain defects associated with that condition. However, in a bizarre turn of history, these early discoveries were apparently regarded as proof of therapeutic *efficacy* rather than *neurotoxicity*.

For example, when Mackiewicz and Gershon found that clinically relevant doses of oral chlorpromazine (18 mg/kg per day for 1 to 3 months) induced degenerative changes in the cerebellum and brainstem of guinea pigs, their tone was generally dispassionate. In 1964, their published comments were completely lacking in terms of any discussion about drug dangers or concern for human subjects:

> "Chronic alternations of neurons were found to occur quite extensively. The most prominent chronic alterations of the nerve cells in the form of vacuolization of the cell body, neuronophagy and concomitant glial reaction are found especially in the reticular formation...
>
> "Active glial response of the **degenerative process of the neurons** was confirmed by the presence of an active proliferative process of the glial elements between the **damaged neurons** and especially in the form of mobilisation of the microglial elements..."[76]   [emphases added]

Today, what is missing from the pages of standard textbooks of psychiatry and pharmacology, and missing equally from medical school and postgraduate curricula, is any mention of this confused and horrific legacy.

In retrospect, it is truly remarkable just *how much* anatomical research was conducted between the 1950s and the late 1970s, and just how strongly and consistently this literature demonstrated the brain damaging effects of neuroleptics. Once again, it is important to appreciate the scholarship of Dr. Kurt Jellinger who, in the aforementioned publication (1977), provided a generous overview of these early results:

Neuropathology Observed in Animals
(adapted from Jellinger) [77]

CPZ = chlorpromazine, RES = reserpine, HAL = haloperidol

1958	Kemali et al	rabbit	CPZ 7.5 mg/kg x 12 days
			RES 7.5 mg/kg x 12 days
		pathology:	neuronal lesion, gliosis
		location:	basal ganglia, hypothalamus
1958	Roizin et al	rat, monkey	CPZ 12.5 mg/kg x 8 months
		pathology:	neuronophagia, chromatolysis, gliosis
		location:	diffuse
1962	Guyeniseman	rat	CPZ 5.0 mg/kg x 30 days
		pathology:	neuronal lesion
		location:	diffuse
1964	Mackiewicz & Gershon	guinea pig	CPZ 10 mg/kg x 4-14 weeks
		pathology:	neuronal lesion, gliosis
		location:	brainstem (reticular formation)
1966	Cazzullo et al	rabbit	CPZ clinical dose x 12 months
		pathology:	neuronal swelling
		location:	brainstem
1967	Dom	rat	HAL therapeutic dose x 4 months
		pathology:	gliosis
		location:	limbic system
1970	Sommer & Quandt	rabbit	CPZ 3-16.7 mg/kg x 6 months
		pathology:	neuronal loss, gliosis
		location:	cortex, hippocampus, globus pallidus
1973	Koizumi et al	rabbit	CPZ 2.0 mg/kg x 2-5 months
		pathology:	axonal and dendritic degeneration
		location:	hypothalamus, hippocampus
1975	Hackenberg et al	rat	CPZ 10-15 mg/kg x 6-8 weeks
		pathology:	neuronal degeneration, neuronal loss
		location:	superior olive (also: diffuse gliosis)

Note: An unfortunate limitation of the Jellinger review was a failure to describe routes of administration (oral, IP, IV). Nevertheless, these studies involved doses that were therapeutically relevant for humans on a dose per weight basis (mg/kg).

Regrettably, as soon as biological reductionism and drug therapy became the prevailing psychiatric approaches in many industrialized societies following World War II (progressing steadily from the 1950s, but exploding in the 1980s and beyond), efforts to define neuroleptic brain hazards all but disappeared. It is in the context of this troubled history that the efforts of a few contemporary research teams have been particularly valuable and audacious.

Study #1   The Birmingham Study   -   Haloperidol in Rats

Despite the longstanding recognition of haloperidol (Haldol) as a precipitant of Parkinson's disease and other dyskinetic syndromes, few – if any – scientists had investigated the basis of this toxicity in the basal ganglia.[78] In 2002, researchers from the University of Birmingham (U.K.) filled this research void when they reported the effects of this neuroleptic upon the striatum (caudate/putamen) and the midbrain (substantia nigra) of rats.

Experimental Protocol

The study involved the use of adult male Sprague-Dawley rats. First, researchers assigned the animals to four different arms of the experiment: no medication vs. low, medium, or high doses of haloperidol. Second, they administered to each animal a fixed dose of medication (1 mg/kg, 4 mg/kg, or 12 mg/kg) via a single (one-time) intraperitoneal injection. Following the administration of haloperidol, the animals were euthanized. Brain tissue was harvested and processed for microscopic examination using two different methods of analysis. One procedure involved the processing of specimens via a technique known as **TUNEL** staining, in order to identify fragmented DNA in dead or dying neurons. The second procedure involved the use of immunocytochemistry (antibody techniques) in order to identify **microglia**. As mentioned in chapter two, microglia are small cells which are activated during times of injury, infection, and/or inflammation in order to clear away dead cells and cell debris.

Findings

Neuronal Death at Clinically Relevant Doses

Cell death was observed in all of the animals. However, **haloperidol induced levels of cell death which were 30 to 80% greater than the unmedicated controls**. These differences were documented in the two brain regions – the striatum and the substantia nigra – which are considered to be the primary anatomical determinants of Parkinson's disease:

mean # of dead neurons per brain section			
		striatum	substantia nigra
no drug	(n=5)	7.6	2.24
1 mg/kg	**(n=4)**	**10.1**	**2.92**
4 mg/kg	(n=5)	13.6	2.73
12 mg/kg	(n=5)	12.2	3.23

Note: Neuron counts reflect that analysis of coronal slices of brain tissue stained with silver methenamine (three to five sections of tissue from each brain region, per animal).

Most importantly, the brain toxicity in this study occurred in response to a *single injection* of haloperidol. When one considers the blood and brain concentrations which have been documented for a 1 mg/kg intraperitoneal injection in rats (blood level = 14 ng/mL, associated brain level = 309 ng/mL), the results observed in the 1 mg/kg subgroup of animals in this experiment were arguably relevant for human patients (human blood levels of haloperidol = 4-56 ng/mL, associated brain levels = 12-1680 ng/mL).[79-81]

Activation of Microglia

The U.K. researchers also confirmed that the drug-induced injury in their experiment was associated with the **activation of microglia** (reactive gliosis). This phenomenon may have been the cause of neuronal death or simply a response to neuronal injury. In either event, it was significant that the scientists identified microglial proliferation in multiple regions of the brain, including the striatum, the substantia nigra, and the hippocampus.

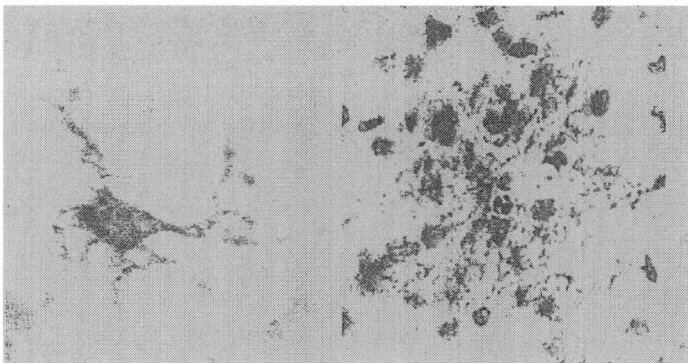

Reprinted from *Neuroscience*, volume 109, issue 1, I.J. Mitchell et al, p. 92, copyright (2002), with permission from Elsevier.

This image depicts two microglia following exposure to haloperidol. Although it is difficult to discern the details, the microglial cell on the right is shown in the process of digesting an apoptotic (dying) neuron. This is an example of reactive gliosis.

> In Sum:
>
> The U.K. team employed multiple advanced techniques which confirmed the neurotoxicity of haloperidol.
>
> The dose which was used (1 mg/kg IP injection) was relevant for humans.
>
> The distribution of pathology was consistent with the anatomic determinants of Parkinson's and other forms of basal ganglia disease.
>
> Even a single injection of the neuroleptic was sufficient to inflict cell death in subcortical structures of the brain.

Study #2   University of Pittsburgh   Monkeys

Curious to know if lab preparation techniques could alter (confound) the results of studies involving the brain effects of neuroleptics, scientists at the University of Pittsburgh performed a series of investigations involving monkeys.[82]

Experimental Procedure

Eighteen adult male macaques (aged 4.5 to 5.3 years) were divided into three groups and were trained to self-administer drug treatments. ***Monkeys received oral doses of haloperidol (27 months), placebo (27 months), or olanzapine (17 months)***. Blood samples were taken periodically and drug doses were adjusted in order to achieve plasma levels equal to, or lower than, those which occur in the treatment of humans (1-1.5 ng/mL for haloperidol; 10-25 ng/mL for olanzapine). At the end of the treatment period, the animals were euthanized. Brains were removed, and brain size (volume and weight) was quantified using several experimental methods (for volume: Archimedes' fluid displacement method and Cavalieri's principle; for weight: calibrated lab scale).

Findings

A variety of behavioral and anatomical effects were observed.

First, all animals appeared to develop an aversion to the taste and/or subjective effects of the medications. This required creative changes in the methods which were used to administer the drug treatments.

Second, a significant number of monkeys became aggressive during the period of study (four of the six monkeys exposed to olanzapine; two of the six monkeys exposed to haloperidol). One monkey, originally placed in the sham treatment group, engaged in self-mutilatory behaviors. A switch to olanzapine resulted in no improvement. However, when the animal was provided with increasing human contact, a doubling of cage space, a decrease in environmental stimuli, and enhanced enrichment, his behavior stabilized.

Third, chronic exposure to the neuroleptics resulted in significant reductions in brain volume and weight, relative to controls (8.8% lower volume for haloperidol, 10.5% lower volume for olanzapine). Regional changes were also significant, with the greatest atrophy identified in the frontal and parietal lobes:

Mean Changes in Fresh Brain Weight (relative to controls)		
	haloperidol	olanzapine
left cerebrum	↓ 8.9%	↓ 10.9%
frontal lobe	↓ 10.1%	↓ 10.4%
temporal lobe	↓ 6.9%	↓ 10.5%
occipital lobe	↓ 9.7%	↓ 9.8%
cerebellum/brainstem	↓ 8.7%	↓ 8.5%
parietal lobe	↓ 11.2%	↓ 13.6%
parietal gray	↓ 11.8%	↓ 15.2%
parietal white	↓ 13.3%	↓ 12.7%

Based upon these results, the researchers concluded that the progressive reductions in brain size which had been historically observed in many studies of schizophrenia "may have reflected the effects of drug treatment." They proposed that further studies be undertaken in order to characterize the precise targets (neurons or glia) and mechanisms of these effects. Subsequently, the same team of scientists published two follow-up reports which described their continuing analyses of drug-related changes in the parietal lobe.

Study #3   **University of Pittsburgh Monkey Study**   (*Follow-Up Study #1*)

Brain tissue from the animals in the previous experiment was subjected to further inspection.[83]   The left cerebrum was divided into five anatomical regions: frontal, parietal, temporal, occipital, cerebellum + brainstem. Sections of **parietal lobe tissue** were then taken from each animal, processed with Nissl stain (a specific treatment which labels an intracellular unit, called the endoplasmic reticulum), and analyzed using a light microscope. In this particular experiment, the researchers performed a computer-assisted counting procedure (stereological assessment) in order to identify the number of cells.

Background Information

For the purpose of orientation, it may be helpful to recall that the **parietal lobe (cortex)** is a brain region which coordinates many functions. In humans, these include speech, mathematical activity, visual perception, spatial orientation, and the integration of sensory information. The picture below shows the location of the parietal lobe in the human brain:

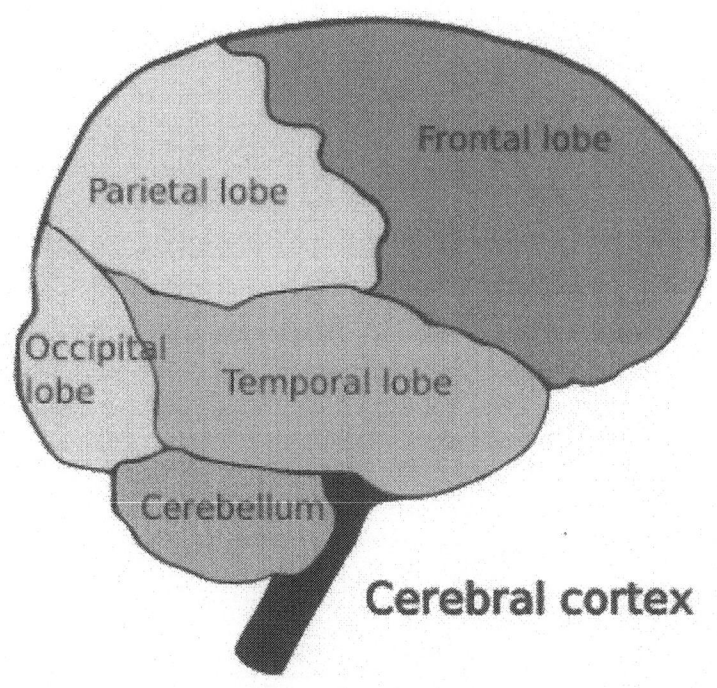

Source: Wikimedia commons. http://en.wikipedia.org/wiki/File: Brainlobes.svg

---

In comparison to the unmedicated monkeys, *the animals who received haloperidol or olanzapine experienced a 6 to 20% reduction in the number of brain cells within the left parietal lobe*:

Left Parietal Lobe			
	mean # of cells x 10^6	cell number vs. drug-free control	
	control	haloperidol	olanzapine
total cells	460	↓ 10.6%	↓ 7.4%
neurons	202.1	↓ 6.3 %	↓ 5.5%
glia	175.3	↓ 18.1%	↓ 10.3%
endothelial	64.7	↓ 7.1%	↓ 5.2%

Study #4   University of Pittsburgh Monkey Study (Follow-Up Study #2)

In a second follow-up study, the Pittsburgh researchers employed two special lab techniques (immunocytochemical labeling and fluorescent microscopy) in order to explore drug effects upon specific populations of glial cells.[84] This time, they detected a drug-induced reduction in the number of **oligodendrocytes** (the cells which produce the fatty lining, or myelin, which ensheaths axons) and **astrocytes** (the glial cells which assume key roles in brain metabolism, the maintenance of the blood-brain barrier, and neurotransmission).

Relative to the unmedicated monkeys, *prominent reductions were noted in both glial cell types*. These reductions were essentially equal for both neuroleptics:

	haloperidol	olanzapine
oligodendrocyte number	↓ 13.9%	↓ 11.8%
astrocyte number	↓ 20.4%	↓ 20.5%

In Sum:

The University of Pittsburgh team employed a number of experimental procedures which confirmed the toxicity of haloperidol *and* olanzapine.

Both drugs resulted in prominent reductions of brain volume and brain weight. These changes were particularly significant in the frontal and parietal lobes.

Follow-up studies revealed a 6 to 20% reduction in the number of neurons *and* glia (astrocytes and oligodendrocytes) within the left parietal cortex.

Chapter Six

Anxiolytics (Anti-Anxiety Drugs)

Until the invention of discrete anxiety disorders by authoritative decree, as evidenced in the 1980 publication of the American Psychiatric Association's diagnostic handbook (*Diagnostic and Statistical Manual of Mental Disorders, 3rd Ed.*), there was no such thing as a specific "anti-anxiety" drug. Rather, throughout the 19th and early 20th centuries, neurologists, psychiatrists, and anesthetists experimented with a series of chemicals which were known to induce sedation and unconsciousness. Based upon their behavioral effects, these medications became known as the *__minor tranquilizers.__[1-2]

Unfortunately for patients, however, there existed in the 1800s and early 1900s no regulatory requirements, no professional motivation, and no substantial scientific capacity to identify the brain toxicities of the minor tranquilizers. Thus, despite the fact that each new pharmaceutical eventually replicated the problems of its predecessors – by inducing cognitive impairment, addiction, and lethality – physicians were only too glad to welcome each novel potion based upon expectations of superior utility. Ultimately, the enthusiastic reception of **meprobamate and the barbiturates resulted in an epidemic of tranquilizer dependence and – through accidental and intentional overdoses – thousands of cases of respiratory suppression, coma, and death.

It was in the context of a long and toxic legacy that the discovery of the benzodiazepines was heralded as a monumental achievement in the history of medicinal chemistry. Appearing on the American market in 1960, the latest tranquilizers were perceived as almost "too good," and it was their *efficacy* which sparked dichotomous and ambivalent reactions. On the one hand, benzodiazepine enthusiasts greeted the drugs as magical elixirs which promised "joy, peace, loving-kindness, and beauty in a pill."[3] On the other hand, some physicians and philosophers worried about the implications of pharmaceuticals which would blunt the very psychic reactions that make human existence meaningful, and which make human survival possible. Meanwhile, social commentators worried that the drugs would be used as a tool of oppression, subduing the responses of victims and diverting attention away from redressing the root causes of anxiety via political, economic, and other kinds of reform.

---

* The term **major tranquilizer** was reserved for the dopamine-blocking agents, such as chlorpromazine. This was replaced by the word "neuroleptic" in the late 1950s, in deference to the tendency of those chemicals to cause not only sedation, but also (or especially) the symptoms of Parkinson's disease.

** meprobamate – a carbamate derivative first marketed in the U.S.A. in 1955 as Miltown; the muscle relaxant, carisoprodol (Soma), derives its potential addictiveness and lethality from the fact that it is a pro-drug which is metabolized to meprobamate

	A Short History of Psychiatric Tranquilizers and Anti-Anxiety Drugs [4-9]		
	discovery	initial use	contemporary examples
paraldehyde	1829	1882	no longer used
bromides*	1857	1857	no longer used by psychiatrists
barbiturates	1864	1904	pentobarbital, phenobarbital, thiopental
meprobamate	1954	1954	metabolite of carisoprodol, aka Soma
MAOIs	1951	1952	isocarboxazid, tranylcypromine
benzodiazepines	1955	1960	chlordiazepoxide, diazepam, clonazepam
TCAs	1955	1958	imipramine, clomipramine, amitriptyline
SSRIs	1971	**1982	fluoxetine, sertraline, paroxetine
azapirones	1980s	1986	buspirone
SNRIs	1980s	1993	venlafaxine, duloxetine

MAOIs = monoamine oxidase inhibitors
TCAs = tricyclic antidepressants
SSRIs = selective serotonin reuptake inhibitors
SNRIs = serotonin norepinephrine reuptake inhibitors

* Bromides are still used in some countries as treatments for epilepsy; highly toxic to the brain, it was bromide which more than likely contributed to the psychosis, mania, and eventual suicide of the famous painter, Vincent Van Gogh.

** SSRIs: The first SSRI, zimelidine, was launched in Europe in 1982. It was withdrawn from the market shortly after its release due to its association with the neurological disorder known as Guillain-Barré syndrome.

Note: Although the MAOIs, SSRIs, SNRIs, and other modulators of monoamines are all used to treat the symptoms of anxiety, they are classified first and foremost as *antidepressants*. The toxicities of the SSRIs are addressed in detail in chapter four.

Eventually, the hazards of benzodiazepines came to the fore. According to historian, Dr. Susan Speaker, the U.S. Congress convened no fewer than twelve investigations between 1958 and 1979, questioning the conduct of the pharmaceutical industry, in general; and probing the "over-use" and "misuse" of minor tranquilizers, in particular. Only with the passage of time did it become clear to members of the medical community that the benzodiazepines could be just as problematic as meprobamate and the barbiturates, in terms of their mind-impairing properties and their capacity to cause physiological and psychological dependence. Today, the risks of tranquilizers have been largely repudiated by revisionist voices and marketing forces which, in America, extol the virtues of lifelong pharmaceuticals for many problems. To the detriment of patients, few clinicians heed the explicit warnings which appear on the labels of benzodiazepines, explicitly encouraging only *short durations* of exposure. Fewer still – by word or by deed – appreciate the potential of these drugs to induce long-lasting cognitive dysfunction.

Neuroimaging Studies of Humans

The story of the benzodiazepines is a tale of many failures. First, there was a failure on the part of the medical profession to address the problem of tachyphylaxis (tolerance = diminished efficacy over time). Second, there was a failure of clinicians to anticipate the problems of drug withdrawal and addiction. Third, there was a failure of toxicologists and regulators to identify the neuropathological consequences of these drugs.

But for the time-limited attention of a few European investigators more than two decades ago, it is possible that these failures would be even worse. Understanding the neurotoxicities which arise from the chronic or excessive use of alcohol, and appreciating the chemical similarities between alcohol and the benzodiazepines (all of which enhance the transmission of GABA), a handful of researchers in the 1980s and early 1990s decided to explore the possibility that the anxiolytics might alter the structure of the human brain.

Background Information

In the 1970s and 1980s, the primary neuroimaging technology was **c**omputerized axial **t**omography or "C.T." Based upon differences in the uptake of radiation, this diagnostic tool could provide an inside view of the skull and its contents. When researchers employed C.T. in the early studies of benzodiazepine effects upon the brain, two different assessments were frequently made: 1) a determination of the VBR (ventricle-to-brain ratio), and 2) a determination of cortical atrophy. Investigators would begin by *measuring the size of the largest fluid-filled spaces inside the brain, known as the lateral ventricles* (shown below):

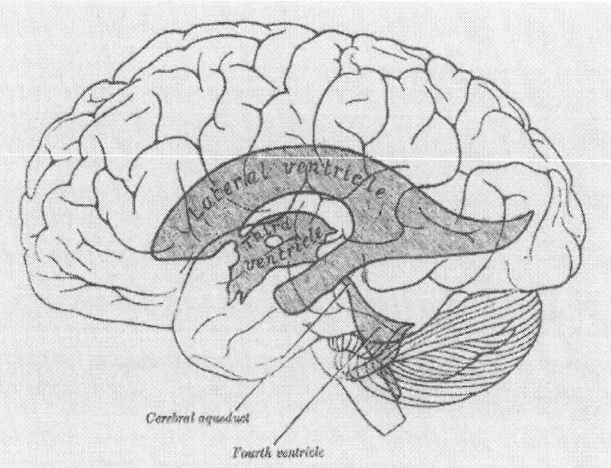

Source: Wikipedia Commons: Image from *Gray's Anatomy, 20th Ed.*, now in public domain.

Next, the cross-sectional area of the ventricular cavity would be divided by the cross-sectional area of the brain around it. The resulting ratio was known as the **ventricle-to-brain ratio or VBR**. The significance of this measurement was based upon the assumption of ratio-related pathology: the larger the VBR, the greater the extent of brain injury or neurodegeneration.

For the purpose of orientation, the anatomy of the brain can be examined according to three different planes or orientations:

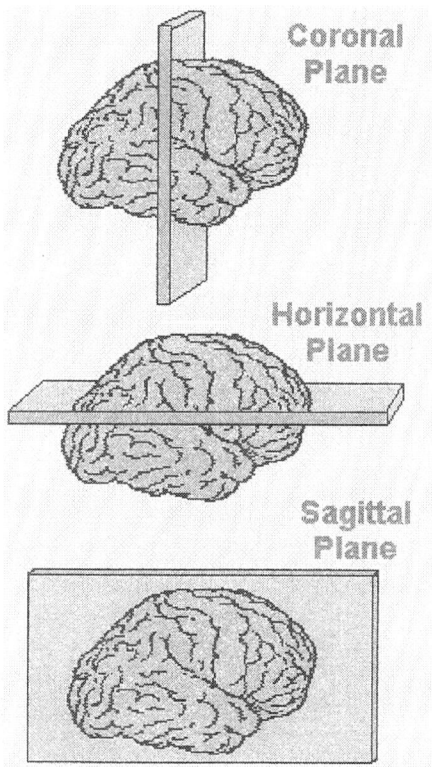

Source: Eric H. Chudler, PhD, http://faculty.washington.edu/chudler/slice.html

coronal: think of inspecting the brain "head on" (front to back)

horizontal: think of inspecting the brain from bottom to top (or top to bottom)

sagittal: think of inspecting the brain after slicing it into two halves (left vs. right)

In studies of benzodiazepine users, the brain images that became the focus of concern were **horizontal sections**, as depicted below.

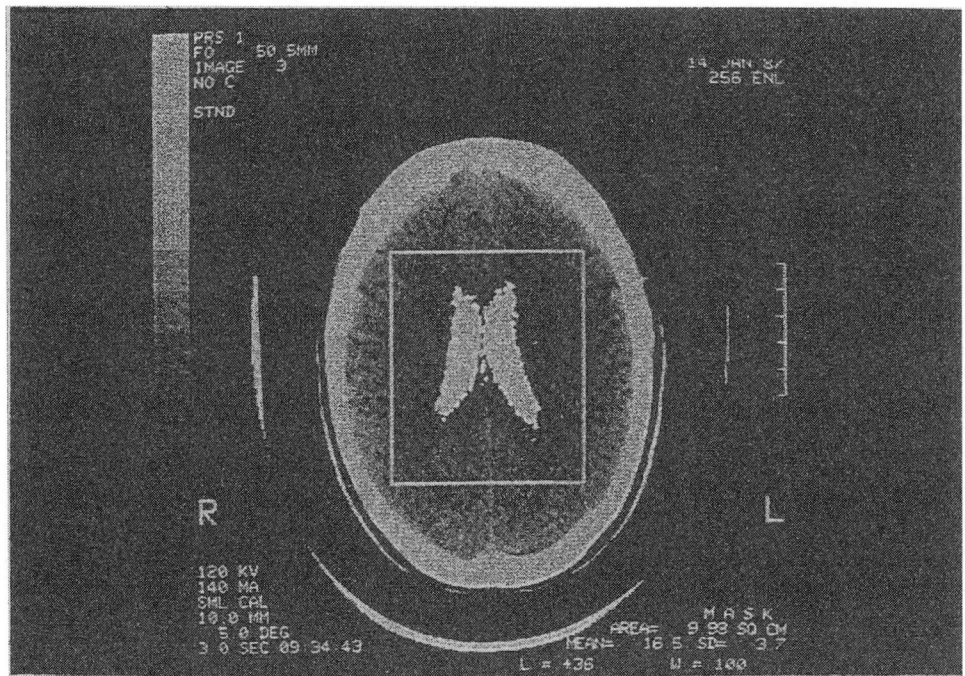

Source: C. Schmauss and J.C. Krieg, "Enlargement of cerebrospinal fluid spaces in long-term benzodiazepine abusers," *Psychological Medicine* 17 (1987): 870, reprinted with permission of Cambridge University Press.

This image is a C.T. scan obtained from a benzodiazepine consumer.

The framed space shows the **lateral ventricles** (right ventricle is on the left; left ventricle is on the right) and surrounding brain. [By convention, radiologists display horizontal slices of the body as though the patient is lying face-up on a table. This perspective requires the viewer to imagine that he or she is viewing the brain from the bottom of the table (i.e., looking up at the brain from the feet).]

In this photo, the white region represents cerebrospinal fluid within the lateral ventricles. In the experiment from which this picture was obtained, all pixels of a certain density (white) were added together and reported in square centimeters. This became the "V" (ventricular portion) of the corresponding **VBR**. To this dimension, the area of the surrounding brain tissue was added in order to provide the "B" portion of the VBR.

The second kind of analysis in benzodiazepine brain scan studies focused upon the extent of **cortical atrophy**. To perform this assessment, researchers would calculate an entity known as the *cortical score*. To understand this rating, it is important to appreciate the appearance of the brain's outer surface or cortex (see chapter two).

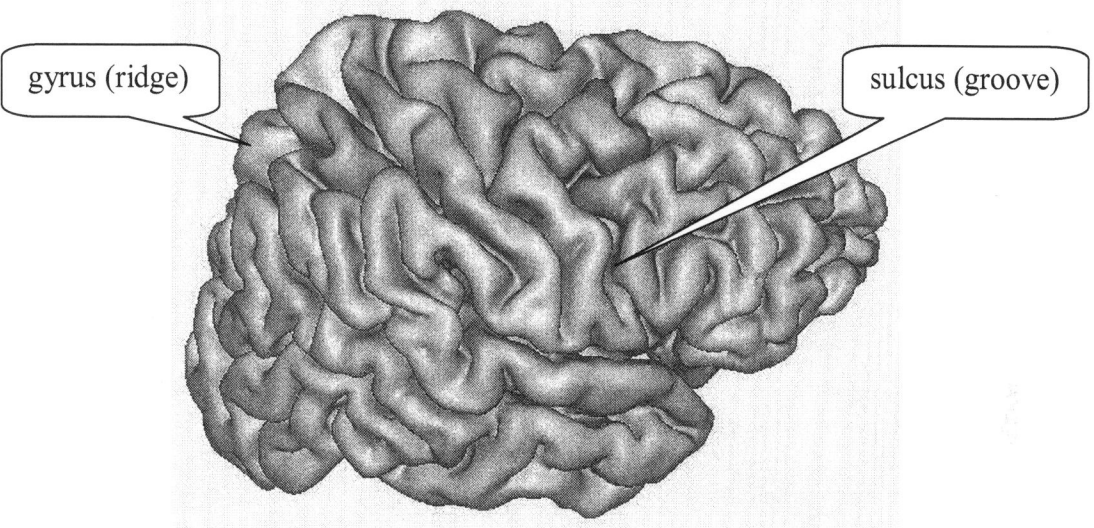

Source: Dr. Jean-Francois Mangin, http://brainvisa.info/doc/brainvisa/images/compute8.jpg (labels added by author).

The cortex of the human brain is organized into a series of folds which consist of ridges and grooves (in medical terminology: gyri and sulci).

With ageing or injury, changes to the cortex occur. These include the flattening or narrowing of the ridges (gyri), and the progressive enlargement of the grooves (sulcal widening). When summed together, the progression of these changes can provide a rough estimate of the severity of neurodegeneration.

Study #1   *Lader et al (1984)*

In 1984, Lader et al reported the results of a C.T. study which compared the brains of 20 former or current users of benzodiazepines (clobazam, diazepam, lorazepam) with an equal number of age- and gender-matched alcoholics and "normal" controls.[10] Unique to this study was the inclusion of individuals who had consumed benzodiazepines for an extremely long period of time (*mean duration: ~ 10 ½ years*). Also unique was a careful review of medical records to exclude patients who were abusers of alcohol or other drugs.

Demographics

	benzo users n = 20	alcoholics n = 19	controls n = 19
Demographics			
mean age:	43.3 years	41.4	41.5
mean duration of drug use:	10.6 yrs	not provided	n/a
current users:	11	n/a	n/a
former users:	9	n/a	n/a
Benzodiazepine Drug Use			
lorazepam	n = 11	dose: 2 to 7.5 mg per day	
diazepam	n = 8	dose: 6 to 20 mg per day	
clobazam	n = 2	dose: 30 to 60 mg per day	

Experimental Procedure

Subjects underwent neuroimaging by C.T. The **ventricle-to-brain ratio** (VBR) was derived from two contiguous pairs of slices (using a technique called mechanical planimetry to estimate the maximal ventricular area). A **cortical score** was also calculated by summing estimates of sulcal and fissure widening on a scale of 0 to 6 (6 = maximal widening).

Findings

The VBR and cortical scores of the benzodiazepine users were intermediate between the abusers of alcohol and normal subjects. No significant differences were observed between former and current users, nor according to duration of use. However, when compared specifically to non-alcoholic controls, ***chronic consumers of anxiolytic drugs demonstrated statistically significant higher brain scores (20 to 30% larger), implying a substantial process of cortical and subcortical atrophy***:

	benzodiazepine	alcoholics	controls
V/B ratio (%)	7.09	9.22	5.77
*cortical score	1.8	2.8	1.4
% of subjects with cortical score of 1 or more	45%	** just over 50%	25%

* cortical score = 0 to 6, with 6 equal to maximal atrophy
** authors of publication did not divulge specific percentage for this subgroup

Study #2   *Perera et al (1987)*

Shortly after the appearance of the report by Lader's team, other researchers attempted to replicate their findings.[11] Investigators in Sheffield, England subsequently performed neuroimaging in a study which included nineteen chronic users of benzodiazepines and an equal number of age- and gender-matched *medical* controls.

Experimental Procedure

The benzodiazepine subgroup consisted of patients with diagnoses of anxiety and/or minor tranquilizer dependence (addiction). A control group of patients was selected from general practice clinics. All of the control group members received neuroimaging for diagnostic reasons (to rule out a structural cause of vertigo or headache), and none of them had consumed benzodiazepines chronically or within the most recent three months. After obtaining C.T. scans of all subjects, the researchers determined a cortical score on the basis of sulcal and fissure widening, and the ventricle-to-brain ratio (VBR) based on measurements of contiguous brain slices.

Limitations

This particular protocol failed to control for the amount or duration of alcohol or other brain-altering substances, failed to match subjects according to skull size or height, and failed to confirm pharmaceutical histories based upon a review of medical records. Furthermore, the researchers failed to exclude controls on the basis of past head injury or past exposure to benzodiazepines. Among the migraine and vertigo sufferers, two individuals had received prescriptions (diazepam, clorazepate) within the past year, and an undisclosed number may have received benzodiazepines historically. Despite these limitations, however, anatomic differences were *still* observed between the users of benzodiazepines and the neurological controls:

Findings

	benzodiazepine recipients n = 19	controls n = 19
mean age	48	"age matched"
mean duration of drug use	6.21 years	not divulged
VBR	5.23%	5.79%
cortical score	**0.95**	**0.53**

Although the previous finding of VBR enlargement was not replicated in this investigation, the *researchers did confirm more pronounced atrophy among chronic users of benzodiazepines* (based upon comparative cortical scores).

Study #3 *Schmauss and Krieg (1987)*

Reacting to the phenomenon of unexpected and, in some cases, protracted psychological impairment following benzodiazepine *abuse*, researchers in Munich, Germany performed a neuroimaging study of seventeen benzodiazepine recipients and twenty-two psychiatric controls.[12]

Experimental Procedure

The investigation enrolled seventeen inpatients (16 female, 1 male) admitted for *benzodiazepine withdrawal therapy*. None of these individuals had histories of any other form of chemical dependency, including alcohol abuse. A control group was comprised of currently unmedicated patients who had been admitted for the treatment of anxiety (77%) or personality disorders (23%). None of the subjects in the control group were believed to be suffering from organic (non-psychogenic) illnesses.

C.T. scans were obtained from all subjects within the first week of hospitalization. In this particular study, images were analyzed using methods to derive various estimates of internal fluid-filled spaces (including VBR) and external fluid-filled spaces (surface cortical dimensions). A unique aspect of this experiment was the comparison of brain features according to "high" or "low" doses of anxiolytic drugs:

high dose = daily benzodiazepine $\geq$ 50 mg diazepam equivalents per day
low dose = daily benzodiazepine $<$ 50 mg diazepam equivalents per day

Limitations

The researchers failed to obtain (or failed to report) the medication histories of all subjects. It is unclear how many members of the *control group* had received prior treatment with benzodiazepines acutely or chronically.

Findings

	benzodiazepine users		drug-free controls (currently not medicated)
	low dose $n = 0$	high dose $n = 8$	$n = 22$
mean age	33	37	37
mean duration of drug therapy	4.8 years	6.6 years	not divulged
VBR	**4.47%**	**6.81%**	2.91%
External Spaces:			
maximal width of cortical sulci	2.39 mm	**3.08 mm**	2.42 mm
width of anterior interhemispheric fissure	**3.87 mm**	**4.54 mm**	2.64 mm
width of insular cistern	**2.70 mm**	**3.56 mm**	2.41 mm
Note: "normal" reference standard for external spaces = 3 mm			

Relative to age- and gender-matched controls, ***the benzodiazepine-dependent patients demonstrated significantly higher VBRs.*** This difference was positively associated with drug dose, but not to age or duration of therapy. ***Similarly, the benzodiazepine-dependent patients demonstrated moderate enlargements of the external spaces of the brain.*** Changes in the size of sulci and fissures were positively associated with dose (the higher the dose, the greater the size of these spaces). Cortical atrophy also tended to be associated with increasing age and duration of treatment.

Study #4 *Moodley et al (1993)*

In what appears to be the last C.T. study of its kind, researchers in the U.K. performed neuroimaging among benzodiazepine users and healthy controls.[13] In this protocol, brain images were compared according to chronicity of past and current drug use, and according to the kinds of medication (lorazepam, diazepam) consumed.

Experimental Procedure

Neuroimaging was performed by C.T. Two contiguous slices from each brain scan were evaluated by visual inspection, and the VBR was derived by mechanical planimetry (a method of estimating a two-dimensional area within the scan). Next, the researchers employed a method known as **densitometry** to compare the intensity of radiation absorption in several regions of the brain.

[Note: Radiation absorption varies according to tissue composition. Because fluid absorbs less radiation than myelin and brain cells, changes in radiation absorption may provide estimates of edema, deterioration, and/or non-regeneration after injury.]

Exclusion Criteria

Specifically excluded from this study were individuals with any history of epilepsy, brain injury, stroke, psychotic condition, major affective or eating disorder, or cardiovascular disease; any history of past treatment with antipsychotic drugs or monoamine oxidase inhibitors; and any history of illicit drug use or excessive consumption of alcohol.

Findings

When chronic users of benzodiazepines were compared to the non-drug-exposed, no significant differences in the ventricle-to-brain ratios were revealed. ***However, based upon changes in radiation absorption, benzodiazepine use was associated with a universal reduction in brain tissue.*** These changes approached or attained statistical significance in five regions: the frontal lobes, the left caudate, and both occipital lobes:

	VBRs and C.T. Density Values (adjusted for age/gender/alcohol/VBR/skull size)				
	* chronic benzo users n = 18		non-benzo n = 25		
VBR	7.7%		9.2%		
	density values		density values		
left frontal	**34.5**	n = 17	36.2	n = 24	$p < 0.08$
right frontal	**34.2**	n = 17	35.9	n = 24	$p < 0.003$
left caudate	**38.5**	n = 17	40.0	n = 24	$p < 0.08$
right caudate	37.4	n = 17	39.9	n = 24	
left thalamus	39.5	n = 17	40.5	n = 24	
right thalamus	38.8	n = 17	40.0	n = 24	
left occipital	**42.1**	n = 17	46.7	n = 12	$p < 0.02$
right occipital	**41.9**	n = 17	44.9	n = 12	$p < 0.001$

* chronic benzodiazepine use = continuous use for 3 or more years

A similar pattern was detected when the research team evaluated a subgroup of patients according to specific drug use (≥ 2 years of continuous lorazepam or diazepam). These differences were *statistically significant* in the left frontal lobe (lorazepam), the right frontal lobe (lorazepam), the right caudate (lorazepam), and the left occipital lobe (lorazepam and diazepam):

	VBR and C.T. Absorption Density Values (adjusted for age, gender, alcohol intake, VBR, and skull size)		
	lorazepam only n = 9	diazepam only n = 6	non-benzo exposed n = 25
mean duration of use	5 ½ years	11 years	n/a
VBR	7.5%	8.2%	9.4%
	density values	density values	density values
left frontal	**33.5**	36.0	36.1
right frontal	**33.0**	36.1	35.7
right caudate	**36.7**	39.2	39.8
left occipital	**42.7**	42.9	46.9

Although the ventricle-to-brain ratios of the benzodiazepine users in this study were *smaller* than those of the healthy controls, the results of C.T. densitometry revealed consistent decrements in the cortex (frontal, occipital lobes) and in one subcortical structure, as well (caudate). ***The interpretation of these findings requires a critical examination of the study design.***

First, it is important to appreciate the limitations of C.T. planimetry and VBR. The former technique (as applied here and in the previously mentioned studies) can provide only a two-dimensional analysis of select brain slices, rather than a volumetric assessment of the whole brain.

Second, the VBR is inherently ambiguous by virtue of the fact that it is a *ratio*. For example, the significance of a dramatic enlargement of the ventricles could be masked by an equal or disproportionate increase of the surrounding brain tissue. In such a scenario, the VBR could result in the erroneous conclusion that the brain had not changed when in fact, the *absolute size* of the numerator and denominator – if disclosed by researchers – might convey the existence of significant pathology (such as brain edema with hydrocephalus).

Ultimately, the clinical significance of the investigation by Moodley et al depends upon the **fidelity of the brain density** findings. The researchers were careful to minimize several technical variables which could have compromised the results. To minimize variation in absorption density caused by *machinery* (a problem known as *scanner drift*), some of the analyses (chart #2, above) were limited to scans that were obtained within a discrete time period. To minimize absorption variance due to *bone* effects (bone reduces the absorption value for the underlying brain tissue), findings were adjusted for skull size. To minimize absorption variance caused by the trapping of fluid and tissue within a C.T. pixel (a problem known as *partial volume artifact*), the researchers employed an automated method to identify brain regions precisely. **All of these precautions reduced the likelihood that the absorption values in this experiment were the result of *technical* difficulties.**

In the final analysis, what did the findings of the Moodley study imply? Given the fact that reduced absorption density is associated with *cell degeneration*, the *loss of myelin*, and an *increase in fluid-filled spaces*, ***the observations made in this experiment were consistent with brain damage and atrophy***. Although a cross-sectional study of this kind could not definitively prove that benzodiazepines had been the cause of neuronal or glial shrinkage in these patients, the results strongly suggested that the benzodiazepines had *either* instigated, enhanced, or allowed the sustained expression of a neurodegenerative process.

Postmortem Studies of Animals

Despite the fact that benzodiazepines replicate many of the chemical, physiological, cognitive, and behavioral effects of alcohol – and despite the fact that alcohol is an undisputed cause of neurodegeneration – there has been an appalling void of research with regard to the **structural** brain effects of these medications.

Although investigators in the United States have recently addressed the subject of benzodiazepine toxicity within a few narrow areas of concern (e.g., the prenatal exposure of embryos and fetuses who receive benzodiazepines via their mothers; and the exposure of infants and young children who receive benzodiazepine anesthesia during surgery), they have long ignored the anatomical consequences of the same drugs in adults. Perhaps this failure stems from the fact that the benzodiazepines quickly became psychiatry's preferred substitute for alcohol and other street drugs, both in the management of acute detoxification, and in the management of chronic psychological distress. Similar to the situation with neuroimaging studies, it is to non-American research teams that one must turn if one seeks to explore the neuropathological effects of these drugs.

Study #1 Carassiti et al (1998)   *Midazolam Effects Upon the Hippocampus*

The benzodiazepine, midazolam (Versed), is well known to physicians as an anesthetic agent used for surgical and medical procedures. In these contexts, the drug is selected for the purpose of producing three major effects: relaxation, sedation, and amnesia. This latter phenomenon – known more specifically as **anterograde amnesia** – refers to the capacity of midazolam to prevent the conscious recollection of events which occur shortly *before and after* the drug is given. Although the precise mechanism which leads to anterograde amnesia has been a longstanding subject of inquiry and debate, most textbooks of pharmacology have traditionally invoked a *transient physiological process* in the memory centers of the brain. However, the work of Spanish scientists suggests that an *anatomical process* is also involved.

Responding to the increasing tendency of some patients to consume midazolam chronically (e.g., as a sleeping pill), researchers from the University of Navarra (Spain) were motivated to explore the long-term impact of this drug upon cell viability in the hippocampus (as explained in chapter four, a key component of the temporal lobe involved in learning and memory).[14]

Experimental Procedure

Young (2 months old) and old (24 months old) Wistar rats were assigned to three different subgroups for chronic treatment with midazolam or saline. Treatments were administered by means of a feeding tube for four months. The purpose of this experiment was to compare two different drug doses (1 mg/kg, 3 mg/kg) and to examine potential differences in drug effects according to age.

	young rats	old rats
saline	n = 10	n = 10
1 mg/kg	n = 20	n = 20
3 mg/kg	n = 20	n = 20

Note: Based upon the timing and duration of developmental processes and lifespan in rodent species, these rat ages were roughly equivalent to human puberty and adulthood.

Following 120 days of exposure, the animals were euthanized. The hippocampal tissue from each animal was processed with two fixatives, stained, and examined by light microscopy. In this specific experiment, manual counts of cell numbers and the size of neuronal nuclei were made in each region of the hippocampus. Interestingly, although the research team considered their doses to be "much higher" than those used clinically, this conclusion was only partly correct. While it is true that the concentrations which the animals received were higher than those given to humans on a *dose per weight* (mg/kg) basis, one must consider inter-species differences in terms of drug absorption and metabolism. When one reviews the *dose-related blood concentrations of midazolam* which have been measured in both species, the drug regimens used in this experiment were far *lower* than those which occur in humans when the same medication is prescribed for sedation or anesthesia.

dose vs. midazolam concentration in peripheral bloodstream [15-18]

	dose	C_{max}	elimination half-life
human	oral dose of 15 mg (0.21 mg/kg)	70-120 ng/mL	1.5 to 2.5 hours
rat	oral dose of 1 mg/kg	2 ng/mL	0.5 hours

oral bioavailability
human 24-46%
rat   2-5%

* **bioavailability** = fraction of unchanged drug that reaches the systemic circulation after an administered dose (influenced by intestinal absorption, metabolism, etc.)

Anxiolytics

Background Information

The hippocampus of the rat (as in humans) is divided into several regions: CA1, CA2, CA3, CA4, dentate gyrus (and subiculum). The Spanish researchers analyzed changes in the number and appearance of neurons within the first five areas.

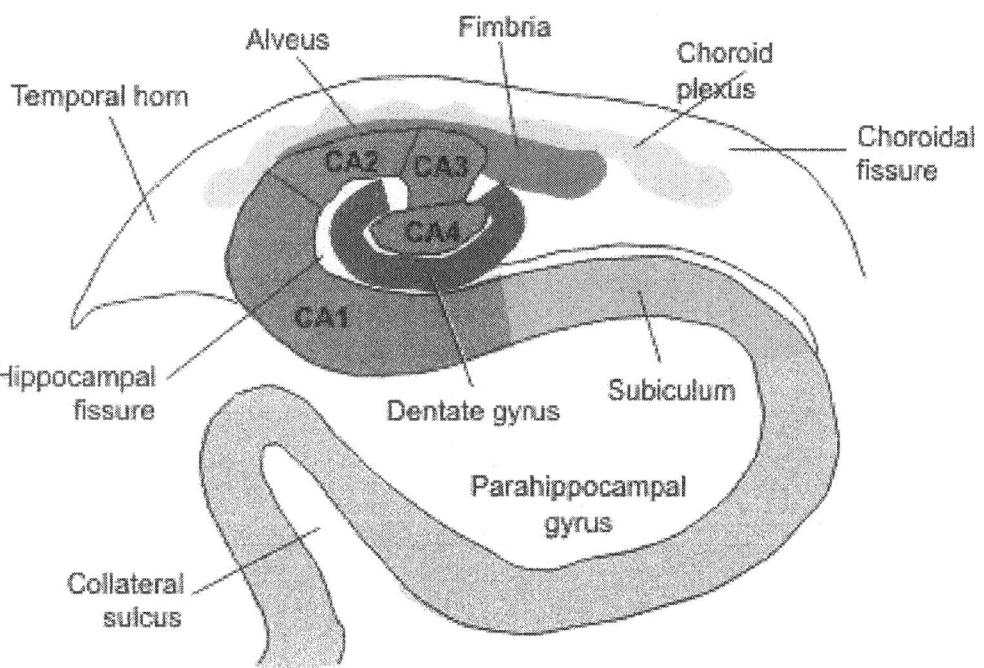

Source: R.R. Edelman, J.R. Hesselink, M.B. Zlatkin, and J. Crues, Ed. *Clinical Magnetic Resonance Imaging, 3rd Ed.* (Philadelphia: Elsevier-Saunders, 2006), figure 47-4. Reprinted with permission of Dr. John Hesselink and Elsevier.

Findings

Chronic midazolam resulted in age-specific changes in brain tissue relative to unmedicated controls. ***The researchers identified diffuse reductions in both the number (2 to 14% fewer) and size (2 to 15% smaller) of neurons in the older rats***:

Adult Rats - Total # of Neurons by Hippocampal Region					
	CA1	CA2	CA3	CA4	Dentate Gyrus
controls	78656	30541	17076	21298	300884
1 mg/kg	70221	26376	14743	19197	265909
3 mg/kg	72801	28348	16358	20873	272509
Adult Rats – Nuclear Size by Region (um^2)					
	CA1	CA2	CA3	CA4	Dentate Gyrus
controls	82.93	107.99	97.89	94.44	48.5
1 mg/kg	80.55	105.42	96.01	87.16	46.21
3 mg/kg	77.09	94.42	94.80	80.63	42.86

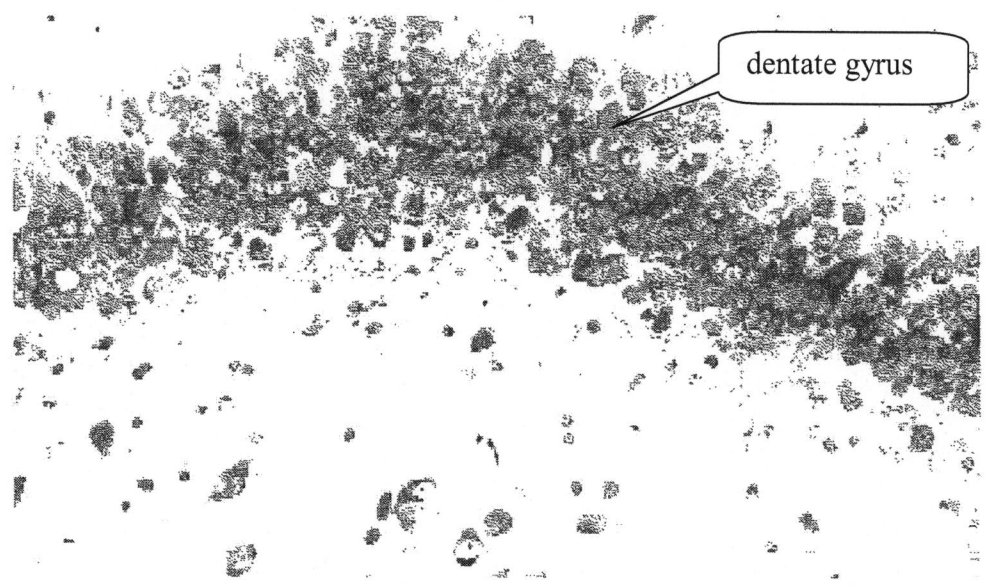

Source: Reproduced with permission, M. Carassiti, A. Ysasi, and L.M. Gonzalo, "Efectos de administracion cronica de Midazolam sobre el hipocampo de ratas," *Revista de Medicina de la Universidad de Navarra* 1-3 (1998): 23.

Above: This image depicts normal neuron features in the **dentate gyrus** of an older rat (saline control).

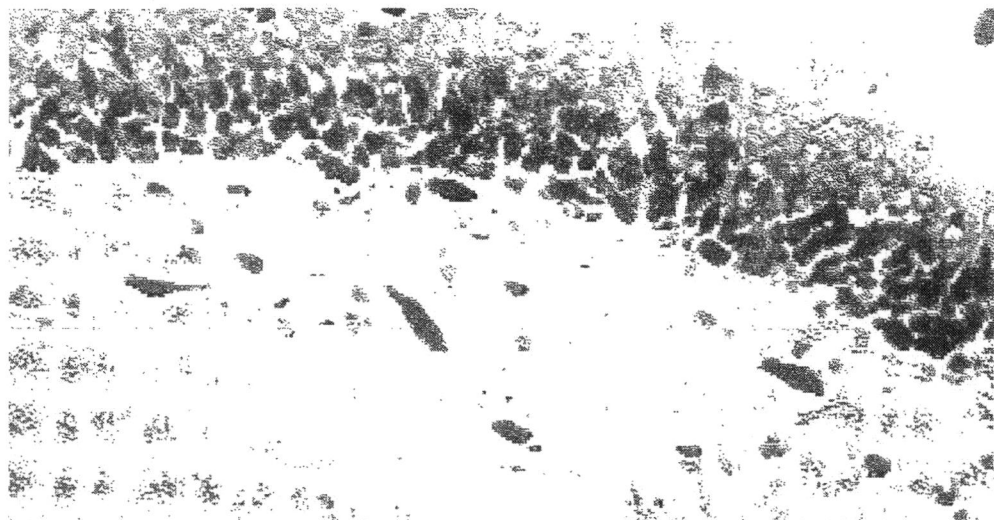

Source: Reproduced with permission, M. Carassiti, A. Ysasi, and L.M. Gonzalo, "Efectos de administracion cronica de Midazolam sobre el hipocampo de ratas," *Revista de Medicina de la Universidad de Navarra* 1-3 (1998): 23.

Above: Following 120 days of daily treatment with midazolam (3 mg/kg per day), ***neurons in the dentate gyrus of older rats were reduced in number and size***.

In adult rats, chronic exposure to midazolam also resulted in the proliferation of **dark neurons** – a tissue change consistent with early cell death. [Although dark neurons can appear in brain specimens as a result of improper fixation techniques, the methods and pattern of results in this study suggested that they were truly reflective of neuropathology.][19-23]

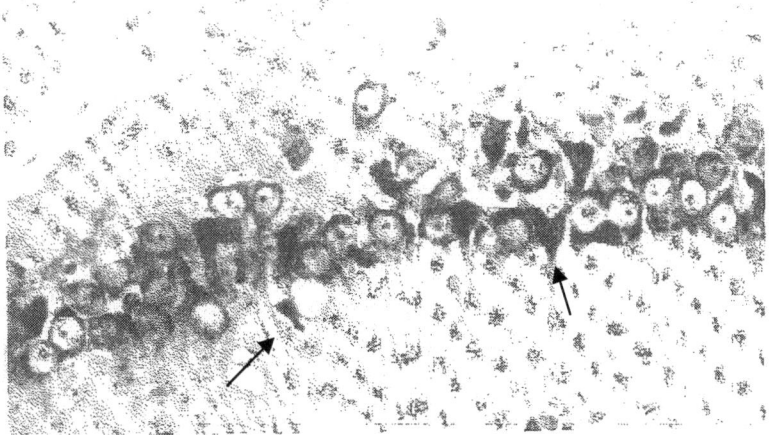

Source: Reproduced with permission, M. Carassiti, A. Ysasi, and L.M. Gonzalo, "Efectos de la a administracion cronica de Midazolam sobre el hipocampo de ratas," *Revista de Medicina de la Universidad de Navarra* 1-3 (1998): 24.

This image depicts dark neurons in the CA1 region of the hippocampus. These cell were obtained from one of the older rats following treatment with midazolam (3 mg/kg per day for 120 days).

In Sum:

Spanish investigators employed a research protocol which involved the administration of midazolam for a protracted period of time (four months). This made their experiment unique among published animal studies, which have generally been limited to the study of acute drug effects.

The procedure involved a dosing regimen which was relevant for humans on the basis of equivalent drug concentrations in the bloodstream.

Chronic midazolam was associated with hippocampal atrophy and neuronal death in the adult brain.

Given the importance of the hippocampus in the acquisition and retrieval of memory, it is quite likely that the anatomic abnormalities observed in this experiment contribute to the amnestic properties of midazolam.

Other Postmortem Studies of Benzodiazepines

While it could be argued that the results of the previous experiment are irrelevant for the residents of some countries (in other words, wherever midazolam is *not* used chronically), the results of another series of investigations are not so easily dismissed.

Background Information

All living organisms are comprised of three kinds of chemicals: fats (lipids), sugars (carbohydrates), and proteins. The term "glycolipid" refers to any molecule which contains both fats and sugars. ***Glycolipids*** are important in biology because they are essential components of cell membranes:

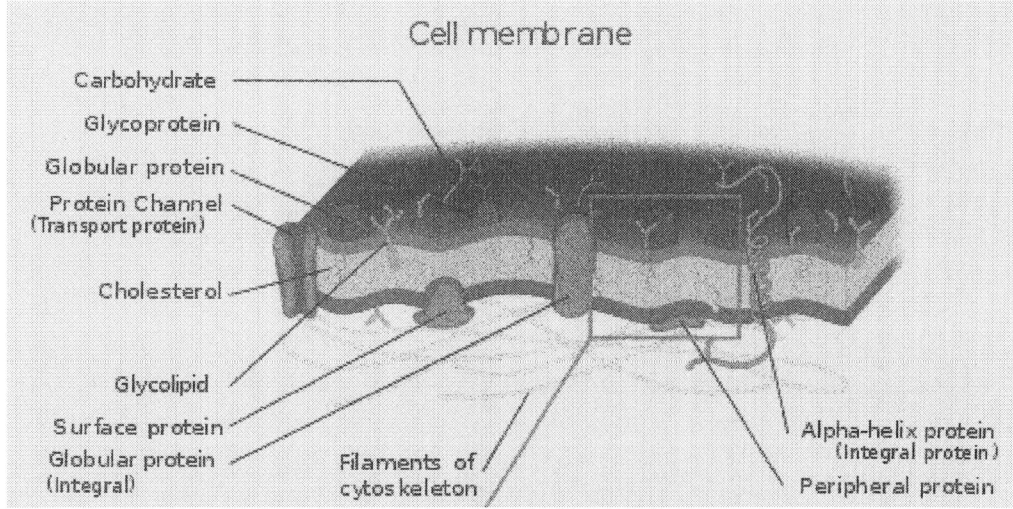

Source: Wikipedia Commons. Cell_membrane_detailed_diagram_4.svg

Within humans, *a specific kind of glycolipid* – called a ***ganglioside*** – accounts for 6% of all brain lipids by weight, and 25% of each neuron's surface membrane. Particularly within the central nervous system, gangliosides fulfill a number of crucial functions which are believed to include: [24]

* the response of immune defense systems
  (binding with toxins, viruses, bacteria, and cytokines)

* the maintenance of cell integrity (i.e., myelin)

* the regulation of cell growth and differentiation, and

* the modulation of apoptosis (programmed cell death)

Gangliosides have complex structures (as shown below). For the technically curious, the skeleton of a ganglioside always consists of three parts: 1) a ceramide backbone; 2) at least one unit of sialic acid (aka, N-acetylneuraminic acid); and 3) three or more sugar residues:

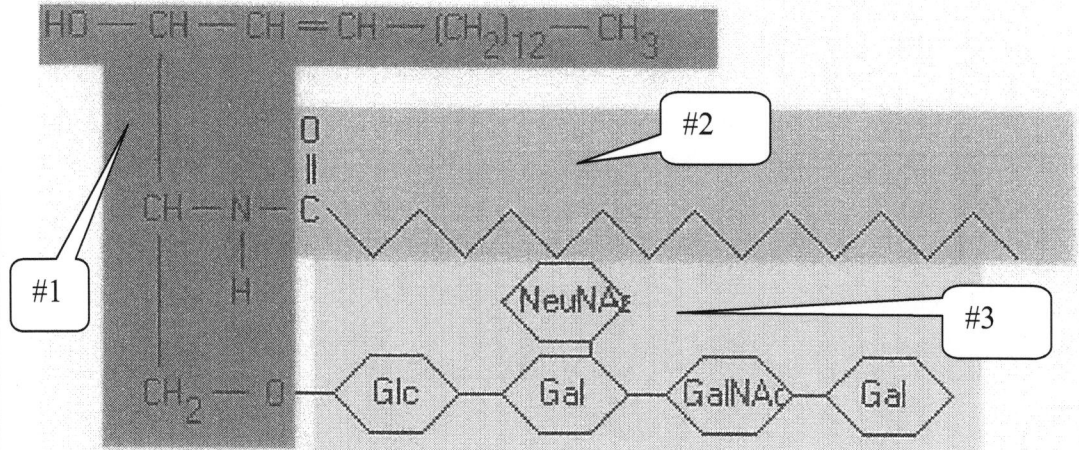

Reprinted with permission of Ines Niehaus. http://www.endotoxin.gmx.home.de/artikel3.html

This is a chemical shorthand description of the ganglioside known as **GM$_1$**.

ceramide backbone = #1 (sphingosine) + #2 (fatty acid)
sialic acid = #3 component (NeuNAc)
sugar residues = #3 component (Glc-Gal-GalNac-Gal)

There are more than 40 different kinds of gangliosides. Based upon the complexity of their structures, these compounds are classified by a distinct nomenclature (GM, GD, GT where M = 1 sialic acid unit, D = 2 sialic acid units, T = 3 sialic acid units). They are further organized into four primary series ("0", "a", "b", and "c"):

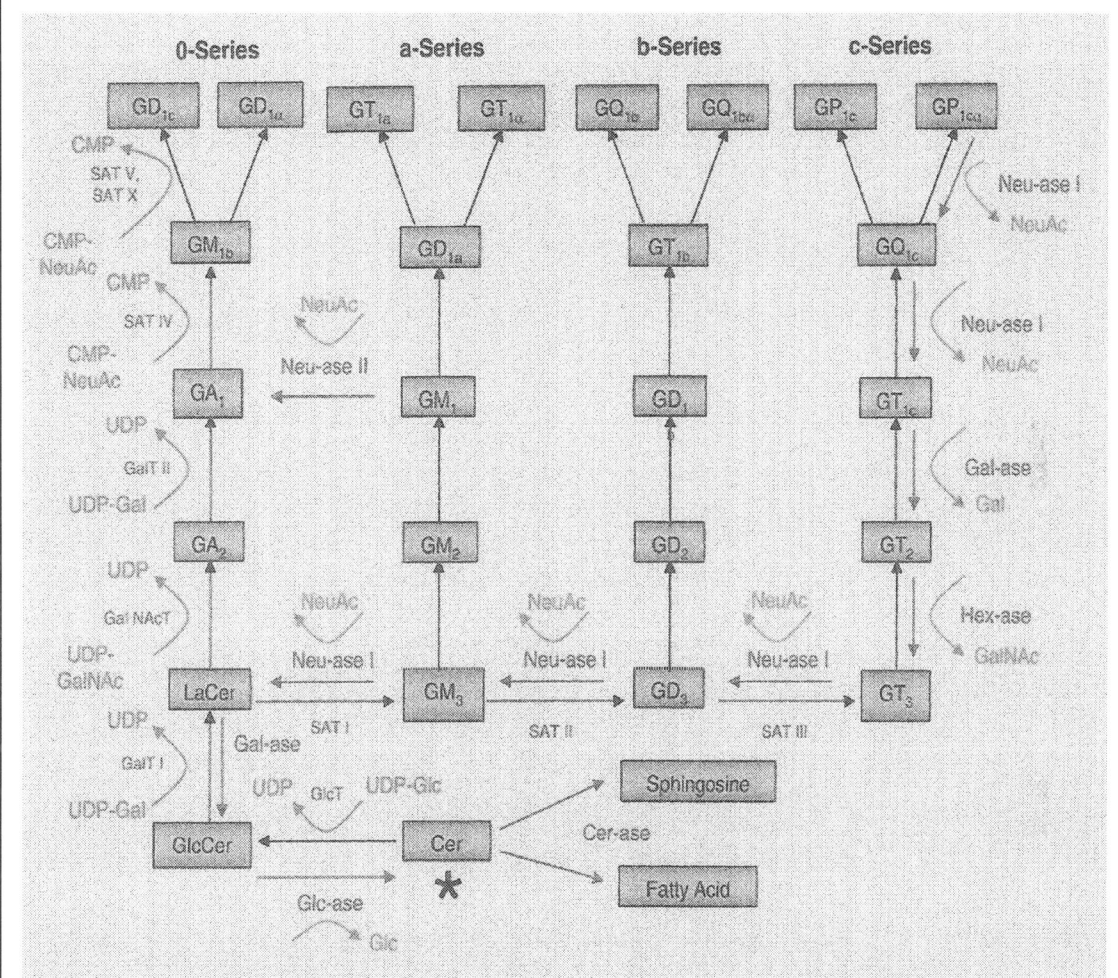

Reprinted by permission from Macmillan Publishers Ltd: *Cell Death and Differentiation*, volume 13, d'Azzo et al, "Gangliosides as apoptotic signals in ER stress response," page 406, copyright.

Why is any of this important?

The gangliosides play vital roles in a number of genetic, infectious, and neurological diseases.[25-27] When specific populations of gangliosides become too numerous or too sparse, the results can be disabling or lethal. For example, a number of genetic conditions, known as **lysosomal storage diseases** (see chapter two) involve the excessive accumulation of gangliosides in various organs of the body. This results in mental retardation, movement abnormalities, blindness, and early death. Conversely, the loss of complex gangliosides, or the inhibition of their function (as occurs when the immune system makes antibodies against them) can result in the degeneration of myelin, axons, or whole neurons within the brain and spinal cord.

Studies #2 and #3   *De Luka et al (1999, 2002)*   *Diazepam and Gangliosides*

Motivated by a desire to understand adaptive changes elicited by chronic drug therapy, researchers at the University of Belgrade conducted a series of experiments which explored the effects of diazepam (Valium) upon brain lipids.

In the first of two published reports, the team described the results of an investigation which was limited to drug effects upon the *cerebellum* (below) – the hindbrain structure which modulates balance, gait, motor dexterity, and cognition.[28] In 2002, they divulged the results of an expanded study which addressed the effects of diazepam upon lipids *throughout the brain*.[29]

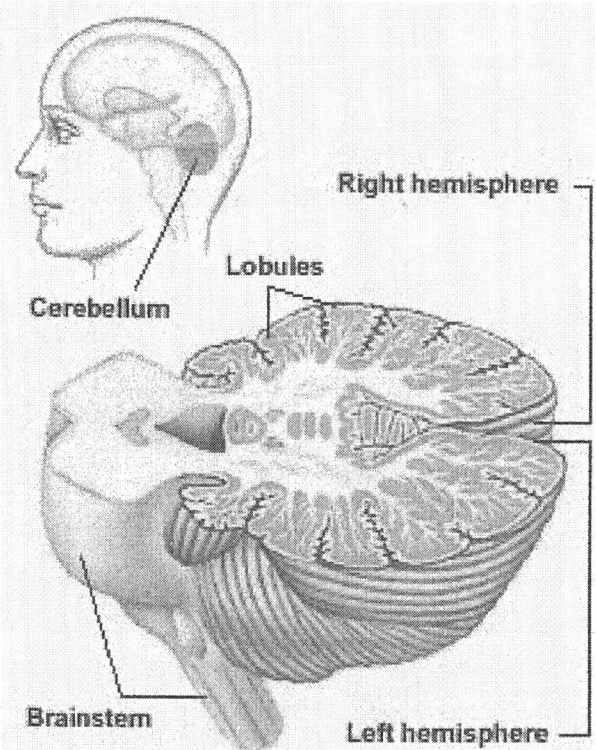

Source: McGill University. The Brain from Top to Bottom.

Experimental Procedures

Experiment #1

Young male Wistar rats (aged 2 months) were divided into four subgroups. Animals received a daily regimen of diazepam (10 mg/kg by diet) for three, five, or six months. One group of animals received the drug for five months, followed by a one-month period of drug withdrawal. At the end of these treatment periods, the rats were euthanized. Brains were removed and processed chemically in order to quantify the concentration of total gangliosides and specific ganglioside subtypes.

Experiment #2

Two-month-old, male Wistar rats were exposed to an oral daily regimen of diazepam (10 mg/kg by diet) for a period of six months. Following this period of treatment, the subjects were euthanized as before (CO_2 inhalation), brains were excised, and tissue was processed for total ganglioside content. In addition, the researchers analyzed changes in specific ganglioside subtypes (series "a" and series "b").

Dosing [30-34]

Differences in the absorption and elimination of diazepam make it difficult to extrapolate pharmacokinetic data from rats to humans. However, animal data from dosing studies suggest that this experiment approximated the physiological effects (blood levels) which one would see during the use of moderate to high doses of diazepam in humans:

Diazepam Pharmacokinetics

	dose	blood level	elimination half-life
human	2-55 mg per day (oral)	16-1412 ng/mL	32.9 hours
rat	5 mg/kg IV	200 to 700 ng/mL	1 hour
	5 mg/kg IP	150-700 ng/mL	0.88 hr
	4 mg/kg p.o. (oral)	277-309 ng/mL	1.10 hr

Findings

In the 1999 cerebellum study, **chronic diazepam was associated with statistically significant reductions in total gangliosides after five and six months** (20% and 46% reductions, respectively). A short period of drug withdrawal resulted in partial but *incomplete recovery* of brain ganglioside content:

Total Cerebellar Gangliosides (umol/g)			
	control n = 6 per group	diazepam (10 mg/kg by mouth) n = 6 per group	
3 months	1.767	1.836	
5 months	1.809	**1.421**	**p < 0.05**
6 months	1.836	**0.996**	**p < 0.01**
5 months of drug + 1 month drug withdrawal	1.836	**1.553**	**p < 0.01**

In 2002, these findings were replicated in ***multiple regions of the brain***. This time, groups of six animals were exposed to diazepam or tap water (control) for a period of six months. Total ganglioside content was measured in six areas. ***Statistically significant reductions were observed in the cerebral cortex, the hippocampus, and the cerebellum***:

Total Brain Ganglioside Content After Six Months (umol/g)			
	control n = 6	diazepam (10 mg/kg by mouth) n = 6	
cerebral cortex	4.18	**2.90 (↓ 30.6%)**	**p < 0.01**
hypothalamus	3.93	3.75	
caudate	3.40	3.74	
thalamus	1.47	1.67	
hippocampus	3.12	**2.74 (↓ 12%)**	**p < 0.05**
cerebellum	1.84	**0.99 (↓ 46%)**	**p < 0.01**

When changes in specific lipid fractions were reviewed, the researchers found regional differences in drug effects. ***In three discrete brain regions, the concentrations of GM_3 (known as a "simple" ganglioside) was significantly increased***:

Changes in GM_3 After Six Months			
	control n = 6	diazepam (10 mg/kg by mouth) n = 6	
cortex	11.13	9.54	
hypothalamus	1.74	**6.46 (↑ 370%)**	**p < 0.01**
caudate	5.66	4.63	
thalamus	1.74	**5.56 (↑ 320%)**	**p < 0.01**
hippocampus	2.18	**6.58 (↑ 302%)**	**p < 0.01**
cerebellum	10.35	5.21 (↓ 50%)	p < 0.05

The clinical significance of these (and other observations) depends upon their correlation to changes which have been documented in neurological disease.[35-39] For example, autopsies of patients with Alzheimer's and Creutzfeldt-Jakob disease – two different forms of dementia – have demonstrated a lower *total* ganglioside content and elevated concentrations of *simple* gangliosides, such as GM_3.

Furthermore, a causal link between lipid changes and neurodegeneration has been suggested by research involving genetically modified animals. In those studies, the *depletion of complex gangliosides* has been shown to result in the loss of myelin and axons within the central and peripheral divisions of the nervous system. Similarly, multiple research groups have confirmed the *hazards of elevated GM_3* via mechanisms which include the induction of beta-amyloid deposits in the brain, and glutamate-mediated cell death.

In Sum:

In young adult Wistar rats, daily exposure to diazepam for 5 to 6 months resulted in a 20 to 50% reduction of the total ganglioside content of the brain.

The regimen which was used in these experiments approximated human therapy on the basis of dose-associated drug levels in the peripheral bloodstream.

Statistically significant reductions in total gangliosides occurred in multiple brain regions, including the cerebral cortex, the hippocampus, and the cerebellum.

A specific analysis of ganglioside fractions revealed lipid abnormalities that were qualitatively similar to those which occur in Alzheimer's and Creutzfeldt-Jakob disease. An important body of research evidence suggests that membrane disruption of this kind can be a cause, as well as effect, of brain cell death.

Chapter Seven

Mood Stabilizers

The term *mood stabilizer* first emerged in the 1990s as a description for specific chemicals (i.e., lithium and numerous anticonvulsant drugs) which have found use in the treatment of manic-depression – otherwise known as bipolar disorder.[1-3] Like all other medicines which are used in psychiatry, the story of these substances is a tale of serendipitous discovery, misguided theories about disease causation, experimentation, and longstanding denial of drug-induced harm.

Lithium

The alkali metal known as lithium is one of the simplest elements in nature. When conjoined with other molecules to form a salt – such as lithium bromide, lithium carbonate, or lithium citrate – the chemical has been used throughout history as a magical elixir for numerous ailments, including gout (mid-1800s); rheumatism, headaches, obesity, constipation, and asthma (early 1900s); and even as a salt substitute (1940s).

Although lithium was used sporadically in the 1870s by physicians who attributed mood disturbances to 'gout in the brain,' the formal adoption of lithium *within psychiatry* originated in the late 1940s. At that time, experiments by the Australian physician, Dr. John Cade, suggested a role for the drug in the treatment of mania – a condition marked by euphoria, racing thoughts, rapid speech, sleeplessness, irritability, and erratic behavior. Believing that manic symptoms must be caused by the circulation of an endogenous toxin, Cade sought to isolate a noxious compound from the urine of his patients. In the course of testing the effects of urine components, Cade employed lithium salts to dissolve uric acid. When he injected guinea pigs with various concoctions – including lithium carbonate with and without uric acid – the animals became calm. This discovery inspired Cade to test lithium carbonate directly in manic patients. Despite the fact that a series of tests were successful, the widespread adoption of the drug was delayed by problems related to production and marketing (the lack of patent exclusivity) ***and*** by the drug's toxicities.

Ultimately, the Danish psychiatrist, Dr. Mogens Schou, confirmed Cade's observations. In the 1950s, it was Schou's experiments which instigated the use of lithium in Europe and North America as a treatment for mania. In 1967, his further discoveries sparked enthusiasm for the drug as a means of preventing *recurrent episodes of high *and* low mood. Since that time, the popularity of lithium has waxed and waned, depending upon the availability of alternative treatments, and depending upon the medical community's selective attention to the salt's limitations and lethality.

---

* There is now abundant evidence that lithium *provokes*, rather than prevents, mood instability. For a discussion of this problem, see chapter references #4-10 at the end of this book.

Anticonvulsant Drugs

According to one account of pharmaceutical history, lithium salts were not widely available in Western Europe or Asian nations in the early or mid-20th century. Consequently, the classification and testing of anti-manic therapies came to focus upon anticonvulsant drugs. How this occurred is a reflection of two unique developments: first, the prolonged warehousing of epileptic patients in asylums alongside the mentally ill; second, the close relationship between neurology and psychiatry in some countries which resulted in a strong biomedical approach to institutionalized patients.[11-13]

Valproic acid / Valpromide / Divalproex Sodium (Depakote)

On the European continent, the first non-lithium treatment for mania grew out of epilepsy research. Interestingly, it was a female American scientist, Dr. Beverly Burton, who synthesized valproic acid (valproate) while working in Germany in 1882. Subsequent to her discovery, valproic acid was widely used in the field of chemistry as an organic solvent. Due to its aliphatic ("fat-like") chemical structure, valproic acid was even used during World War II as a butter substitute. However, the **neuropsychiatric** applications of valproic acid and several related compounds originated in 1963 when researchers in France discovered the anticonvulsant properties of these drugs. Eventually, the anti-seizure, sedative, and behavioral effects of these chemicals led to their use as treatments for mania in France (1970s), Germany (1980s), and North America (1990s).

Carbamazepine (Tegretol)

Carbamazepine – a three-ring or "tricyclic" compound related to the first generation of antidepressants – was invented in 1964. Similar to the history of valproate, researchers quickly came to appreciate the fact that this compound possessed anticonvulsant and sedative properties. Within Japan, the treatment of seizure disorders fell within the domain of psychiatry, and for this reason, carbamazepine was first tested in that nation within asylums. In the 1970s, Japanese clinicians confirmed the efficacy of carbamazepine as an alternative to neuroleptics and lithium in subduing mania and aggression.

Within the United States, the psychiatric use of carbamazepine was not accepted until the 1980s. First, it was necessary for physicians and patients to perceive the limitations and hazards of lithium. Second, it was helpful (though not essential) for physicians to embrace a new philosophical framework, according to which antiepileptic drugs could be viewed as *specific* treatments for bipolar disorder.

Responding to both needs, researchers at the National Institute of Mental Health provided the perfect construct. According to the *kindling theory of mood disorders*, advanced by Drs. Susan Weiss and Robert Post, the physiological underpinnings of recurrent mania and depression were alleged to be analogous to the substrates of recurrent seizures.[14-15] The theory was based upon experiments in animals which showed that repeated electrical stimuli lowered the threshold for each triggered seizure. **The term "kindling" referred to the fact that convulsions would, over time, respond to smaller and smaller amounts of applied current.** For drug advocates, the kindling theory was attractive because it supported the view of mood disorders as a product of malfunctioning brain circuits. Here was a model which justified lifelong and aggressive treatment with pharmaceuticals, based upon the belief that unmedicated neurons – like the embers of a smoldering fire – were just waiting to burst forth into behavioral and emotional conflagrations.

Backed by this new dogma, the therapeutic efficacy of carbamazepine and all other anticonvulsant drugs was attributed to the normalization of *assumed* aberrant neural activity. Consequently, many antiepileptic drugs were imbued with imaginary "mood stabilizing" properties. Over time, this trend resulted in the uncritical acceptance of these drugs for approved and experimental (off-label) uses:

Drug	**FDA Approval for Epilepsy**	**Status of FDA Approval for Mania/Bipolar Disorder**
carbamazepine (Tegretol)	3/11/1968	12/10/2004 (Equetro)
divalproex sodium (Depakote)	3/10/1983	5/1995
felbamate (Felbatol)	7/29/1993	phase II trial done (results pending)
gabapentin (Neurontin)	12/30/1993	many trials done, more in progress
lamotrigine (Lamictal)	12/27/1994	6/20/2003
levetiracetam (Keppra)	11/30/1999	trials in progress
*pregabalin (Lyrica)	12/30/2004	trials in progress
tiagabine (Gabitril)	9/30/1997	pilot studies done
topiramate (Topamax)	12/24/1996	phase III trial done
**valproate (Depakene)	2/28/1978	5/1995
zonisamide (Zonegran)	3/27/2000	pilot studies done

* Pregabalin was approved in Europe in 2006 as a treatment for anxiety. Currently, the drug is under investigation in the U.S. as a treatment for anxiety. A federal patent was issued in March 2002 for the drug's use in the treatment of mania.

** Valproic acid (valproate) is the generic term for the American drug known as *Depakene*. Many bipolar and epileptic patients receive divalproex sodium (aka, *Depakote*) which *combines* sodium valproate and valproic acid. An injectable form is known as *Depacon* (valproate sodium).

Background Information

Problems of the Kindling Theory of Epilepsy and Mood Disorders

For those clinicians who hold rigidly to the view that neuroscience, rather than wishful thinking, should justify the use of pharmaceuticals, the alarming fact is that the kindling theory has been invalidated by several realities.

First, no scientist has *ever* confirmed a pre-existing firestorm (kindling) in the brains of depressed or manic patients. While the kindling model is heuristically appealing, there has never been any objective marker or test which confirms a hyper-excitable state in the case of mood disorders.

Second, the kindling theory has failed to predict the real-world pattern of neurological *and* psychiatric phenomena. According to Weiss and Post, the kindling phenomenon requires that symptoms become more intense, more frequent, and ultimately spontaneous (i.e., independent of environmental triggers). However, animals exposed to repeated electrical stimuli generally do *not* progress to spontaneous convulsions.[16-17] Many seizure conditions in humans remit naturally without the use of medication, and even when seizures *do* recur, the majority of humans experience symptoms which remain stable in their intensity or which diminish over time.[18-20]

Third, if the kindling theory of mood disorders were correct, one would expect to see little variation between different antiepileptic drugs in terms of their capacity to "stabilize" the brain. Interestingly, several recent investigations (topiramate, zonisamide) have found *no superior efficacy for the newer agents when compared to placebo*, and in one small study (tiagabine), exposure to the anticonvulsant drug resulted either in no change, or in the symptomatic deterioration of 77% of the bipolar patients.[21-23]

Arguably the most concerning aspect of the kindling theory has been the medical community's unwillingness to appreciate its contributions to iatrogenic harm. By virtue of the very mechanisms which reduce the frequency and intensity of seizures (the blockade of various ion channels in cell membranes), and by virtue of the collateral damage which these drugs inflict upon the endocrine and nervous systems, all of the mood stabilizers have proven to be inherently *destabilizing*. For some treatments, these effects have been unambiguously identified as brain damage and dementia.

What is Lithium?

Many patients are under the impression that lithium is "just a salt." Presumably, this perception originates in discussions or educational materials distributed by prescribers, drug makers, and support groups, all of whom imply that lithium is perfectly safe "as long as blood levels of the drug remain within a therapeutic range." Sadly, these assumptions betray the fact that lithium is a potent nerve toxin whose properties have been repeatedly mischaracterized for decades.

For example, in the late 1800s and early 1900s, lithium-spiked beverages, such as lithiated spring water and lithium-laced cola, were briefly hailed as tonics for numerous afflictions. These claims did not endure. In 1949, the Food and Drug Administration banned lithium salt from the food supply due to its toxicity. Two decades later, lithium compounds won FDA approval for psychiatry.

In the early 1980s, neurologists explored the use of lithium as a treatment for a variety of conditions, including Huntington's chorea, Parkinson's disease, tardive dyskinesia, and epilepsy. None of these uses found evidentiary support.[24-25] Most recently, advances in molecular biology have led to the claim that lithium has neuroprotective and neuroregenerative properties. According to the latest view, lithium can prevent, reverse, or mitigate a variety of neuropathologies, including stroke; Canavan disease (a genetic disorder which results in the loss of myelin and in the degeneration of white and gray matter within the brain); amyotrophic lateral sclerosis (Lou Gehrig's disease); and several other neurodegenerative conditions (Parkinson's, Alzheimer's, and Huntington's disease).[26-31] Unfortunately, the scientists who now promote lithium as a neurological "cure-all" have ignored substantial evidence to the contrary.

Not Just Another Salt ...

In contrast to dietary table salt (sodium chloride) which is extracted from bedrock or seawater, the lithium salts which are used in psychiatry are specifically synthesized by compounding the alkali metal with acids. The resulting salts (lithium carbonate, lithium orotate) commonly exert deleterious effects upon multiple organ systems, including the thyroid, the parathyroid, the kidney, and the brain. The characteristics and prevalence of some (but by no means all) lithium toxicities are summarized below:

Lithium Toxicities [32-41]	
Disease & Associated Features	*Prevalence*
hyperparathyroidism – excessive blood levels of calcium (caused by insensitivity of the parathyroid to calcium feedback, possible over-secretion of parathyroid hormone)	6-50%
symptoms: renal stones, bone resorption and bone pain, gastrointestinal distress, psychiatric distress (changes in mood, cognition, perception)	
hypothyroidism – dysfunction of thyroid gland (caused by impaired uptake of iodide, impaired release and synthesis of T3 and T4)	30-40%
symptoms: weight gain, fatigue, weakness, dry skin, hair loss, depression, psychosis, neuropathy	
kidney disease –	
impaired ability to produce concentrated urine	50%
full-blown diabetes insipidus (nephrogenic)	10%
tubular atrophy, interstitial fibrosis (CITN)	unknown
chronic kidney disease (reduced GFR)	40-80%
symptoms: excessive thirst and urination, high blood pressure, heart dysrhythmias	
lithium intoxication	$\geq 30\%$
SILENT (see below)	$\geq 5\%$
tremor – dysfunction / deterioration of brain regions which modulate movement	40-80%
symptoms: rhythmic, involuntary oscillatory movements (hands, limbs, head) at rest, when resisting gravity, and/or with targeted movement	

Lithium Intoxication: Brain Damage in the Acute and Chronic Setting

Presumably related to excessive levels of medication within the brain, the syndrome of **lithium intoxication** refers to a cluster of predominantly neuropsychiatric symptoms which include the following:

Symptoms of Lithium Intoxication [42-44]

Cognitive	*Motor/Sensory	Psychiatric
clouded consciousness	muscular weakness	apathy
drowsiness	slow movement	confusion
slow cognition	slurred speech	depression
coma	visual / ocular disturbances	
death	unstable gait (ataxia)	
	tingling/numbness	
	tremor	
	seizures	

* may include features of Parkinson's disease (extrapyramidal symptoms) and/or other movement abnormalities (myoclonus, chorea, dystonia, dyskinesias)

According to the timing and context of the inciting event, three major patterns of intoxication are recognized. [45-46] *Acute poisoning* occurs in patients who were not previously taking lithium. *Chronic poisoning* refers to toxicity which occurs unexpectedly in the context of ordinary or usual doses. *Acute-on-chronic poisoning* refers to an accidental or intentional overdose of lithium during maintenance therapy (either by swallowing too much lithium, or by failing to eliminate normal doses in the usual fashion).

Patterns of Lithium Exposure Resulting In Poisoning

Acute	no previous lithium
Acute-on-Chronic	previous lithium + accidental or intentional overdose
Chronic	previous lithium, no change in dose (treatment as usual)
Results:	excessive brain levels of lithium due to acute or chronic drug accumulation (affected by changes in diet, other medications, renal or thyroid disease, infection, fever, and dehydration)

Consistent with several previous studies, a recent investigation conducted by authorities in Scotland revealed that the ***most serious symptoms of lithium poisoning were associated with chronic treatment***.[47] Furthermore, the Scottish study also confirmed that it was the pattern of drug exposure – rather than age, dose, or serum drug level – which predicted the severity of symptoms.

The importance of these findings cannot be overstated. Although health care providers are generally well prepared to identify and respond to acute overdose, ***they are usually not trained to identify the consequences of chronic toxicity***. In this regard, *standard textbooks* have been inappropriately glib when characterizing lithium as a treatment with a low therapeutic index (low safety margin) and implying that harmful side effects might be avoided as long as blood levels of the drug remain within the so-called "therapeutic" range (0.8 to 1.5 mEq/L for acute mania, 0.5 to 1.2 mEq/L for "mood stability").[48]

In reality, history has repeatedly demonstrated that the problem with lithium is not a therapeutic index which is narrow, so much as it is a therapeutic index which is arguably *non-existent*. Because of the fact that physicians have no way of predicting when or if lithium intoxication will emerge in the context of maintenance therapy, and because of the fact that *blood levels* of lithium commonly fail to predict *brain levels* or *brain consequences* of exposure, all lithium prescribers and consumers engage in the pharmaceutical equivalent of Russian roulette. Most critically, the significance of lithium intoxication arises not only from the potential lethality of an *acute* event, but also from the capacity of an intoxication episode to induce *delayed or persistent dysfunction*.

SILENT – Syndrome of Irreversible Lithium Effectuated Neurotoxicity

The propensity of lithium salts to inflict **permanent brain damage** was first recognized in the early 1970s.[49-51] Subsequently, professional fervor about lithium as a so-called anti-manic deflected critical discussions about long-term harm. Only at rare intervals have astute clinicians interrupted this legacy of neglect. In 1987, one such individual (a physician self-identified only as Adityanjee) coined the term **SILENT** to remind other health care professionals about the perpetually ignored, post-intoxication **Syndrome of Irreversible Lithium Effectuated Neurotoxicity**.[52]

Adityanjee's mnemonic has not yet found its way into lecture halls, textbooks, or continuing education materials. Nevertheless, the subject matter of SILENT has become more accessible to discerning researchers in recent years due to a proliferation of clinical anecdotes and occasional reviews in the medical literature.[53-55] ***Common to all of these reports has been the failure of drug levels, drug monitoring, and even*** <u>***drug***</u> ***withdrawal to predict or prevent destruction of the brainstem and/or cerebellum***. The disabling effects of SILENT may include any or all features of lithium intoxication, as well as the symptoms of various neurodegenerative conditions: Parkinson's disease, Lewy body dementia, Creutzfeld-Jakob disease, or Alzheimer's disease.

What About Antiepileptic Drugs (AEDs)?

In the words of the late British physician, Sir Edward Henry Sieveking, "there is scarcely a substance in the world capable of passing through the gullet of a man that has not at one time or another enjoyed the reputation of being anti-epileptic."[56] With its long record of "trial and error" treatments fueled by irrational theories about the cause of seizures, no specialty of medicine has surpassed the capacity of neurology to ignore, deny, and/or minimize the problems of target organ toxicity.[57] Unfortunately, when psychiatry embraced the kindling theory to justify the use of anticonvulsants as quenchers of mood disorders, it emulated this distressing trend.

Today, few physicians in the United State appear to be trained to recognize the pathophysiological mechanisms through which antiepileptic drugs harm the brain. For example, leading textbooks of pharmacology, psychiatry, and neurology routinely fail to mention the potential of America's leading anti-seizure drug to induce a syndrome which has been variably referred to as *dementia, encephalopathy*, and *pseudo-atrophy* of the brain.[58-61] Perhaps this failure is partly explained by the continuing ambiguity which surrounds the use of each one of these terms.

Background Information

Even the most authoritative sources do not discriminate clearly between dementia and encephalopathy. This semantic confusion can be well appreciated by consulting the pages of a medical dictionary in which these two conditions are only vaguely defined:[62]

dementia:	organic loss of intellectual function
encephalopathy:	any degenerative brain disease

The website of the National Institute of Neurological Disorders and Stroke reflects a similar degree of linguistic murkiness:[63-64]

dementia:	a collection of symptoms that can be caused by a number of disorders that affect the brain; impaired intellectual functioning that interferes with normal activities and relationships
encephalopathy:	any diffuse disease of the brain that alters brain function or structure; the hallmark is an altered mental state; symptoms may include dementia

---

> Textbooks of neurology typically distinguish these conditions in terms of temporality and reversibility, reserving the term dementia for any degenerative condition which involves the gradual and usually irreversible loss of two or more cognitive functions (such as memory, language, or planning ability). Meanwhile, encephalopathy is reserved for any transient syndrome which involves *cognitive deficits along with an altered mental state* (drowsiness, lethargy, disorientation, and/or loss of consciousness). In reality, the signs and symptoms of dementia and encephalopathy frequently overlap, and in deciding between two possible diagnostic labels, physicians are often influenced by patient age (older patients = demented; younger patients = encephalopathic).
>
> Finally, the word "pseudo-atrophy" appears in several publications which describe the effects of valproate and other AEDs, just as the word "pseudo-Parkinsonism" appears in many papers about neuroleptics. Originating in the Latin term for something false or fraudulent, the compound nouns which begin with ***pseudo*** have been habitually used in medicine to imply that drug-induced changes are less than real.[65] This is a dangerous anachronism, based upon the outdated expectation that any *organic condition should not reverse. As data from pathology and neuroimaging research have repeatedly revealed, there is nothing artificial or pretentious about the toxicities of valproate (or any other psychiatric drug), and there remains no valid defense for linguistic contrivances which conceal the true nature of iatrogenic disease.
>
> * organic = caused by a fundamental abnormality in brain function or structure,
>   as opposed to being caused by emotions, the psyche, or the soul

Depakote Dementia

The phenomenon of ***Depakote dementia*** first came to the attention of this writer incidentally, for it was not an FDA advisory, nor a Dear Doctor Letter, nor a Continuing Medical Education program which served as the conduit for this vital information. Rather, the source of my education was a piece of non-medical journalism which appeared in the pages of a monthly woman's periodical. The story concerned an eight-year-old girl and her first experiences with treatment for epilepsy.[66] Under the influence of Depakote, this young patient developed severe changes in behavior, cognition, and mood. Although the prescribing clinician eventually recognized the emergent problems as iatrogenic and successfully tapered the medication – with immediately beneficial results – the family suffered needlessly for an entire school year because they were not informed about the likely hazards of treatment. Recalling the events as they unfolded in 2000 and 2001, the mother described the comments of her daughter's doctor and her own reactions:

>...*" 'well, I told you about **Depakote dementia**,' the physician said as he rolled his eyes. Actually, no, he hadn't. I went home and researched the drug online and was floored by my findings. Buried deep amid reams of pharmacological information, under "adverse reactions," it read:
>
>> 'Emotional upset, depression, psychosis, aggression, hyperactivity, hostility and behavioral deterioration.'
>
> Everything I had pointed out to the experts for months. Yet they shrugged it off. My daughter may have lost a year of her life because no one was reading the fine print." [67]

In actuality, not even the fine print of the 2008 Depakote product label could have given this family an *explicit* warning about dementia.[68] Ironically, although a Black Box Warning does exist, it alerts physicians to the potential hazards of pancreatitis, liver failure, and birth defects. ***Notably, there is no mention about the chemical's inherent brain toxicity***. This omission is all the more remarkable when one considers the fact that the National Library of Medicine's own search engine, PubMed, now references more than 130 publications on this topic. Included among these papers are at least 180 cases of valproate-induced encephalopathy (with and without high blood levels of ammonia) and more than 85 cases of drug-related Parkinsonism – some of them, fatal. At this juncture, one cannot help but wonder how much suffering and disability might have been avoided had standard textbooks and training programs paid attention to history, and had regulatory officials called for the characterization of **target organ toxicity**.

Depakote Dementia According to PubMed		
# of publications (unique articles)	VHE/VIE # of cases	Parkinsonism # of cases
1980s 14	41	6
1990s 41	26	39
2000-2008 76	117	42

VHE = valproic hepatic encephalopathy and/or "reversible dementia"
VIE = valproic-induced encephalopathy and/or "reversible dementia"

Features of Depakote Dementia and Encephalopathy

	dementia	encephalopathy
can appear at any time during course of therapy	+	+
occurs despite "normal" serum levels of medication	+	+
agitation/irritability	+	++
impaired memory	+	+
impaired cognition	+	+
lethargy/drowsiness/stupor	-	+/-
loss of consciousness	-	+/-
impaired activities of daily living	++	+
high serum ammonia	+	++
Parkinsonism	++	+
with hearing loss	++	+
other neurological signs		
ataxia	-	+
asterixis	-	+
dysarthria	-	+
dysphagia	-	+
myoclonus	-	+
tremor	++	+
visual disturbance	-	+
abnormal brain scan	+/-	++
gastrointestinal distress (nausea/vomiting)	-	+
rapid resolution of symptoms upon drug withdrawal	+/-	+
complete resolution of symptoms upon drug withdrawal	+/-	+

key: + common finding   - rare or absent   +/- variable finding

Despite abundant evidence to the contrary, many researchers have claimed that Depakote dementia is extremely rare. This impression appears to reflect wishful thinking and a superficial consideration of the published record. As far as this writer is aware, no nation maintains accurate statistics on the true incidence and prevalence of this phenomenon. However, the epidemiological evidence which *is* available suggests a problem that may be far more extensive than clinicians would like to believe.

For example, more than a decade ago, physicians associated with the Veterans Administration Medical Center became concerned about **hearing loss** as a possible reaction to Depakote. Following the detection of this side effect in the second of two patients, Armon et al decided to review the records of all Depakote-exposed subjects within a North Carolina epilepsy clinic.[69] The results revealed an alarmingly high prevalence of neurological dysfunction, including treatment-emergent Parkinsonism, cognitive impairment, and a variety of other neurological signs and symptoms. The probability that these conditions were, indeed, drug-related was suggested by a trial taper of the anticonvulsant in 35 of the 36 subjects. This resulted in significant improvements in the condition of most patients, including no change in the frequency or severity of seizures:

Depakote Dysfunction: Veterans Administration (Armon et al, 1996)		
Demographics		
# of patients	36	
age range	22 to 74	(mean age: 51.5 years)
male	35 (97%)	
duration of Depakote	1 to 11 years	(median: 3 years)
monotherapy	64%	
+ other AEDs	36%	
	% With Problems During Depakote Therapy	% Improving After Depakote Withdrawal
cognitive impairment	86%	72%
Parkinsonism	75%	96%
memory dysfunction	69%	63%
progression of brain atrophy	69%	100% (n = 2, repeat CT scan)
progressive hearing loss (via audiometry)	83%	48%

Other teams of investigators have denied that Depakote dementia is a widespread problem, but their studies have often been compromised by serious methodological confounders. Easterford et al (2004) analyzed the experiences of 50 epilepsy patients who were recruited from English clinics between May and November 2001.[70] The research protocol excluded anyone who had stopped taking Depakote within the previous five years, as well as anyone with current exposure to Depakote lasting less than one year. Thus, a serious selection bias omitted data from individuals who may have developed drug-induced dementia or encephalopathy leading to withdrawal of the offending agent ("sick quitter" effect). Despite these design flaws, and despite the relative youth of this sample (mean age: 43 years), the study still managed to detect a high prevalence of tremor (22% of the subjects) and a surprisingly high rate of Parkinsonism (6% of the subjects, based upon a Unified Parkinson Disease Rating Scale score of 30 or more).

In a similar study which consisted of 364 subjects, recruited over a nine-year period, Ristic et al (2006) claimed that only 1% of their Depakote cohort developed Parkinsonism.[71] However, the findings of this investigation were necessarily influenced by a younger cohort (median age: 25 years) and a relatively short duration of Depakote exposure (3 years, 4 months) relative to other studies. Furthermore, a selection bias again clouded the results. The protocol excluded data from more than 400 patients who had been unable to tolerate Depakote for at least one year. In other words, the exclusion criteria prevented the consideration of any patient who might have experienced drug-induced dementia or encephalopathy, followed by Depakote withdrawal, within the first year of therapy.

Given the fact that Depakote, like all AEDs, is an inherent suppressor of brain activity (essentially encephalopathogenic); and given the fact that Depakote - like all AEDs – exerts multiple pathological effects upon cell metabolism (see chapter two), there is good reason to doubt the claim that Depakote dementia is a rare, avoidable, or unexpectable event.[72-74]

Postmortem Studies of Animals

The relevance of animal research for human patients depends largely upon the use of treatment protocols which employ therapeutically relevant doses. The following table compares the blood and brain levels of lithium which have been measured in humans and rats:

Pharmacokinetics of Lithium [75-81]		
	Human	Rat
dose		
oral	600 mg to 2400 mg per day 16.2 mmol to 65 mmol per day ~ 9 mg/kg to 30 mg/kg per day ~ 0.23 to 0.93 mmol/kg per day [8.12 mmol per 300 mg]	55 mmol/kg
intraperitoneal	n/a	5 mmol/kg
subcutaneous	n/a	0.9 mmol/kg
peripheral blood level	0.5 to 1.5 mmol/L	0.44 mmol/L (oral)* 0.1 to 10 mmol/L (IP)** 0.7 to 1 mmol/L (subQ)***
brain level	0.5 to 1.95 mmol/L	0.482 mmol/kg (oral)* 0.5 to 1.5 mmol/kg (IP)** 0.08 to 1.0 mmol/kg (subQ)***
time to max. blood level	3 to 8 hrs	8 hrs
time to max. brain level	5 to 10 hrs	12 hrs
elimination half-life		
blood	24 to 30 hrs	5 to 7 hrs
brain tissue	48 hrs	8 to 12.7 hrs

*   oral = steady-state levels after four days of **oral** dosing
**   IP = measured 0 to 50 hrs after single **intraperitoneal** injection
***   subQ = 0 to 50 hrs after 2 days of **subcutaneous** injections given every 6 hours

Study #1   Dethy et al (1997)   *Lithium Destroys White Matter in the Cerebellum*

Appreciating the clinical significance of lithium intoxication, researchers in Belgium designed an animal experiment with the goal of confirming the anatomic consequences of a **single high-dose exposure** to the medicinal salt.[82]

Experimental Protocol

The investigation proceeded in two phases. In phase I, five male Wistar rats were implanted with microdialysis probes in the cerebellum. Following a single injection of lithium chloride (250 mg/kg intraperitoneal injection = 6.77 mmol/kg), serial measurements of several neurotransmitters were made by sampling fluid via these probes. Five hours after treatment, a blood sample was withdrawn from the carotid artery in order to verify the post-treatment level of the medication.

In phase II, twenty-nine male Wistar rats received a single intraperitoneal injection of lithium chloride (again at the dose of 250 mg/kg). Fourteen animals were sacrificed five hours after treatment, and the remaining subjects were sacrificed following a two-week period of recovery. Brain tissue from both groups of animals was excised and examined using standard staining techniques for light microscopy.

Findings

Five hours after injection with lithium chloride, the average serum drug level was 1.5 mmol/L (upper limit of the therapeutic reference range in humans). At this level of exposure, *__fifty percent of the rats developed severe pathology within the white matter of the cerebellum and mild damage to the brainstem__*. In order to appreciate the appearance and implications of the resulting injury, it is important to understand the normal structure and functions of the cerebellum.

Background Information

Derived from Latin, the word cerebellum ("little brain") refers to a region of the central nervous system which is associated with the control of balance, coordination, eye movements, and cognition. For reasons which remain unknown, the cerebellum is particularly sensitive to the toxic effects of exogenous chemicals (including alcohol, lithium, anesthetics, and all of the anticonvulsants) and to injury by fever and infectious agents.

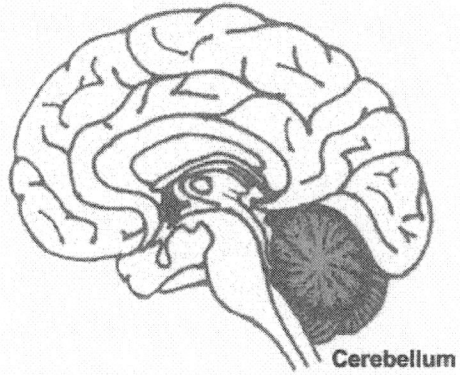

Source: McGill University, The Brain from Top to Bottom

When sliced and examined under a microscope, cerebellar tissue demonstrates a precise arrangement of component cell types (neurons and glia), according to which neuroscientists have identified three major layers. Starting at the cortical surface and working towards the inner core, these are known as **the molecular layer**, the **Purkinje cell layer**, and the **granule cell layer**. The **white matter** of the cerebellum lies beneath all three zones:

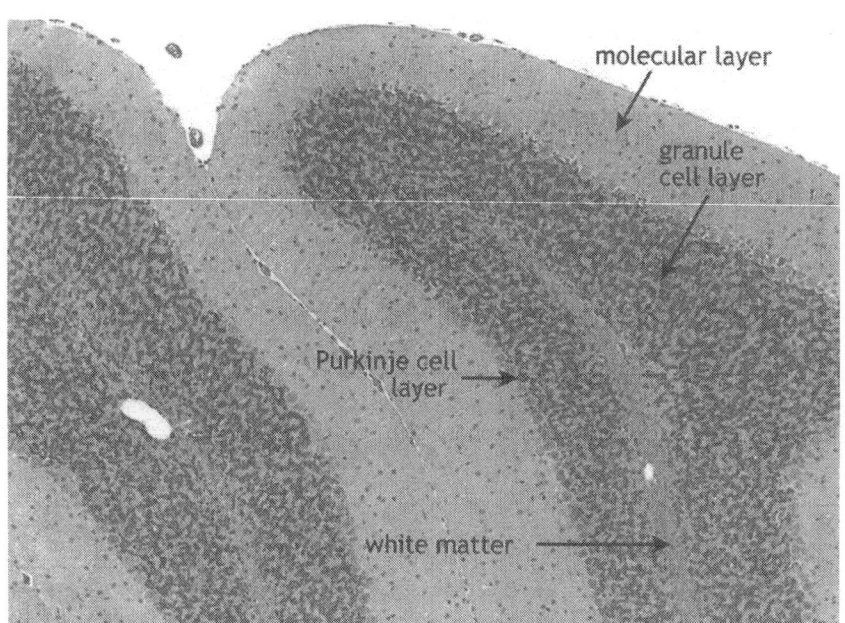

Source: Deltagen, Inc. (San Mateo, CA), reproduced with permission.

Returning to the Belgian experiment, half of the lithium-exposed animals experienced the onset of extensive vacuolization within the myelinated fibers of the cerebellum. (Vacuoles refer to membrane-bound vesicles or sacs which form inside brain cells in the context of several neurodegenerative conditions.) Because they impart a Swiss cheese or sponge-like appearance to the affected tissue, the corresponding abnormalities are referred to by pathologists as *spongiform* changes:

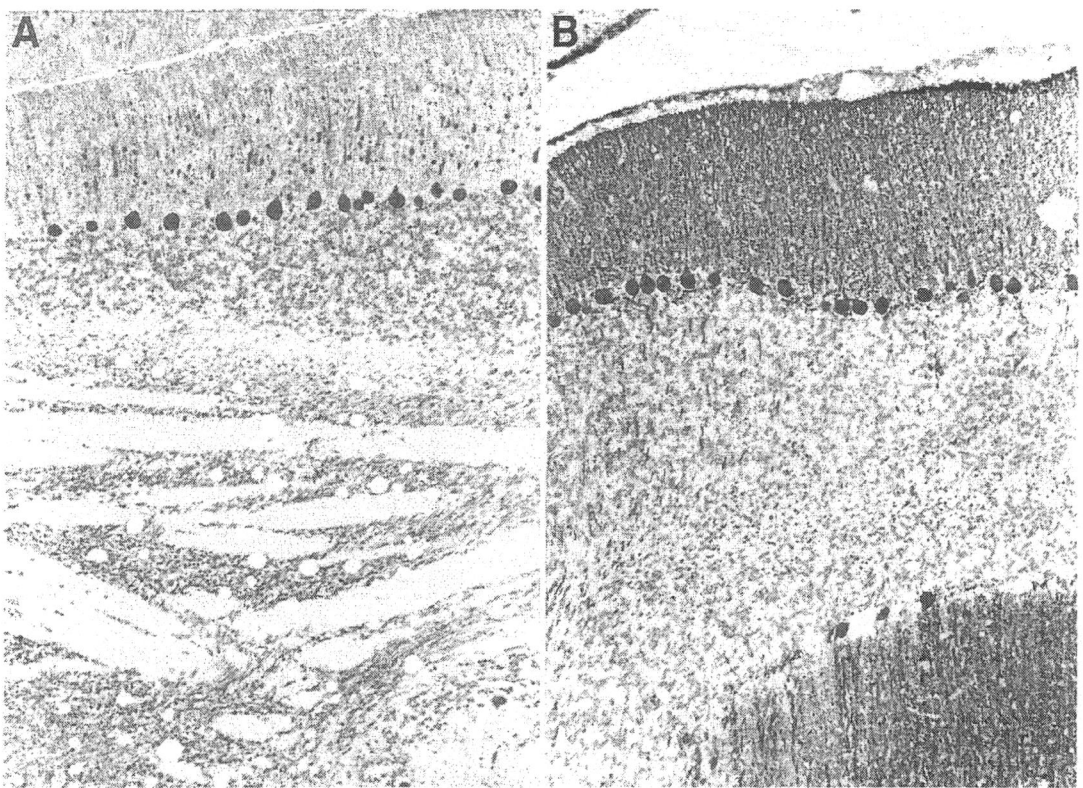

Reprinted from *Neuroscience Letters*, volume 224, issue 1, Dethy et al, "Cerebellar spongiform degeneration induced by acute lithium intoxication in the rat," p. 27, copyright (1997), with permission from Elsevier.

Left: Appearance of rat cerebellum five hours after injection with lithium salt

Spongiform changes (white holes = vacuoles) can be seen in the deepest layer of this specimen, consistent with the location of the white matter.

Right: Appearance of normal cerebellum and normal white matter from a non-exposed (control) rat

This particular experiment was significant not only because it replicated the location of major drug effects, as seen in humans (i.e., *cerebellar* damage), but also because the *type* of damage which it confirmed was consistent with another important kind of human pathology.

Within the mainstream medical literature, several clinicians have published case reports about lithium users who developed a clinical syndrome identical to Creutzfeld-Jakob disease or CJD. This neurodegenerative condition (mentioned briefly in chapter two) is characterized by dementia, myoclonus (intense muscle jerks), and electrophysiological abnormalities (periodic sharp-wave complexes and generalized slowing).[83-89] Pertinent to this discussion, however, is the fact that **CJD and lithium intoxication share several anatomical features**. These include spongiform changes, neuronal death, and gliosis (see below) in the cerebellum, and damage to axons in the peripheral nervous system.[90-94]

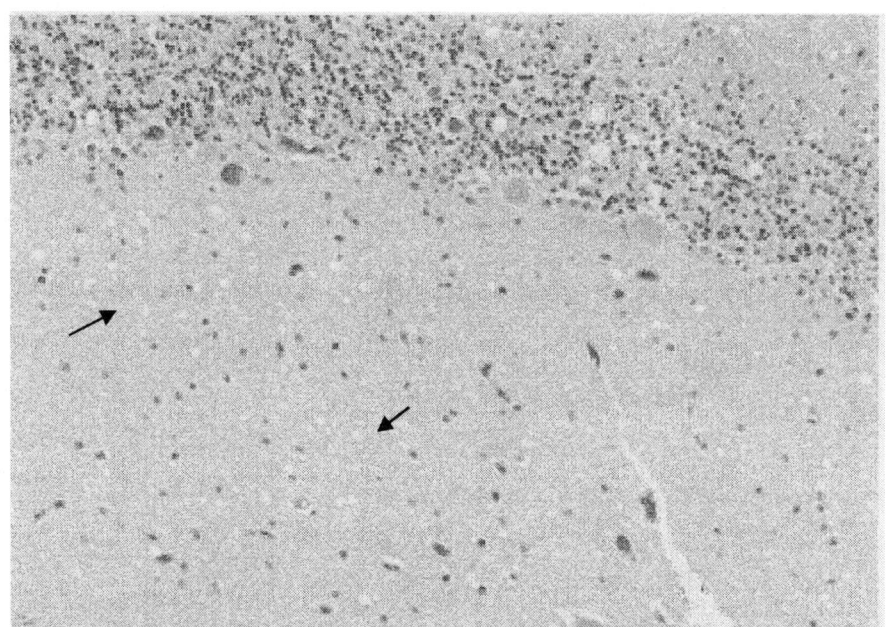

Source of image: National CJD Surveillance Unit, University of Edinburgh – Western General Hospital, Edinburgh UK. Reproduced with permission of Professor James W. Ironside.

Above: This image displays spongiform changes (small vacuoles = holes) in the **molecular layer** of the cerebellum. This specimen was obtained from a patient with **Creutzfeldt-Jakob disease**.

The neurotoxic effects of lithium extend beyond the cerebellum. Animal experiments have also revealed drug-induced injury in the ***hippocampus***. Before examining this body of research evidence, it is important to appreciate the significance of one of the major anatomical markers which toxicologists associate with chemical harm.

Background Information

In response to mechanical trauma, disease, and/or chemicals, microglia and astrocytes expand in number and activity. This proliferation of glial cells, known as **reactive gliosis**, is important in the discussion of pharmaceuticals for several reasons.

First, **reactive gliosis** *can serve as a sensitive and early signal of brain damage* because it frequently precedes the disintegration of nerve terminals, the loss of synapses, and the death of neurons.[95-97] Thus, by identifying gliosis with appropriate lab techniques, scientists can confirm harmful changes which might otherwise be overlooked:

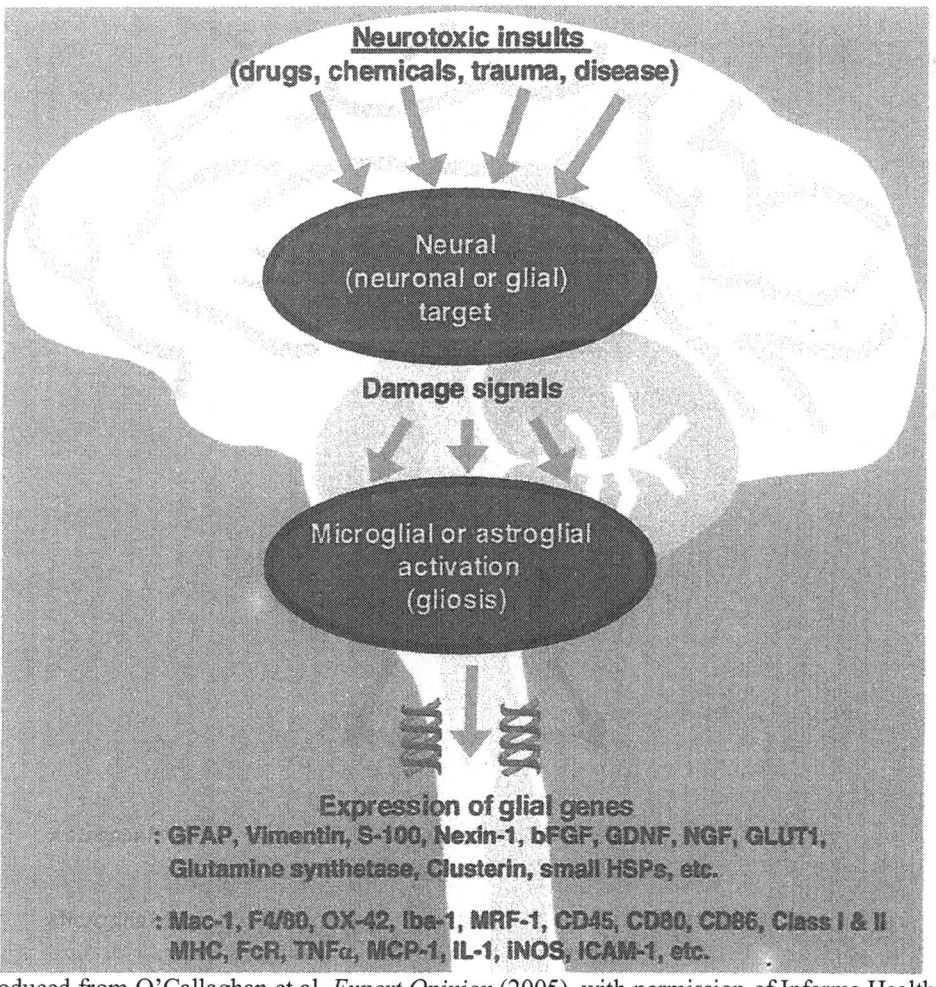

Reproduced from O'Callaghan et al, *Expert Opinion* (2005), with permission of Informa Healthcare

Second, the phenomenon of reactive gliosis is important for neurologists and psychiatrists because it can have positive *and* negative effects.[98-99] Here, it is important to appreciate the function of astrocytes and microglia after injury. Both kinds of cells produce and secrete chemicals such as cytokines (and, in the case of astrocytes, chondroitin sulfate proteoglycans). *Depending upon the timing and intensity of their release, these substances can impair the regeneration of neurons and promote additional and potentially long-lasting dysfunction.*

Third, autopsy analyses of brain tissue from human subjects have demonstrated a positive association between the severity of gliosis, and the symptoms and pathologies of dementia. For example, in a recent study performed by researchers in Hawaii, the severity of antemortem cognitive dysfunction and intensity of postmortem abnormalities (plaques, tangles) were both associated with high levels of astrocytic proliferation within the temporal, parietal, and occipital lobes.[100]

In sum, the experimental detection of *reactive gliosis* provides valuable information about the presence of degenerative changes, whether genetically or environmentally provoked. In order to confirm gliosis, scientists often employ a sophisticated method (known as immunohistochemistry, or IHC) which coats brain specimens with antibodies targeting glia-specific proteins. Subsequent procedures permit the researchers to "find" these antibodies and to stain the cells of interest. (Otherwise, these cells would not be visible with light microscopy.) One of the key antibody targets which is used to identify astrocytes is **glial fibrillary acidic protein** or **GFAP**. Excessive levels of this protein have been found in the brains of patients with dementia, and in the brains of experimental animals following exposure to various kinds of neurotoxicants.

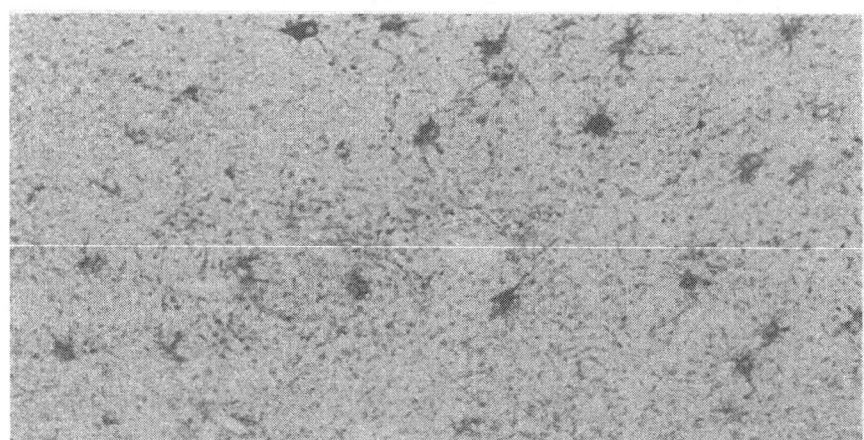

Source: University of Oklahoma Health Science Center, with permission of Dr. Kar Ming-Fung

Above: This image demonstrates **reactive astrocytes** from a patient who developed an inflammatory demyelinating disease of the brain. The dark appearance of these astrocytes (star-shaped cells, above) has been caused by the staining of glial fibrillary acidic protein or **GFAP**.

Mood Stabilizers

Studies #2-3   Rocha et al (1994, 1998)   *Lithium Damages the Hippocampus*

In two studies of increasing complexity, researchers in Brazil exposed animals to clinically relevant, *oral* doses of lithium chloride for a period of four weeks.[101-102] Following treatment, brain specimens were processed using immunocytochemical techniques which permitted the study of glia.

Experimental Protocol #1

In the first experiment, an undisclosed number of adult male Wistar rats were administered a normal diet or a regimen supplemented by daily lithium (lithium chloride, 60 mmol/kg x four weeks). At the end of this treatment interval, the animals were sacrificed, brains were removed, and samples of tissue from the **hippocampus** and **caudate** were processed using biochemical procedures (immunoblotting and radiographic densitometry) to identify the content of **GFAP**.

Experimental Protocol #2

In the second experiment, the researchers employed the same initial methodology. Following the four-week period of normal or lithium-enhanced diet, an undisclosed number of male Wistar rats were surgically euthanized. Brains were removed and slices from the **hippocampus** were prepared using immunocytochemical methods to **label GFAP**. Following the staining procedure, the specimens were viewed using a special photomicroscope. Automated and hand-drawn images (*camera lucida **drawings**) of hippocampal glia were produced.

---

* **camera lucida** – an optical device which superimposes a viewed object (in this case, tissue viewed through a microscope) onto the surface upon which an artist is drawing; until the advent of technologies which permitted inexpensive reproductions of photomicrographs, many textbooks in science and medicine featured camera lucida drawings because of their clarity[103]

Findings

Experiment #1

The results of the first study revealed *serum lithium levels which corresponded to the concentrations observed in the treatment of humans* (0.6 to 1.2 mmol/L). Following four weeks of exposure to the medication, the animals experienced a **34% increase in GFAP** in the **hippocampus** (a temporal lobe structure which mediates learning and memory). Similar changes were observed in tissue samples obtained from the **caudate** (a basal ganglia structure associated with the control of involuntary movement and cognition).

Experiment #2

As before, the scientists found that lithium doses associated with typical serum levels (0.6 to 1.2 mmol/L) were sufficient to induce a ***35% increase in the GFAP content throughout all regions of the hippocampus***:

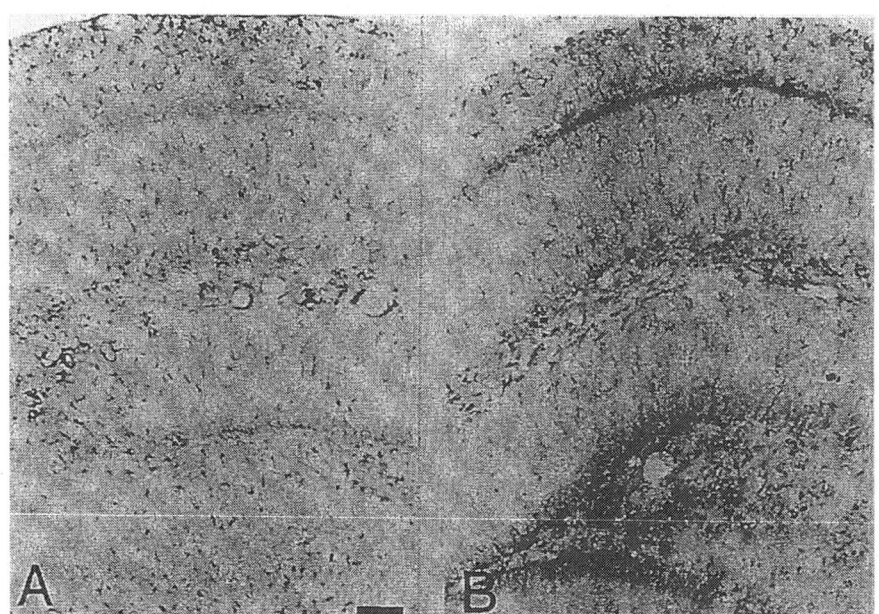

Source: E. Rocha, *NeuroReport*, volume 9, issue 17, "Lithium treatment causes gliosis and modifies the morphology of hippocampal astrocytes in rats," p. 3972, (1998), reprinted with permission of Wolters Kluwer / LWW.

Left: Appearance of normal hippocampus from a drug-free control

Right: Darkened cells (GFAP+) indicative of **gliosis** in the hippocampus following four weeks of lithium chloride

In addition to observing anatomical changes consistent with gliosis (either more astrocytes, or larger astrocytes, or both), the Brazilian research team also documented a profound alteration in glial alignment. Under healthy conditions (control animals), the processes or "arms" of each astrocyte assumed an orderly, perpendicular orientation within several of the hippocampal layers. *Under the influence of lithium, however, this cellular arrangement became markedly disrupted*.

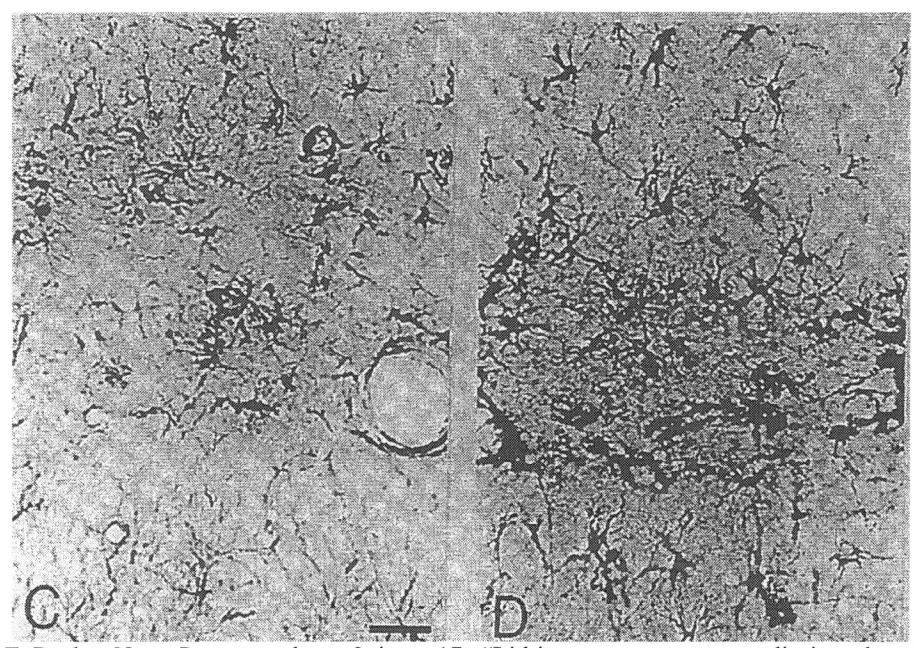

Source: E. Rocha, *NeuroReport*, volume 9, issue 17, "Lithium treatment causes gliosis and modifies the morphology of hippocampal astrocytes in rats," p. 3972, (1998), reprinted with permission of Wolters Kluwer / LWW.

    Left: This slide depicts hippocampal astrocytes from a drug-free control.

    Right: Following four weeks of lithium, astrocytes changed in number and size, and their alignment became chaotic.

The loss of perpendicular alignment was captured well by camera lucida drawings of the pertinent brain tissue:

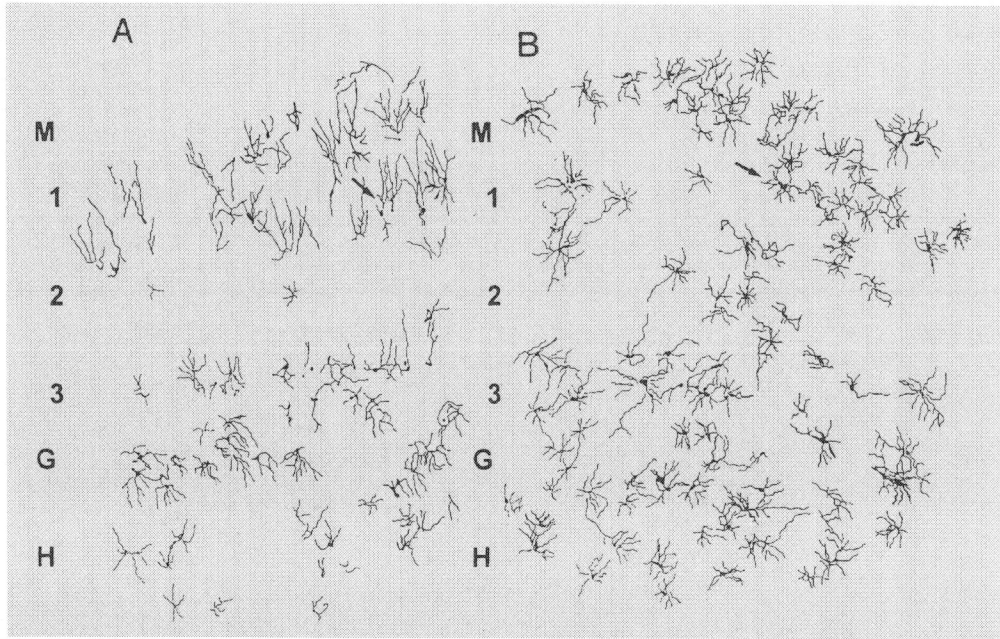

Source: E. Rocha, *NeuroReport*, volume 9, issue 17, "Lithium treatment causes gliosis and modifies the morphology of hippocampal astrocytes in rats," p. 3973, (1998), reprinted with permission of Wolters Kluwer / LWW.

Left (A): Perpendicular alignment of astrocytic processes in the layers of the hippocampus (unmedicated control)

Right (B): Disorganized pattern of astrocytic processes following lithium exposure

Although the Brazilian team did not comment upon the implications of their observations with respect to drug-induced toxicity, more recent studies by other teams have demonstrated ***how and why gliosis – particularly, within the hippocampus – can result in cognitive impairment***.

For example, a series of investigations conducted by researchers in Czechoslavakia have explored the role of glial cells in regulating the flow of ions, water, and neurochemicals throughout the brain.[104-105] When glial cells expand in size or number, and when they assume a disorganized orientation – as occurs under the influence of illness, injuries, or ageing – they impede function by forming barriers to the diffusion (movement) of substances through the brain.

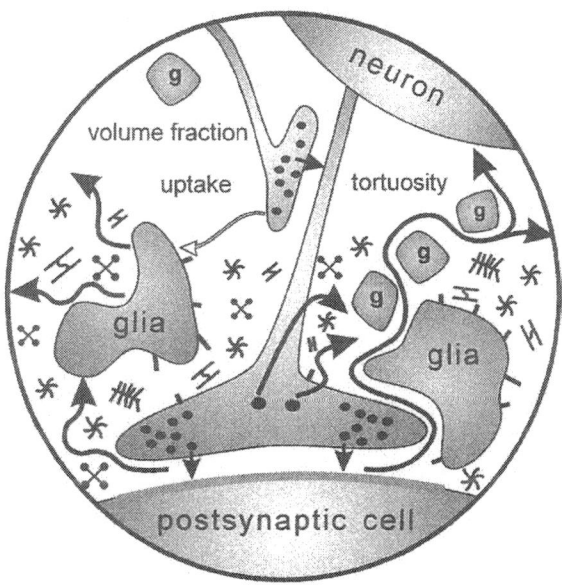

diffusion barriers

Source: Reprinted from *Neurochemistry International*, volume 45, issue 4, Syokova, copyright (2004), with permission from Elsevier.

This cartoon demonstrates how gliosis and glial misalignment disrupt the flow of chemicals between neural cells.

The implications of this research are important for the users of lithium. Assuming that humans experience the same kinds of changes which have been observed in rats, one would reasonably expect lithium to impair cell-to-cell communications within the hippocampus (and wherever else gliosis occurs). These changes would predictably impair learning, memory, and other neurobehavioral processes.

Study #4   Phatak et al (2006)   *Lithium Disrupts Water Balance Within the Brain*

Understanding the well established link between lithium and changes in water balance within the soma (non-brain), an international team of investigators was curious to know if lithium might exert similar physiological effects upon the brain.[106]

Experimental Protocol

Adult male Sprague-Dawley rats were randomly divided into two experimental treatment groups for exposure to regular food and water, or lithium-supplemented rat chow and fluid. Feedings occurred orally for **11 days** (acute therapy) or **5 weeks** (chronic therapy). At the end of each study period, lithium and sodium concentrations were measured via carotid artery sampling. The animals were sacrificed, brains were excised, and the content of water and inositol* was determined in three different regions using pertinent lab techniques [for water: wet-dry method, 48 hours of oven drying; for inositol: gas chromatography and spectrometry].

Findings

No significant differences in water content were detected following 11 days of drug exposure. ***However, chronic exposure to dietary lithium at clinically relevant doses (mean serum lithium level = 0.81 mmol/L) resulted in a statistically significant increase in the water content (3%) of the frontal cortex.*** Chronic lithium also resulted in reduced levels of inositol throughout the brain (~ 25% reduction in the frontal cortex and hippocampus):

% Brain Tissue Water Content			
% Tissue Water = [(wet tissue weight – dry tissue wt) ÷ (wet tissue weight)] x 100			
	lithium	control	p-value
frontal cortex	**86.7**	**83.6**	**0.05**
hippocampus	**83.3**	**80.2**	**0.13**
cerebellum	79.3	78.5	082

---

* inositol - an intracellular chemical and osmolyte; as fluid levels *outside* of neurons decline, neurons must transport or synthesize inositol in order to prevent the escape of water to the surrounding environment

Implications

Whereas numerous investigators have recently proposed that lithium exerts a *neuroprotective* role upon the brain, based upon MRI findings of regional enlargement after drug therapy, the study by Phatak et al provides an important alternative explanation. In other words, the apparent expansion of gray matter which has been documented in several radiological studies may reflect changes in water content (and/or gliosis), rather than the presumed proliferation of new neurons.[107-110] Although scientists have not yet determined the mechanisms or consequences of lithium-induced changes in brain hydration, it is important to consider the biological coherence between the current rat study, and the pathophysiological foundations of two conditions which have emerged in human consumers of lithium.

Background Information

Aquaporins, Lithium, and the Brain [111-121]

In 1991, a team of chemists from Johns Hopkins University identified a new class of membrane proteins known as *aquaporins*. As the term implies (aqua *pore*), these cell components were named for their capacity to function as water channels which regulate the entry and exit of fluid. Although scientists have not completed the task of characterizing the identities and properties of these proteins (*to date, thirteen different aquaporins have been identified in mammals – three of them, within the brain*), research has already revealed several findings which are clinically relevant.

First, the connection between lithium and polyuria (excessive urination) is now known to be mediated by a drug-induced reduction of aquaporin 2 (AQP2). Without adequate levels of AQP2 in renal cells, the kidney cannot respond to endocrine signals which regulate the reabsorption of water.

Second, at least two teams of researchers have confirmed an association between increased levels of a major *brain* aquaporin (AQP4, located on astrocytes) and the progression and severity of Creutzfeldt-Jakob Disease. Given the fact that lithium has been known to induce the features of CJD, it is plausible to suspect that lithium may exert this effect by disrupting AQP4 levels within the brain.

Third, the medical literature has revealed sporadic reports about another fluid-related lithium effect. The clinical syndrome known as **pseudotumor cerebri** (PC) refers to a cluster of features which mimic the effects of a brain tumor when, in fact, no tumor is present. Signs and symptoms include severe headache, elevated intracranial pressure, eye pain, blurred vision, and sensory changes. Although lithium, vitamin A, and tetracycline-related antibiotics have all been linked to this condition for more than 20 years, the underlying biology has long been obscure. However, with the discovery of aquaporins on the glia which line the ventricular cavities of the brain [e.g., AQP1 on ependymal cells which absorb and secrete cerebrospinal fluid], scientists have obtained a new target for clarifying thr mechanisms which give rise to this disabling condition.

Studies #5-6   Sobaniec-Lotowska   *Valproate Damages Multiple Regions of the Brain*

Motivated by the desire to understand the anatomic consequences of chronic anticonvulsant therapy, a pathologist in Bialystok, Poland (Dr. Maria Sobaniec-Lotowska) has performed a series of animal experiments involving ***clinically relevant doses*** of valproate. A particular goal of this work has been the characterization of drug-induced changes which might account for the phenomenon of valproate-induced encephalopathy (aka, Depakene / Depakote dementia).

Although this section will report findings from three of Dr. Sobaniec-Lotowska's many papers and their respective concerns, it is important to appreciate the unique work of this scientist in performing numerous electron microscope studies of valproate effects.[122-127] All of these experiments have confirmed ***diffuse drug-induced pathology*** involving the capillary cells, neurons, and glia (astrocytes and microglia). ***Degenerative changes have been observed in the neocortex, the brainstem, the cerebellum, and the hippocampus.***

Experimental Protocol

Study #5   Effects of Valproate Upon the Cerebellum [128]

In this study, adult male Wistar rats were divided into three different groups.

Group I – Chronic Treatment Followed by Immediate Brain Examination
Thirty animals were assigned to separate subgroups for varying durations of drug treatment. Via an intragastric feeding tube, each animal received valproate (dissolved in saline) at the dose of 200 mg/kg per day for treatment periods which consisted of 1, 3, 6, 9, or 12 months (doses resulted in serum drug levels = 60 to 135 ug/mL). Twenty-four hours after the last dose, these animals were euthanized. Brain specimens from the **cerebellum** were processed and examined using light and electron microscopy.

Group II – Chronic Treatment Followed by Brief Periods of Post-Drug Recovery
Twelve animals were exposed to valproate (same regimen) for a period of twelve months. Subjects were sacrificed following one-month or three-month recovery periods, and brain specimens were processed and examined as above.

Group III – Control Group
Fourteen rats were matched to the other animals on the basis of age and body mass. Via intragastric feeding tubes, these subjects received the same amount of saline as the other rats. Brain specimens were examined following six or twelve months of observation.

Study #6   *Effects of Valproate Upon the Hippocampus* [129-130]

In this investigation, the same treatment protocol was performed as described above. Postmortem exams focused upon the analysis of drug-induced changes in astrocytes and microglia obtained from specimens of the **hippocampus**.

Background Information

A distinctive feature of the Sobaniec-Lotowska studies was the use of the transmission electron microscope to provide **ultrastructural analyses** of the brain. As the name implies, the term ultrastructure refers to the detection of detailed units within cells. By focusing a beam of electrons upon tissue, the electron microscope can magnify objects up to 2 million times their normal size (1000X more powerful than the most sophisticated light microscope).[131]

Another rare feature of these studies was the use of dosing regimens designed to simulate human treatment conditions [for the technically curious, this author has constructed an inter-species comparison of valproate metabolism, below]:

Pharmacokinetics of Valproate [132-142]

	Human	Rat
oral dose	15 to 60 mg/kg per day	200-400 mg/kg q.d., b.i.d., t.i.d.
peripheral blood level	50 to 180 ug/mL	15 to 135 ug/mL

[Note: Blood levels after oral doses in rodents vary widely, depending upon rat strain.]

	Human	Rat
brain level	4.57 ug/g based on surgical biopsies	1 to 2.4 ug/g (300 mg/kg, t.i.d.)
time to max. blood level	4 to 8 hrs	< 30 minutes
elimination half-life (blood)	9 to 16 hrs	< 30 minutes
ammonia levels (venous blood):		
during VPA therapy	30 to 132 umol/L	* 323 umol/L
** during VIE	≥ 100-180 umol/L	not applicable
normal levels (adults)	5 to 50 umol/L	not available

* 323 umol/L = ammonia level after 8 weeks of oral dosing 300 mg/kg per day
** VIE = valproate induced encephalopathy

Findings

Cerebellum

Following nine and twelve months of daily exposure to valproate, neurons in the cerebellum (**Purkinje cell layer**) were severely damaged.

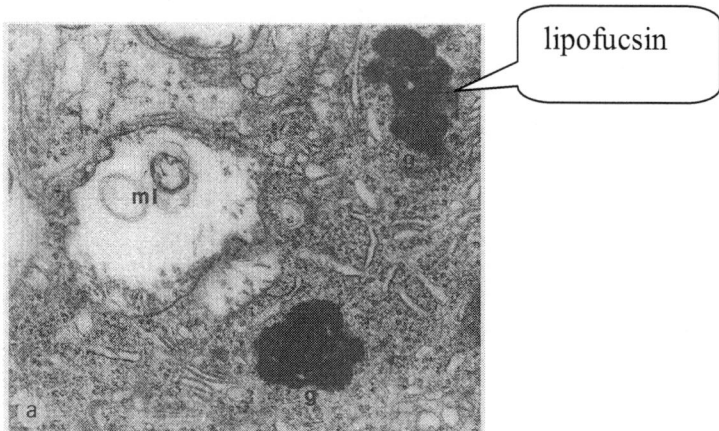

Source: M. Sobaniec-Lotowska, *International Journal of Experimental Pathology*, volume 82, page 341, copyright (2001), reprinted with permission of Blackwell Publishing Ltd.

This slide demonstrates **neuronal degeneration** following nine months of exposure to valproate. All of the neuron's components are swollen and disorganized. Prominent abnormalities in this image include **lipofucsin granules** (byproducts of membrane breakdown) and a dying **mitochondrion** (**ml**) which contains a myelin body in its center. The normal architecture (mitochondrial cristae) can no longer be seen.

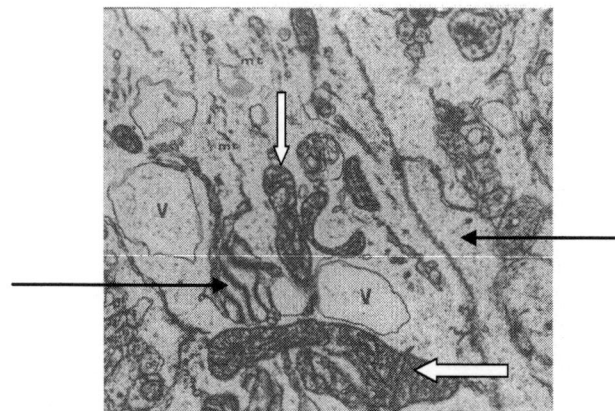

Source: M. Sobaniec-Lotowska, *International Journal of Experimental Pathology*, volume 82, page 343, copyright (2001), reprinted with permission of Blackwell Publishing Ltd.

This slide depicts **persistent damage** in a **cerebellar neuron** (12 months of valproate, followed by 3 months of drug withdrawal). The internal structure (cytoskeleton) has collapsed. Swollen vacuoles (**v**), atrophic mitochondria (thick arrows), and ballooned organelles (thin arrows) reflect necrotic changes due to severe chemical injury

Hippocampus

Chronic exposure to valproate resulted in **reactive gliosis** in the hippocampus. This represented another response to severe chemical injury.

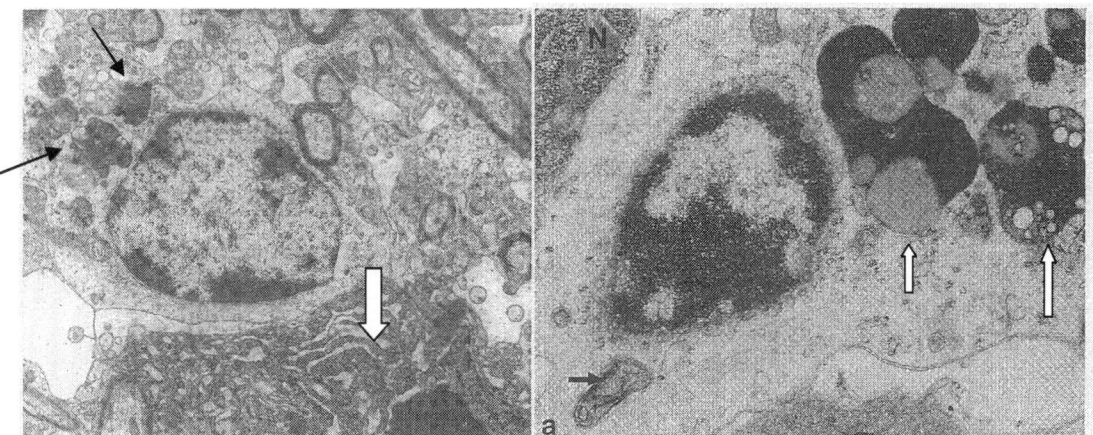

Source: [left] M. Sobaniec-Lotowska, *International Journal of Experimental Pathology*, vol. 84, p. 119, copyright (2003); and [right] volume 86, page 93, copyright (2005), reprinted with permission of Blackwell Publishing Ltd.

Gliosis Following Chronic Valproate

Left: Following 12 months of exposure to valproate, **astrocytes** within the hippocampus displayed evidence of phagocytosis (cell digestion). This image shows a swollen astrocyte adjacent to a dark, dying neuron (white arrow). Two storage sacs, called lysosomes, (thin arrows) contain partly degraded material within the astrocyte.

Right: This image shows a **microglial cell body** (soma) following 9 months of exposure to valproate. The nucleus appears in the center of the slide (dark material = condensed DNA). Like astrocytes, microglia also become activated in response to injury. Within this microglial cell, one can see a multi-walled vesicle (small black arrow, lower left) and several dark storage sacs known as phagosomes (white arrows). All of these units are used to collect and metabolize dead or degenerating debris.

Neuroimaging Studies of Humans

Three major findings have occurred consistently in the neuroimaging research of psychiatric patients exposed to therapy with mood stabilizers: 1) brain tissue lesions (MRI hyperintensities reflecting edema, ischemia, gliosis, and/or loss of myelin); 2) atrophy (e.g., shrinkage of the cerebellum); and 3) enlargement of fluid-filled spaces (ventricles, cisterns).[143-145]

Although the medical literature has generally been limited by the absence of large *longitudinal* studies (showing "before and after" drug effects in bipolar patients), case reports of individual patients have provided evidence which clearly contradicts the premise that mood stabilizers are either neuroprotective or neuroregenerative.

Lithium

Studies #1-2 *Cerebellar Atrophy Associated with SILENT*

In 2007, clinicians affiliated with Columbia University described the long-lasting symptoms which developed in a 52-year-old female who had consumed lithium for more than twenty years.[146] Following an episode of lithium intoxication (serum level = 3.5 mmol/L), the patient subsequently experienced **prolonged difficulties with speech, balance, gait, and the control of eye movements** (SILENT = syndrome of irreversible lithium effectuated neurotoxicity). A neuroimaging study (magnetic resonance imaging, or MRI) performed two years after the acute event revealed pronounced atrophy of the cerebellum:

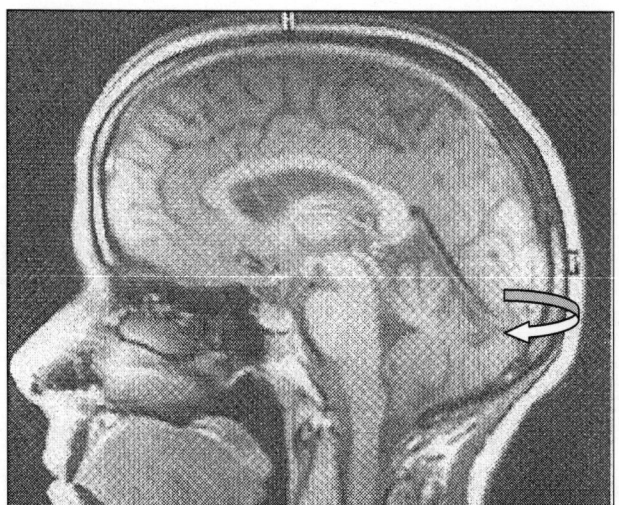

Source: *Movement Disorders*, volume 22, issue 4, M. Niethammer and B. Ford, "Permanent Lithium-Induced Cerebellar Toxicity: Three Cases and Review of the Literature," page 570, copyright (2007), with permission of John Wiley & Sons, Inc.

Similar neuroimaging results were reported in 2008, when clinicians in Brazil described the progression of symptoms in a female lithium consumer.[147] At the age of 26, the subject experienced an episode of drug intoxication in the context of pneumonia (serum lithium level = 1.9 mmol/L). Although she recovered from the respiratory infection, she was unable to walk for six months. Under the influence of a new "mood stabilizer" (valproate), signs of cerebellar dysfunction persisted (abnormal gait, unclear speech, incoordination, and diminished muscle tone). An MRI of the brain obtained at the age of 28 demonstrated pronounced shrinkage of the cerebellum:

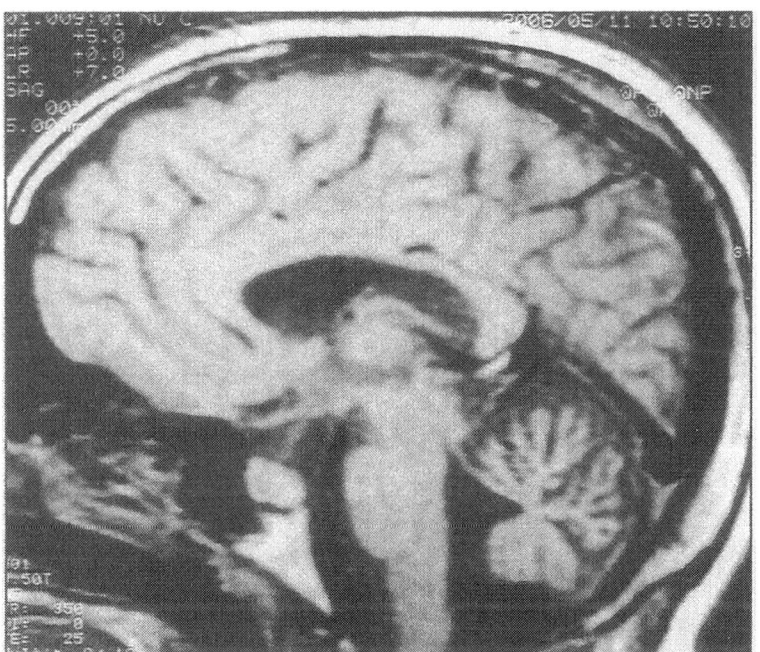

Source: A.C. Rodrigues de Cerqueira, M. Costa dos Reis, F.D. Novis, J.M.F. Bezerra, G. Canedo de Magahlaes, et al, "Cerebellar Degeneration Secondary to Acute Lithium Carbonate Intoxication," *Arquivos de Neuropsiquiatria* 66:3-A (2008): 578, reproduced with permission of Dr. Antonio Spina-Franca.

Cerebellar atrophy is by no means specific for lithium intoxication or the related syndrome of prolonged toxicity. Neuroimaging research involving patients diagnosed with bipolar disorder, migraine headaches, and epilepsy has repeatedly shown damage to the cerebellum in the context of **anticonvulsant drug therapy**.

Background Information

The **anatomic regions of the cerebellum** consist of two **hemispheres** (left and right) and a central zone known as **the vermis**. These areas are further described in terms of their alignment with respect to the front or back of the body (anterior lobe, posterior lobe) and – at the base of the brain – the flocculonodular lobe:

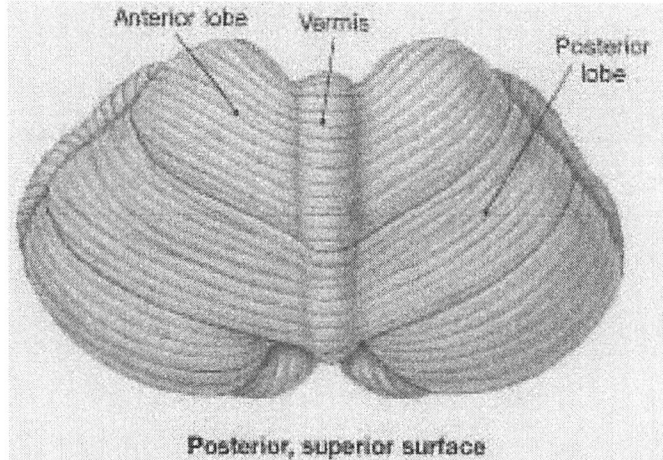

Source: McGill University, The Brain from Top to Bottom, copyleft.

When radiologists examine brain scans of the cerebellum, they commonly employ a classification scheme which identifies three discrete zones and nine subzones within the **vermis**: **V1** (lingula, central lobule, culmen, declive, folium); **V2** (tuber, pyramid); and **V3** (uvula, nodule).

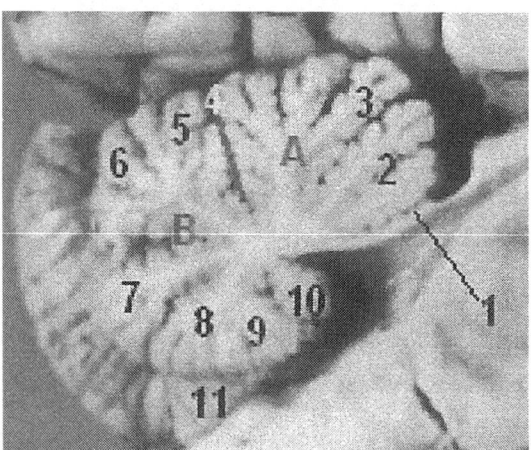

Source: Wikipedia, John A. Beal, PhD, LSU Health Sciences Center, Shreveport.

Regions of the Cerebellum

1 = lingula, 2 = central lobule, 3 = culmen, 5 = declive, 6 = folium vermis,
7 = tuber vermis, 8 = pyramid vermis, 9 = uvula vermis, 10 = flocculonodulus
11 = tonsils of the cerebellum

Studies #3 and 4   *Mood Stabilizers Shrink the Cerebellar Vermis*

Researchers affiliated with the University of Cincinnati have performed two neuroimaging studies of bipolar patients which have detected shrinkage of the cerebellum. In the first of these investigations, they employed magnetic resonance imaging to compare the brain features of individuals who had experienced one (first-episode) or multiple episodes of mania.[148] Although both groups of subjects included a significant number of patients with histories of substance abuse, the patients diverged with respect to exposure to medications:

	First-Episode Mania n = 16	Recurrent Mania n = 14	Controls n = 15
mean age	24	29	27
gender	44% female	36% female	47%
substance abuse history	44%	64%	20%
duration of mental illness	2.8 yrs	8.7 yrs	n/a
months of lithium	**0**	**6**	**0**

Findings

No significant differences were found when the researchers compared the brain size of normal controls and patients without previous exposures to mood stabilizers. In contrast, patients with multiple episodes of mania and long treatment histories displayed an **8% reduction in one specific region of the cerebellar vermis** (V3, Courchesne classification):

	First-Episode Mania n = 16	Recurrent Mania n = 14	Controls n = 15
mean area mm^2			
V1	393.0	396.0	395.5
V2	276.8	261.7	257.4
V3	321.2	**297.1**	322.9

In a brief report of a second brain scan study, the Cincinnati team revised their methodology to include a three-dimensional estimate of changes in regional brain *volume*.[149] The researchers also included more details about the past and current use of medications:

	First-Episode Mania n = 18	Recurrent Mania n = 21	Controls n = 32
mean age	22	25	24
female	39%	62%	50%
duration of illness	2 yrs	10 yrs	n/a
previous use:			
antidepressants	6%	**43%**	none
antipsychotics	0%	**76%**	none
mood stabilizers	**0%**	**81%**	none
current use:			
antidepressants	6%	24%	none
antipsychotics	61%	67%	none
mood stabilizers	**94%**	**81%**	none

Findings

As before, the researchers detected no significant differences in the brain size of first-episode bipolar patients and normal controls. However, ***patients with extensive histories of exposure to mood stabilizers (and other psychiatric drugs) manifested statistically significant shrinkage in two regions of the cerebellar vermis***:

	First-Episode Mania n = 18	Recurrent Mania n = 21	Controls n = 32	
	volume (cm^3)	volume (cm^3)	volume (cm^3)	
V1	1952	1869	1940	
V2	1241	**1048** (↓ 13%)	1203	p = 0.002
V3	1467	**1407** (↓ 7%)	1505	p = 0.03

Valproate (Depakote/Depakene/Depacon)

Despite the existence of several dozen case reports of valproate-induced encephalopathy in the medical literature, there have been relatively few papers with detailed photos of neuroimaging results. Of the publications which have displayed such images, an intriguing variety of abnormalities have been observed in patients, irrespective of age.[150-155]

Background Information

The *presence of excessive fluid in the ventricles or subarachnoid spaces* gives rise to a neurological condition known as **hydrocephalus** (from the Greek words: *hydro*, meaning water, and *kephalos*, for head). Accompanied either by elevated or normal pressure within the skull (the latter condition, known as NPH), hydrocephalus is caused by one of three abnormalities: 1) the excessive production of cerebrospinal fluid by the choroid plexus (the cells which line the ventricles); 2) the obstruction of that fluid's movement as it flows through channels within the brain (a condition known as "non-communicating" hydrocephalus); or 3) the impaired absorption of cerebrospinal fluid when it reaches specialized cells (the arachnoid granulations) atop the surface of the brain.

Despite the existence of numerous case reports of Depakote dementia accompanied by ventricular dilation, these accounts have always omitted the diagnosis of hydrocephalus. This is indeed curious when one considers the definition of this disorder: "a congenital or acquired condition marked by dilatation of the cerebral ventricles...typically, there is ...brain atrophy, mental deterioration, and convulsions." [156]

Pertinent to the discussion of valproate is the documented link between this drug and the disruption of the mitochondria – the energy-producing units of each cell. Valproate has been known to induce or enhance the features of a rare condition known as MELAS (discussed in much greater detail, below). It is interesting to note that several clinicians have observed hydrocephalus in the context of MELAS.[157] Others have noted mitochondrial dysfunction in the context of hydrocephalus.[158-161] Thus, there is good reason to suspect that valproate's toxic effects upon mitochondria contribute to the emergence of hydrocephalus. This would also explain why some valproate patients experience symptoms which mimic the classic triad of **normal pressure hydrocephalus** – those being dementia, gait instability (imbalance), and/or urinary incontinence.

Studies #5-7   *Encephalopathy with Ventricular Enlargement*

Guerrini et al (1998)   *Depakote Encephalopathy in a Young Girl*

A six-year old female developed a seizure disorder for which phenobarbital was initially prescribed. The baseline MRI revealed mild enlargement of the fluid-filled spaces (ventricles) of the brain. Ongoing epileptiform activity triggered a medication change to valproate at age eight. This resulted in the insidious development of cognitive difficulties (emerging at age 9 years 5 months).

By the age of 10 years 8 months, drug therapy consisted of valproate and lamotrigine. Cognitive difficulties had progressed to the point of severe dysfunction. *Signs and symptoms of encephalopathy included an 18 point reduction in intellectual aptitude* (Wechsler Intelligence Scale for Children IQ = 72), accompanied by disorientation, apraxia (inability to perform learned, purposeful movements), abnormal gait, and learning difficulties. An MRI revealed increased dilation of the ventricles and subarachnoid spaces. Valproate was replaced by phenobarbital four months later. Within three months of this change, the patient displayed dramatic improvements in cognition (IQ = 92) and brain anatomy.

Key aspects of this case:
1) serum levels of valproate were occasionally but only mildly elevated
   (101-106 ug/mL, normal reference range for epilepsy being 50 to 100 ug/mL)
2) liver function tests were normal
3) serum ammonia levels were also within normal limits.

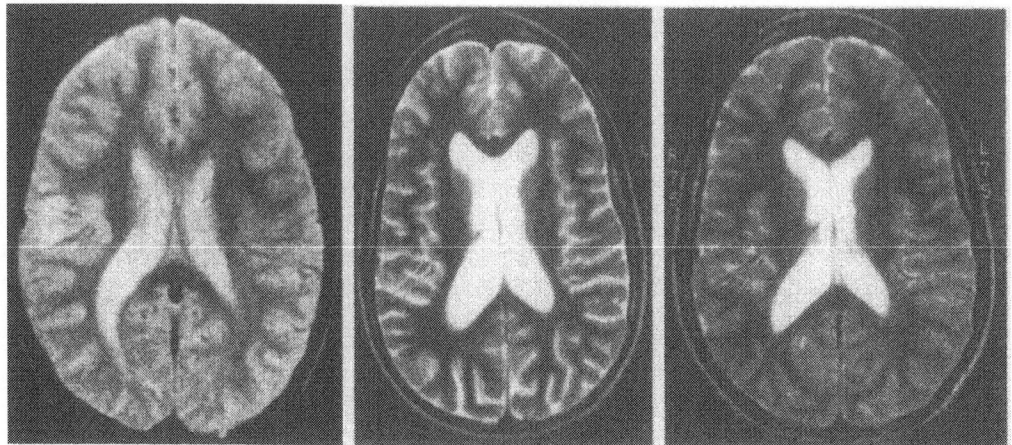

Source: *Epilepsia*, volume 39, issue 1, Guerrini et al, p. 28, copyright (1998), with permission of Wiley-Blackwell.

age 6	**age 10 years 8 months**	**age 11 years 3 months**
pre-antiepileptic drugs	valproate + lamotrigine	phenobarbital
mild dilation of ventricles	expansion of fluid	partial normalization of fluid

Papazian et al (1995) *Encephalopathy and Cerebellar Atrophy in Childhood*

Clinicians in Miami, Florida and Venezuela reported the experiences of two young patients who received valproate for benign rolandic epilepsy. (This summary will be limited to the published case report which included photos from neuroimaging). The patient was a young male who developed partial seizures around the age of six. A series of pharmaceuticals were used over the ensuing 4 ½ years, in an effort to prevent or control recurrent convulsions, hyperactivity, and – presumably – intermittent psychosis. At the age 10 years 6 months – while consuming valproate + thioridazine + benztropine – this child experienced the onset of behavioral regression and a variety of neurological disturbances: ataxia, imbalance, intention tremor, dysmetria, inactivity. By the following year, these problems had progressed to include a profound decline in cognitive functioning (***27 point drop in performance IQ, 12 point drop in full-scale IQ***). An MRI revealed ventricular dilation and marked atrophy of the cerebellum. Valproate was subsequently withdrawn. Over the ensuing 12 months, this patient experienced the gradual reversal of symptoms and the normalization of brain anatomy.

Key aspects of this case:
1) mildly elevated levels of valproate (110-119 mg/dL)
2) moderately high levels of ammonia (53-80 ug/dL; normal reference = 8.5 to 85 ug/dL)
3) dementia without loss of consciousness, delirium, or worsening of seizures

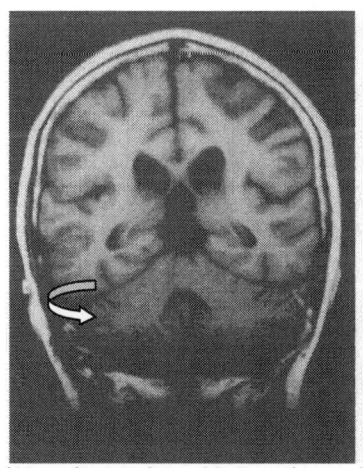 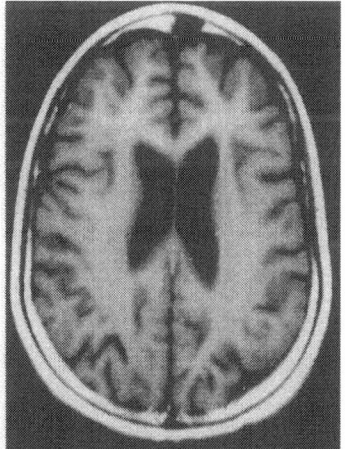

Source: *Annals of Neurology*, volume 38, Papazian et al, "Reversible Dementia and Apparent Brain Atrophy During Valproate Therapy," p. 689, copyright (1995), with permission of John Wiley & Sons, Inc.

MRI of Brain During Depakote Dementia – Patient Age 11 Years, 4 Months

Left: This image shows enlarged ventricles (black = fluid) and cortical atrophy. Prominent pathology is visible in the cerebellum (arrow).

Right: This image depicts cortical shrinkage. The enlarged sulci (grooves) on the outside surface of the brain reflect degeneration of the underlying tissue.

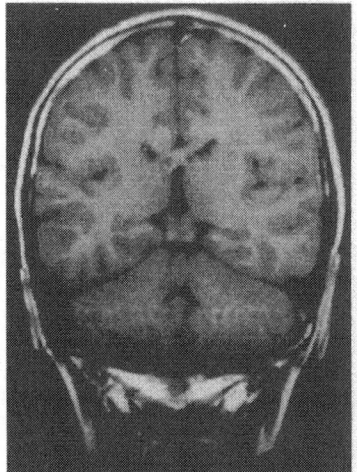
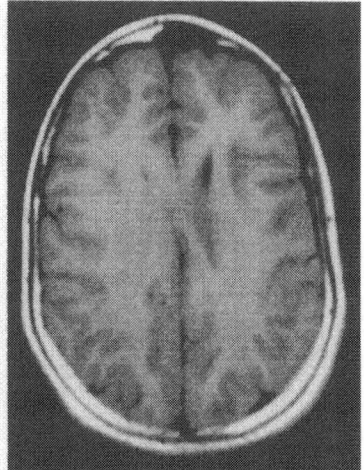

Source: *Annals of Neurology*, volume 38, Papazian et al, "Reversible Dementia and Apparent Brain Atrophy During Valproate Therapy," p. 689, copyright (1995), with permission of John Wiley & Sons, Inc.

MRI After Withdrawal of Depakote – Patient Age 12 Years 6 Months

Reversal of brain atrophy one year after the cessation of valproate

Galimberti et al (2006)   *Valproate Encephalopathy Associated and MELAS*

Clinicians in Pavia, Italy reported the events which involved a four-year-old child who experienced an isolated seizure. This event was followed by mild learning difficulties. At the age of eight, the patient was diagnosed with epilepsy and treatment with valproate was begun. Over the next two years, this child experienced progressive irritability and deterioration of academic and cognitive functioning, ultimately performing three to four years below his chronological age.

Suspecting the presence of a neurodegenerative condition, physicians obtained an MRI of the brain when the patient was ten years old. This examination revealed ventricular dilation and extensive atrophy of the cortex and cerebellum. Valproate was gradually withdrawn, in favor of a brief period of treatment with a benzodiazepine (clobazam). Although a follow-up brain scan four months later revealed normal findings, this child continued to experience moderate deficits in intellectual functioning (performing approximately two to three years below his chronological age).

Further tests and interviews were then performed in an effort to refine the diagnosis. Appreciating the existence of at least one case report which had previously linked valproate to the manifestation of MELAS – a rare condition caused by defects in the mitochondria – the treatment team eventually confirmed a genetic mutation in this patient and in several relatives. ***This led them to speculate that valproate had triggered the expression of MELAS features in their patient, presumably by impairing the integrity and functions of brain cell mitochondria.***

MELAS and Drug-Induced MELAS

Source: R.R. da Fonseca, W.E. Johnson, S.J. O'Brien, M.J. Ramos, and A. Antunes, "The adaptive evolution of the mammalian mitochondrial genome," *BMC Genomics* 9 (2008):119, open access license.

The Mammalian Mitochodrion: Genome and Energy Production

Left: Within mammals, cells contain varying numbers of mitochondria. Using oxygen, these bean-shaped units produce energy through a process known as cellular respiration.

Top: Each mitochondrion contains multiple copies of DNA (mtDNA), organized in the shape of a single-strand loop. Like an architectural blueprint, the genes along this oval code for proteins which are used to produce energy. To date, mutations in six genes have been linked to the syndrome known as MELAS.

Bottom: The inner membrane of the mitochondrion contains a series of electron-accepting units, known as complexes. As mentioned earlier (chapter two), these units are collectively referred to as the respiratory or electron transport chain (ETC). Drugs like valproate directly impair the ETC. They may also hinder the production of energy by mutating the mitochondrial DNA.

Background Information [162-176]

The acronym **MELAS** (**M**itochondrial **E**ncephalomyopathy **L**actic **A**cidosis and **S**troke-Like Episodes) refers to a constellation of symptoms – including dementia, hearing loss, seizures, muscle weakness, diabetes, and stroke-like episodes (e.g., nausea/vomiting/headache/confusion/hemiparesis/hemianopia) – which arise from defects in mitochondrial (or nuclear) DNA.

First described by Pavlakis et al in 1984, MELAS has since been related to several point mutations in the genes for transfer RNA (tRNA). These aberrations disrupt the synthesis of proteins which are essential for cell respiration (see chapter two). The signs and symptoms of MELAS are attributed to the dysfunction or death of tissue, particularly within organs that depend upon high levels of energy: hair cells of the inner ear; muscle cells of the heart and limbs; and neurons, glia, and blood vessels within the brain.

Although most cases of MELAS are believed to be congenital (inherited through the maternal lineage, since mitochondrial DNA is derived almost entirely from the ovum) with symptoms first emerging in childhood, there have been case reports of adult-onset MELAS. Furthermore, and pertinent to the discussion of pharmaceutical toxicity, there have been several case reports of valproate-related encephalopathy in patients who have tested positive for MELAS-associated mutations. In such cases, the writers have always assumed that valproate served as an environmental trigger which unmasked a latent but *pre-existing genetic disorder*.

At this juncture, the writers of case reports have universally ignored the potential for valproate to incite a *de novo* or acquired form of MELAS in patients who lack a family history of the disease. However, it would seem reasonable to suspect the *possibility* of a phenomenon which might be called **Drug-Induced MELAS (DIM)** based upon the following discoveries.

First, evidence from recent research has demonstrated that the process of ageing, itself, involves the accumulation of mitochondrial mutations. Second, scientists have confirmed a link between oxidative stress (the generation of free radicals) and mutations in the mitochondria. Third, it has become increasingly apparent that many pharmaceuticals, including valproate, induce or enhance oxidative stress. Provided that these drug-related processes were able to affect sufficient numbers of the pertinent target genes, one might reasonably predict the transformation of DNA to such an extent that *any* patient might experience the onset of iatrogenic MELAS.

Features of MELAS

signs	symptoms	diagnostic abnormalities
hemianopia	confusion/dementia	brain scan:
hemiparesis	hearing loss	cortical atrophy
short stature	exercise intolerance	ventricular dilation
skin changes	stroke-like episodes	basal ganglia calcification
diabetes	recurrent vomiting	lucencies (infarct/ischemia)
fever	psychosis	
cardiomyopathy	mania/depression	EEG: epileptiform spike discharges
renal failure	fatigue	
rhabdomyolysis	muscle weakness	EKG: conduction abnormalities
deafness	migraine headache	
optic atrophy/	blindness	labs (serum):
retinopathy	numbness/tingling	↑ lactate, ↑ creatine kinase
neuropathy	frequent urination	
hypotonia	thirst	skeletal muscle biopsy:
seizures		deficiency of respiratory chain
tremor		activity (complex I, IV);
myoclonus		presence of red ragged fibers
dystonia		
abnormal gait		genetic analysis (mtDNA):
		mutations of cells obtained from
		blood, skeletal muscle, buccal
		mucosa, hair follicles, urinary
		sediment

(% of mutant mtDNA varies by tissue)

Key aspects of the Italian case of valproate-induced MELAS:
1) declining cognition and mood despite normal (therapeutic) drug levels
2) no evidence of elevated serum ammonia levels

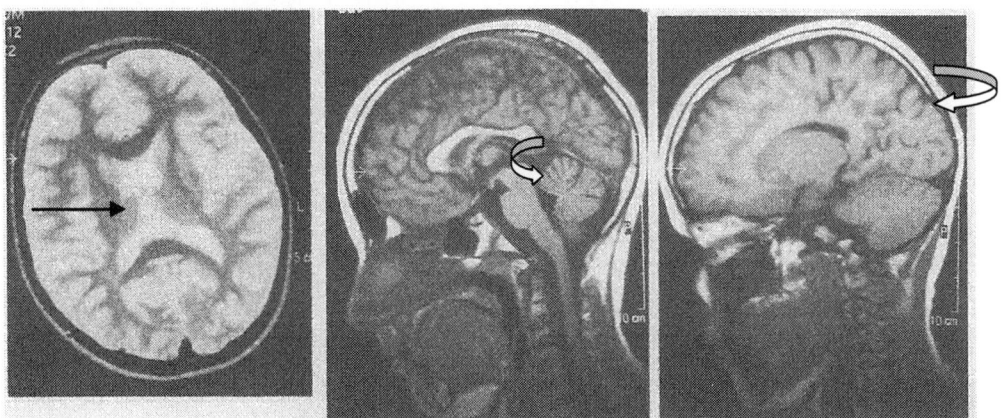

Source: C.A. Galimberti, M. Diegoli, I. Sartori, C. Uggetti, A. Brega, A. Tartara, et al, "Brain pseudoatrophy and mental regression on valproate and a mitochondrial DNA mutation," *Neurology* 67 (2006): 1716, reproduced with permission of Lippincott, Williams, & Wilkins.

Deterioration of Brain While On Valproate – Age 10 Years 1 Month

Left: Arrow shows ventricular dilation.
Middle: Arrow denotes atrophy of the cerebellum
Right: Arrow highlights atrophy of the parieto-occipital cortex.

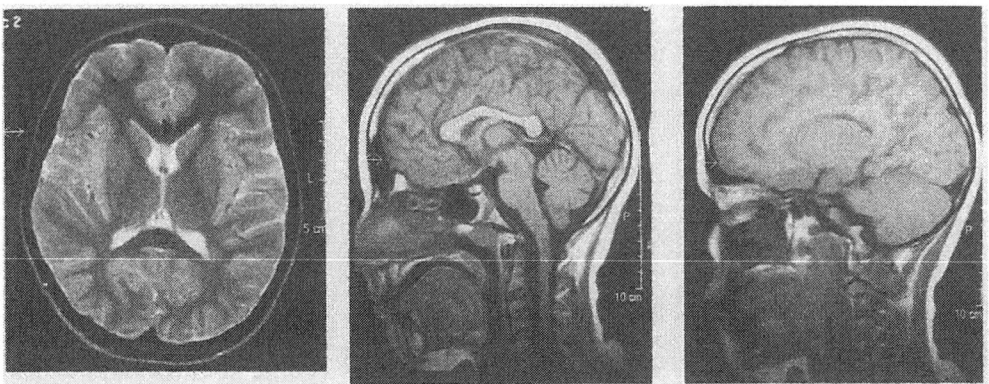

Source: C.A. Galimberti, M. Diegoli, I. Sartori, C. Uggetti, A. Brega, A. Tartara, et al, "Brain pseudoatrophy and mental regression on valproate and a mitochondrial DNA mutation," *Neurology* 67 (2006): 1716, reproduced with permission of Lippincott, Williams, & Wilkins.

Resolution of Brain Atrophy After Drug Withdrawal – Age 10 Years, 8 Months

Mood Stabilizers

Postmortem Studies of Humans

Lithium

Despite the fact that an estimated 15% of lithium consumers experience at least one episode of drug intoxication during treatment, and despite the fact that an even larger number develop insidious brain damage during chronic therapy, there have been remarkably few reports in the medical literature dedicated to the anatomic consequences of lithium. One important exception to this void occurred in 1994 when clinicians in Atlanta, Georgia divulged the autopsy results from a 67-year-old male who had died within 10 weeks of hospitalization for lithium poisoning (serum level of lithium carbonate = 4.04 mmol/L), followed by irreversible neuropsychiatric symptoms.[177]

Although the case was complicated by the subject's history of polypharmacy (three different drugs for high blood pressure and three more drugs for bipolar disorder), hypertension, and renal cancer (the latter, diagnosed posthumously), the neuropathology findings were consistent with **lithium-induced degeneration of the cerebellum**. Specifically, this patient had developed pronounced spongiform changes (white matter of the vermis); neuronal loss (within the Purkinje cell layer and the dentate nucleus); and extensive proliferation of astrocytes (gliosis):

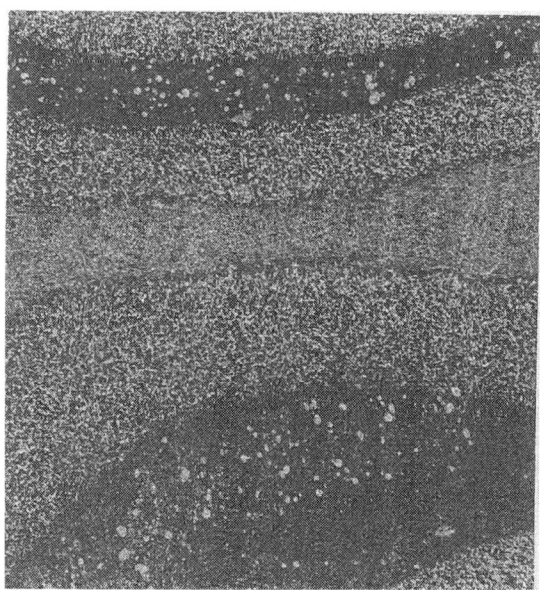

Source: *Annals of Neurology*, volume 36, J.A. Schneider and S.S. Mirra, "Neuropathologic Correlates of Persistent Neurologic Deficit in Lithium Intoxication," page 930, copyright (1994), reproduced with permission of John Wiley & Sons, Inc.

Cerebellar Vermis of Patient Who Died From SILENT

This picture demonstrates abnormal anatomy in the vermis. The vacuoles (white holes) impart a sponge-like appearance to the **white matter layers** of the cerebellum.

The significance of these pathologies extends beyond the phenomena of lithium intoxication and SILENT. It is notable that neuroscientists have recently confirmed a connection between anatomic *abnormalities within the cerebellum* and **essential tremor** – the latter, a movement disorder experienced by 40 to 80% of lithium users which involves the involuntary quivering of the hands, limbs, head, and/or trunk.

For example, in an autopsy analysis of brain tissue obtained from an elderly patient who had suffered from essential tremor, researchers at Columbia University identified structural changes which were *not* present in the brains of Parkinsonian patients or age-matched, neurologically healthy controls.[178] Interestingly, many of the pathologies associated with essential tremor were identical to the changes observed in lithium poisoning: neuronal loss, axonal damage, and gliosis within the Purkinje cell layer; and degeneration of the gray and white matter within the cerebellar dentate nucleus. ***This suggests that lithium-induced tremor, just like essential tremor, is probably not the benign event which most physicians have been trained to believe***, since it appears to reflect degenerative changes within the cerebellum.

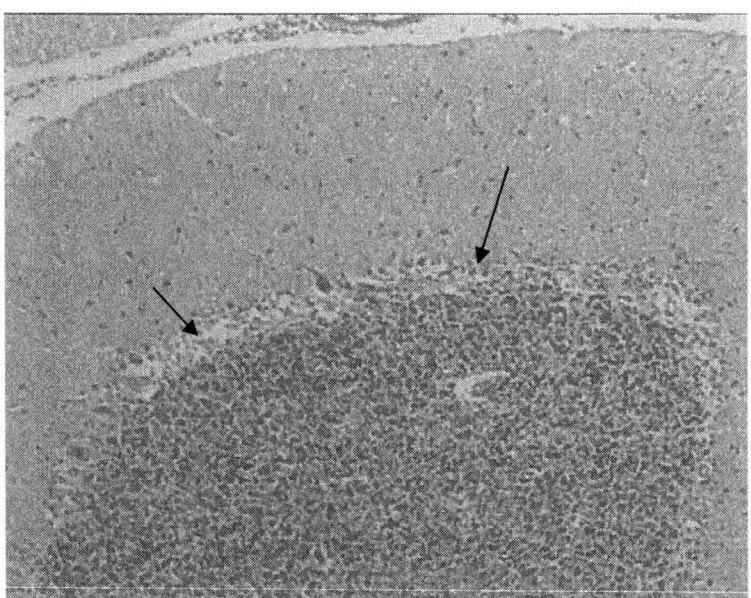

Source: *Archives of Neurology*, volume 63, August, E.D. Louis, J.P.G. Vonsattel, L.S. Honig, A. Lawton, C. Moskowitz, B. Ford, and S. Frucht, "Essential Tremor Associated With Pathologic Changes in the Cerebellum," page 1190, copyright (2006) American Medical Association, all rights reserved.

This slide depicts cerebellar tissue obtained from the tremor patient (described above). The arrows point to neuronal loss (white band) from the Purkinje cell layer. Although essential tremor is an extremely common side effect of lithium, few physicians are trained to believe that this movement disorder reflects overt cell death or anatomical damage to the brain.

Mood Stabilizers

Antiepileptic Drugs   (Anticonvulsants)

With the exception of detailed case reports, the quality of most postmortem studies involving consumers of mood stabilizers has been undermined by a variety of methodological confounders. Despite the existence of more than 200 published studies on the topic of bipolar brains, the integrity of this research has been repeatedly compromised by the following features:

- a failure to provide full information about past exposures to psychiatric drugs (including antidepressants, anxiolytics, neuroleptics, and stimulants)

- a failure to provide full information about past exposures to street drugs, alcohol, and other mind-altering chemicals

- a failure to provide information about past episodes of drug intoxication or other serious adverse drug effects (e.g., neuroleptic malignant syndrome, serotonin syndrome)

- a failure to obtain and examine fresh brain tissue in a timely fashion.

According to one source, eighty percent (80%) of the published studies have relied upon brain specimens obtained from two major brain collections in the U.S.A.[179] Unfortunately, the quality of the resulting research has inevitably been affected by lengthy postmortem delays, extended periods of tissue fixation (storage times), and variable processing techniques prior to microscopic exam.

Each of these variables has made it difficult to draw definitive conclusions about the anatomic effects of lithium and anticonvulsants. However, consistent with the findings from neuroimaging research, postmortem studies solidly contradict the new claim that "mood stabilizers" are heroic agents of neuroprotection. To the contrary, autopsy analyses have habitually revealed a pattern of brain volume reduction (atrophy), neural cell loss, white matter abnormalities, and – most recently – protein changes consistent with the pathology of Alzheimer's disease.[180-190]

Examples of Neuropathology Following Exposure to Mood Stabilizers

Postmortem Study　　　　**Source of Brain Tissue**　　# **of Bipolar Subjects**

Alzheimer's Pathology (ApoE, ApoD)

Thomas et al (2003)
　　　　　Mental Health Research Institute　　　n = 8
　　　　　Victoria (Australia)

　　　key findings: *increased levels of apolipoprotein D in dorsolateral and lateral prefrontal cortex, orbitofrontal cortex, parietal cortex, cingular cortex, caudate, anygdala, and thalamus (16 to 224% increase in apoD, relative to non-psychiatric, non-neurological controls)*

Digney et al (2004)
　　　　　Mental Health Research Institute　　　n = 8
　　　　　Victoria (Australia)

　　　key findings: *increased levels of apolipoprotein E in prefrontal cortex and striatum (caudate, putamen)*

Cortical Pathology

Ongur et al (1998)
　　　　　Harvard Brain Tissue Resource Center　　n = 4
　　　　　postmortem interval: 19 hrs
　　　　　fixation time: 64 months
　　　　　Stanley Foundation　　　　　　　　　n = 14
　　　　　postmortem interval: 31 hrs
　　　　　fixation time: 27 months

　　　key findings: *reduced brain volume (20%), reduced number of neurons and glia in ventral prefrontal cortex*

Benes et al (2001)
　　　　　Harvard Brain Tissue Resource Center　　n = 10
　　　　　postmortem interval: 22 hrs
　　　　　fixation time: 757 days (2.1 years)

　　　key findings: *12% reduction in thickness of layer III; 28% reduction in density of non-pyramidal neurons in layer II of cingulate cortex*

Neuropathology Following Exposure to Mood Stabilizers (continued)

Postmortem Study   **Source of Brain Tissue**   **# of Bipolar Subjects**

Cortical Pathology (continued)

Rajkowska et al (2001)
        Harvard Brain Tissue Resource Center   n = 7
        Case Western Reserve University   n = 3
        postmortem interval: 18 hrs
        fixation time: 14.3 months
        9 with past or continuing use of lithium
        2 with recent use of valproate

    key findings: ***10 to 20% reduction in neuronal density, 20-60% reduction in glial density (small, medium, large cells) in discrete layers of the prefrontal cortex***

Cotter et al (2005)
        Stanley Foundation Brain Consortium   n = 15
        postmortem interval: 32.5 hrs
        fixation time: 620 days (1.7 yrs)
        5 on lithium, 6 on other mood stabilizers at time of death

    key findings: ***reduced neuronal size in all layers of caudal orbitofrontal cortex (statistically significant changes in cortical layers I and V)***

Regenold et al (2007)
        Stanley Medical Research Institute   n = 15
        Bethesda, MD

    key findings: ***4% reduction in brain weight, 14% reduction in staining intensity of deep white matter, 8% reduction in staining intensity of gyral white matter from dorsolateral prefrontal cortex (DLPFC)***
                  [findings suggestive of damage to myelin and/or axons]

Neuropathology Following Exposure to Mood Stabilizers (continued)		
Postmortem Study	<u>**Source of Brain Tissue**</u>	**# of Bipolar Subjects**

Subcortical Pathology

Baumann et al (1999)
       <u>Magdeburg, Germany</u>   n = 2
       postmortem interval: 4 to 72 hrs
       storage time: ≥ 3 months

    key findings: *15 to 32% reduction in volumes of subcortical structures (relative to controls) including the left nucleus accumbens, right putamen, left and right globus pallidus pars externa)*

Beilau et al (2004)
       <u>Magdeburg, Germany</u>   n = 11
       postmortem interval: 4 to 72 hrs
       storage time: ≥ 3 months

    key findings: *4 to 17% reduction in volume of subcortical structures (relative to controls) including the basal ganglia, thalamus, hypothalamus, amygdala, and hippocampus*

Bezchlibnyk et al (2006)
       <u>Stanley Neuropathology Consortium</u>   n = 11
       postmortem interval: 13 to 62 hrs
       formalin fixation: 2 to 16 months

    key findings: *~ 30% decrease in size of neuron bodies (somas) in two discrete sections of the amygdala*

Liu et al (2007)
       <u>Stanley Neuropathology Consortium</u>   n = 14
       postmortem interval: 13 to 62 hrs
       formalin fixation: 2 to 14 months

    key findings: *12% smaller cell size of pyramidal neurons in hippocampus (CA1 region) relative to non-psychiatric controls*

Chapter Eight

Stimulants

Botanical stimulants, such as caffeine, coca leaves, ma huang, and nicotine, have been used for centuries to enhance stamina, prolong wakefulness, and increase mental alertness. However, the creation of *synthetic stimulants* began in the late 19th century with the inception of the organic chemical industry.[1-3]

Stimulants Before World War I

In 1887, a Romanian chemist named Lazar Edeleanu, synthesized the first amphetamine (phenylisopropylamine) while working at the University of Berlin. That same year, Japanese chemist, Nagajosi Nagai, isolated ephedrine from the herb ma huang (*Ephedra vulgaris*). It was Nagai's discovery which gave rise to the manufacture of methamphetamine (1893) and crystallized methamphetamine (the latter, synthesized by Akira Ogata in 1919). Due to the outbreak of the first World War, however, medicinal applications of these new compounds were not actively pursued.

Stimulants in the 1920s

Within the United States, the therapeutic use of synthetic stimulants originally focused upon the capacity of these drugs to relieve nasal congestion and improve breathing. In the 1920s, American-trained chemists (Ku Kuei Chen and Carl F. Schmidt) re-isolated the active ingredient of ma huang. **Ephedrine** was approved for use in America in 1926 and soon gained favor as an oral replacement for epinephrine in the treatment of respiratory conditions (such as asthma, bronchitis, emphysema, and pertussis).

Concurrent with these developments, a team of Californian researchers led by Gordon Alles revived the work of European scientists in producing several forms of amphetamines. The *racemic mixture, known as **Benzedrine**, became a key ingredient in over-the-counter inhalants because of its ability to reduce nasal secretions. The **d-enantiomer of the same drug (**d-amphetamine** or **Dexedrine**) became a treatment for narcolepsy (a sleep disorder) and obesity.

---

* **racemic** – Within nature, certain carbon-containing molecules exist in two different configurations called *enantiomers*. Like a pair of gloves, enantiomers are mirror image structures which cannot be superimposed on each other. A racemic mixture (like Benzedrine) contains equal portions of the "left-" and "right-" hand gloves.

** **d-enantiomer** – Within chemistry, there are many systems for naming molecules. One of these methods identifies compounds according to their optical properties (i.e., how the structure rotates a plane of polarized light). The **d**-enantiomer rotates light in a clockwise direction (*dextrorotatory: to the right*), and the **l**-enantiomer rotates light the other way (*levorotatory: to the left*).

Stimulant Epidemics Worldwide

Over the next three decades, **Benzedrine** and **Dexedrine** were recognized around the world for their euphoric and alerting effects. In the 1930s, college students found that they could remove the paper strips inside Benzedrine nasal inhalers and swallow them to relieve fatigue and enhance mental focus. Soldiers and industrial workers were encouraged to use stimulants during World War II for the same reasons. (In Germany, pilots' rations included chocolate bars which contained methamphetamine, and Hitler's personal physician reportedly injected him with the same substance.) Unfortunately, the post-war experiences around the world – especially in Japan – underscored the addictive potential of amphetamines when an epidemic of drug abuse evolved. Reacting to this problem in 1959, the U.S. Food and Drug Administration banned the Benzedrine inhaler and restricted the availability of amphetamines to prescription use only. Subsequent epidemics resulted in numerous revisions of federal laws, all of which attempted to restrict or control the manufacture and distribution of these drugs.

The Psychiatric Use of Stimulants

The use of stimulants within American psychiatry had a particularly fortuitous beginning. According to one published account, a Rhode Island physician by the name of Charles Bradley first tested Benzedrine in children in 1937. Working with thirty hospitalized subjects who had developed headaches after diagnostic lumbar punctures (more specifically, pneumoencephalography), Bradley hoped that the drug might relieve their pain by "stimulating" the production of spinal fluid. Although Benzedrine failed to do this, fourteen of his patients reported an improvement in their learning capacity and came to refer to the Benzedrine tablets as *math pills*. Bradley continued to use the stimulant in his patients, but his discoveries were overlooked for more than a decade.

In 1944, **methylphenidate (Ritalin)** was created by scientists at Ciba (now Novartis). However, it took ten years for researchers to establish the drug's mechanism of action, and approximately ten more years to test the drug clinically. The landmark study occurred in 1963, when investigators at Johns Hopkins University compared the efficacy of Ritalin and Dexedrine in controlling the behaviors of *institutionalized children*. Four years later, they expanded the scope of their studies to include *schoolchildren* with disruptive classroom behaviors (among the target symptoms: talking loudly, giggling for no reason, and daydreaming).

Aided by research initiatives of the National Institute of Mental Health and major universities, by the birth of the learning disability industry, and by the dominance of the

Stimulants

pharmaceutical companies and its front groups (such as *CHADD and **NAMI), the use of stimulants to restrain youth accelerated in the United States between the 1970s and the new millennium. Throughout this period, the condition known as Attention Deficit Disorder /Attention Deficit Hyperactivity Disorder (ADD/ADHD) worked its way into the popular vernacular and found acceptance within the medical lexicon.

Today, stimulants remain popular as treatments for hyperactivity (in immature or emotionally disturbed children); for narcolepsy (a sleep disturbance characterized by the inability to remain awake); for obesity (due to their capacity to suppress appetite); and for the enhancement of stamina and cognitive functions. Ironically, as the medical profession continues to celebrate the use of prescription stimulants for these purposes, it decries the use of street drugs which have identical chemical and behavioral effects.

According to the prevailing consensus, synthetic stimulants do not, and cannot, replicate the neurotoxicities of cocaine or the modified amphetamines (such as crystal methamphetamine or Ecstasy) when they are consumed *orally* and in the low doses which are commonly prescribed. In other words, the current dogma in American medicine compels physicians to believe that medicinal stimulants simply do not harm the brain. This chapter presents scientific evidence against this claim.

---

* CHADD – Children and Adults with Attention Deficit/Hyperactivity Disorder, a national non-profit organization founded in 1987 for the purposes of education, advocacy, and support; as a conduit of pharmaceutical company advertising and lobbying, CHADD has been criticized by the World Health Organization for violating the 1971 Psychotropic Drugs Convention[4-5]

** NAMI – National Alliance for the Mentally Ill, a non-profit support and advocacy organization founded in the United States in 1979;[6] renamed the National Alliance on Mental Illness in the summer of 2005, the organization is another conduit of pharmaceutical company marketing and lobbying

Postmortem Studies of Animals

Historically, the neurological toxicities of *prescription stimulants* have been ignored in most textbooks of psychopharmacology. In fact, it is only within the past decade that *sporadic studies* have appeared in the medical literature, addressing the potential hazards of clinically relevant (rather than intentionally toxic) doses.

Responding to the fact that catecholamines, such as dopamine and norepinephrine, play a critical role in the formation of the fetal and postnatal brain, and responding equally to the epidemic use of drugs which affect these chemicals, a minority of concerned scientists have felt an urgent need to explore the **developmental effects of early chemical exposures**. The goal of their work has been the detection of delayed and/or persistent drug toxicities using animal models.

As is true of any study involving animals, however, the applicability of this research for humans has depended upon two premises: first, the idea that "developmental windows of vulnerability" can be identified and compared between different species; and second, the idea that equivalent blood levels of medications predict drug effects.

Developmental Windows of Vulnerability

Because rodents reproduce quickly, and because – as mammals – they share a high degree of genetic similarity to humans, rats and mice have become favorite subjects in lab research. Based upon *chronological development in each species (sexual maturity, lifespan)* the following timelines have been used to compare rodents and humans:[7-8]

	rats / mice	humans
gestational period:	21 days	266 to 280 days
neonatal	PND 1 through 7	< 1 month
infant	PND 8 to 10	1 month to 1 year
juvenile	PND 11 to 27	1 year to puberty
adolescent	PND 28 through 45	puberty / teens
young adult	PND 46 through 59	20s
mature adult	PND 60 to death	30s >>

* PND = postnatal day

However, in terms of **anatomical maturation** – a crucial variable when considering psychiatric drug effects – the previous table must be modified to reflect *corresponding levels of* **neurodevelopment**:[9-10]

Stimulants

neurodevelopment	rats / mice	humans
neonatal and infant	**PND 1 through 10**	third trimester
early juvenile	**PND 12 to 13**	day of birth
juvenile	**PND 14 to 27**	infancy to childhood
adolescence	PND 28 through 45	puberty / teens
young adult	PND 46 through 59	20s
mature adult	PND 60 to death	30s >>

* PND = postnatal day

Equivalent Blood Levels of Drugs in Different Species

The second criterion for interpreting animal experiments is the application of therapeutically relevant drug doses. Generally speaking, scientists rely upon the use of treatment protocols in animals which replicate the behavioral and/or physiological effects of the same chemicals when they are administered to humans. One of the methods for achieving this goal is to administer drug doses which result in blood levels or brain levels of pharmaceuticals that are equivalent in human and non-human species.

With respect to methylphenidate (Ritalin), it has always been difficult for scientists to calculate equivalent doses of methylphenidate for rodents and humans due to differences in the absorption, metabolism, and elimination of the drug. However, if one considers both the *dose-related blood levels* and the *duration of peak methylphenidate effects* as they have been measured in both species, the experiments selected and described in this chapter should be relevant for human patients: [11-13]

	methylphenidate dose	elimination half-life	blood level
rat	10 mg/kg po	2 to 3 hrs	40 ng/mL (1 to 3 hrs after dose)
	3.5 mg/kg IP injection	1 hr	62 ng/mL (15 min after dose)
human	0.25 to 1 mg/kg p.o.	2 to 8 hrs	8-40 ng/mL (1 hr after dosing) peak dose is reached in 2 hrs

p.o. = oral dosing

As noted by Gerasimov et al (2000), matching peak plasma levels in humans and rats is considered misleading when choosing clinically relevant doses of this particular drug, largely because rodents (unlike humans) fail to achieve sustained brain levels of methylphenidate. Based upon neurochemical and behavioral effects, a 2 mg/kg IP dose of the stimulant in rats is roughly equivalent to 5 mg/kg delivered orally. Thus, doses of 5 mg/kg IP (as administered by Gray et al, described below) may very well simulate the neurological effects *and* blood levels of human drug regimens, due to the shorter duration of peak serum levels (minutes), and the more rapid elimination of methylphenidate in rats.

Methylphenidate (Ritalin)

Study #1 *Gray et al (2007)   Methylphenidate Reduces Brain Cell Density*

Motivated by the desire to understand the developmental consequences of methylphenidate in the young rat, a team of researchers from universities in New York and Montreal explored the anatomic consequences of early exposure to methylphenidate.[14]

Experimental Procedure

The experiment involved the use of male Sprague-Dawley rat pups. At postnatal day 7, the animals were divided into two groups. The experimental group received twice-a-day, intraperitoneal injections of methylphenidate (5 mg/kg) for 28 days. The control group received an equal volume of saline according to the same schedule. Following the period of drug or saline exposure, some of the animals were immediately euthanized. The goal of this design was to permit a comparison between the immediate effects of drug treatment (PND 7 through 35) during a stage of rat development analogous to human youth. The second phase of the experiment (three-month recovery period) was performed in an effort to detect delayed or persistent drug effects once the animals had become mature adults.

Study Design		
	methylphenidate n = 12	saline n = 12
PND 7 to 35	5 mg/kg IP b.i.d.	equal volume IP b.i.d.
PND 35	½ sacrificed	½ sacrificed
PND 135	remaining animals were sacrificed	
b.i.d. = twice per day		

After the animals were sacrificed, brains were removed, sliced into horizontal sections, and prepared for microscopic examination using a number of advanced staining methods. The primary goal of this experiment was to identify changes in the density of **catecholamine neurons** throughout the brain.

Stimulants

Background Information

Catecholamines refer to chemicals in the body which possess a specific structure known as a *catechol* ring (shown below):

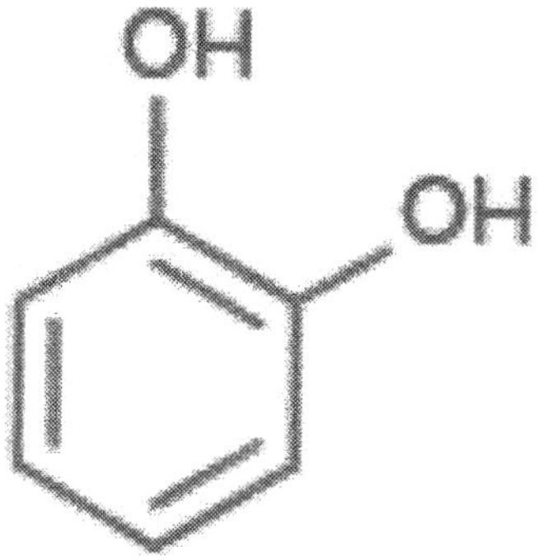

[catechol ring structure: benzene ring with two adjacent OH groups]

and at least one subgroup, known as an *amine* (shown below):

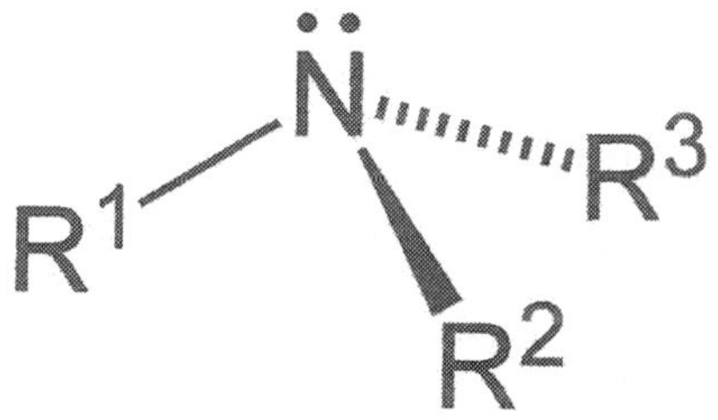

[amine structure: N with lone pair bonded to R¹, R², R³]

Within the human nervous system, the primary catecholamines are **dopamine** and **norepinephrine**. Like all neurotransmitters, these chemicals are synthesized from a series of building blocks. Some of these materials (amino acids and cofactors) are transferred into the brain from the bloodstream. Other materials – called enzymes – are made by the machinery of *specific* cells. This permits specialization within the brain. For example, a serotonin neuron must make its own tryptophan hydroxylase. A dopamine neuron must make its own tyrosine hydroxylase. Lacking the pertinent enzymes, a serotonin neuron will not make dopamine. A dopamine neuron will not make serotonin. ***Thus, the identification of key enzymes within cells allows scientists to trace the anatomy of chemical networks (dopamine network, serotonin network, etc.) throughout the brain.***

In the case of catecholamines, the amino acid known as tyrosine is turned into dopamine, norepinephrine, and epinephrine through the sequence of reactions shown below:

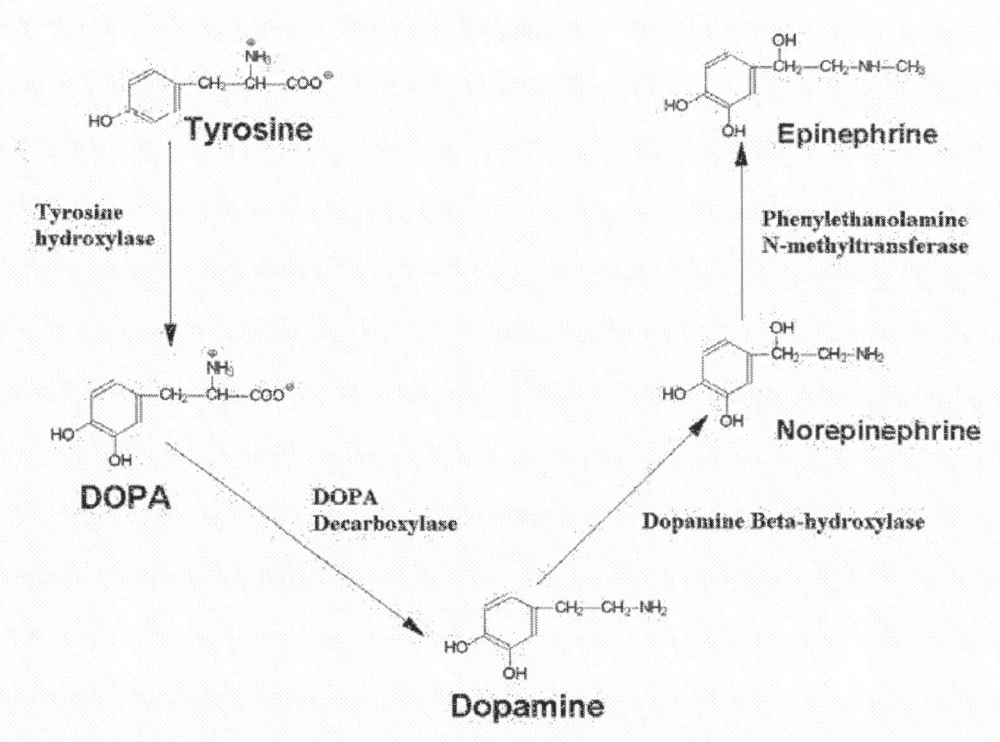

In the methylphenidate (Ritalin) experiment by Gray et al, the scientists employed a variety of lab techniques which **stained brain cells** according to the presence of certain membrane proteins (reuptake pumps or transporters) or enzymes known to be affected by that drug:

target of labeling	significance
tyrosine hydroxylase	labels dopamine *and* norepinephrine neurons
dopamine beta hydroxylase	labels norepinephrine neurons
DAT = dopamine transporter	labels dopamine reuptake pump (located primarily on dopamine neurons)
NET = norepinephrine transporter	labels norepinephrine reuptake pump (located primarily on norepinephrine neurons)
SERT = serotonin reuptake transporter	labels serotonin reuptake pump (located on serotonin neurons and astrocytes)

Findings

Following a four-week period of early exposure to methylphenidate, animals demonstrated several abnormalities in the catecholamine pathways of the brain. *Based upon changes in the density of labeling for enzymes (tyrosine hydroxylase) and reuptake transporters (NET, DAT),* ***methylphenidate appeared to stunt the development of the prefrontal cortex, the hippocampus, and the striatum***:

Key Changes in the Density of Immunological Labeling	
medial prefrontal cortex	**40% decrease in norepinephrine fibers** (based upon staining for NET)
hippocampus	**51% decrease in norepinephrine fibers** (based upon staining for NET)
striatum	**21% decrease in dopamine fibers** (21% ↓ in staining for tyrosine hydroxylase and DAT)

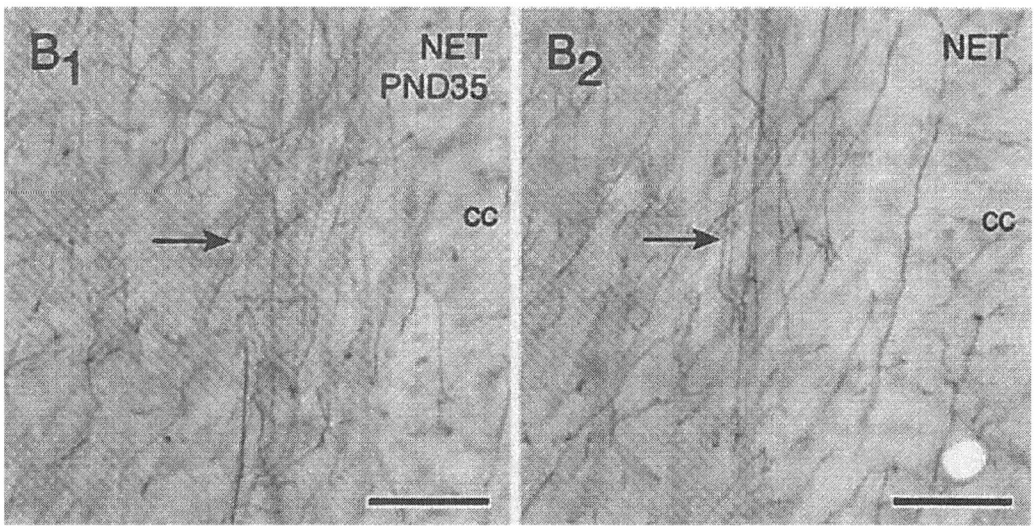

Reprinted with permission, *Journal of Neuroscience*, volume 27, issue 27, Gray et al, "Methylphenidate Administration to Juvenile Rats Alters Brain Areas Involved in Cognition, Motivated Behaviors, Appetite, and Stress," page 7201, copyright (2007).

Prefrontal Cortex: Reduced Norepinephrine Fiber Density After Ritalin

This image depicts **norepinephrine fibers** in the medial prefrontal cortex of rats exposed to saline (left) and methylphenidate (right) between the ages of 7 and 35 days. The image on the right demonstrates the appearance of **reduced fiber density** (neurons) following stimulant medication (fibers have been labeled for norepinephrine transporters).

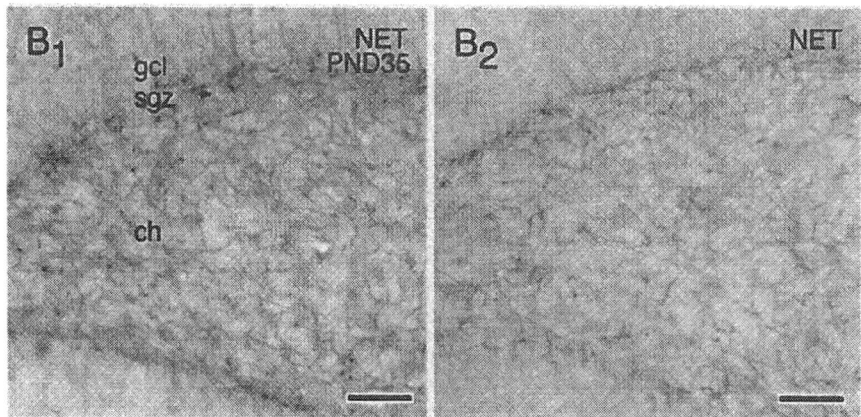

Reprinted with permission, *Journal of Neuroscience*, volume 27, issue 27, Gray et al, "Methylphenidate Administration to Juvenile Rats Alters Brain Areas Involved in Cognition, Motivated Behaviors, Appetite, and Stress," page 7203, copyright (2007).

Hippocampus: Reduced Norepinephrine Fiber Density After Ritalin

Above: This image shows reduced **norepinephrine fiber density** in the hippocampus of animals after four weeks methylphenidate (right) versus drug-free controls (left).

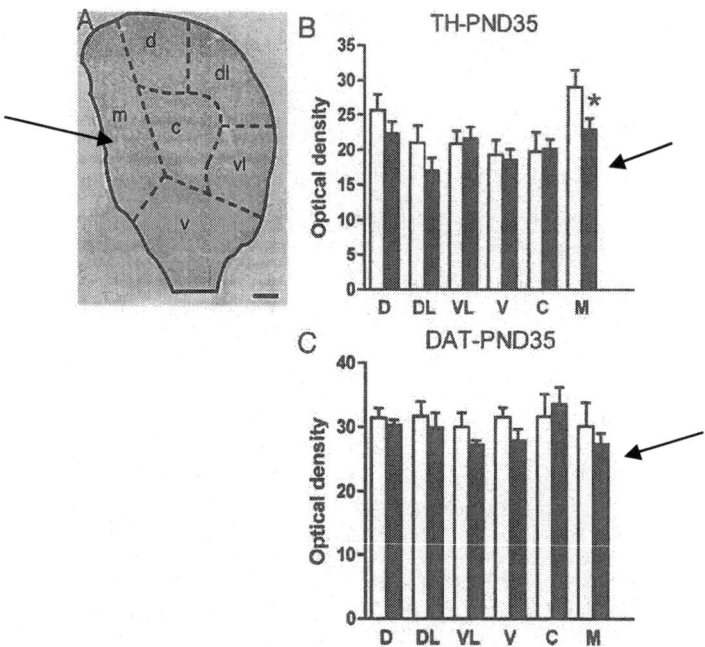

Reprinted with permission, *Journal of Neuroscience*, volume 27, issue 27, Gray et al, "Methylphenidate Administration to Juvenile Rats Alters Brain Areas Involved in Cognition, Motivated Behaviors, Appetite, and Stress," page 7202, copyright (2007).

Striatum: Reduction in Dopamine Neurons After Ritalin

These bar graphs reflect changes in the **density of dopamine neurons** in the **striatum** (**m = medial** part of the striatum) using immunological labeling techniques. As mentioned in previous chapters, the striatum is a brain region whose functions include the regulation of involuntary movement, mood, and cognition. Bar graphs display cell densities based upon staining for tyrosine hydroxylase (TH) and the dopamine transporter (DAT), two markers of dopamine neurons.

> In Sum:
>
> Gray et al administered clinically relevant doses of methylphenidate to young rats between the ages of 7 and 35 days. Although the investigators selected this developmental period as an analog for human youth, it is possible that their model corresponded to fetal, neonatal, *and* juvenile exposure due to inter-species differences in the speed and timing of neurodevelopment.
>
> ***Early exposure to methylphenidate resulted in a 20 to 50% reduction of dopamine and norepinephrine fiber density in the prefrontal cortex, the hippocampus, and the striatum.***
>
> To the extent that methylphenidate induces similar abnormalities in *humans*, the location and intensity of these changes imply a toxic effect. Specifically, the early use of this stimulant medication impairs the growth and/or survival of neurons in brain regions which are considered to be essential for judgment, impulse control, learning, memory, and movement.

Study #2   Husson et al (2004)   *Methylphenidate Kills Brain Cells (Basal Ganglia)*

Responding to the need for further experiments in the field of developmental neurotoxicology, a team of French researchers undertook an analysis of methylphenidate's effects upon neonatal mice.[15] In this investigation, scientists explored the anatomical consequences of the stimulant upon three distinct regions of the brain.

Experimental Procedure

Swiss mouse pups (PND 5) were divided into four groups and exposed to a single intraperitoneal injection of methylphenidate (either 0.5 mg/kg, 5 mg/kg, or 50 mg/kg) or saline (control). The following day, the animals were sacrificed. Brains were removed, sliced into coronal sections, and prepared for examination by light microscopy. In one of several experimental phases, the researchers employed a procedure known as ***TUNEL labeling** in order to identify fragmented DNA within cells that were dead or dying.

---

* **TUNEL** – an acronym for the dUTP nick end labeling technique; this method is used in biological research to detect cells which are dying via apoptosis or late-stage necrosis

Findings

All three doses of methylphenidate (0.5 mg/kg, 5 mg/kg, and 50 mg/kg) resulted in statistically significant increases in neural cell death. These changes were observed in the basal ganglia and in the periventricular (subcortical) white matter of the brain:

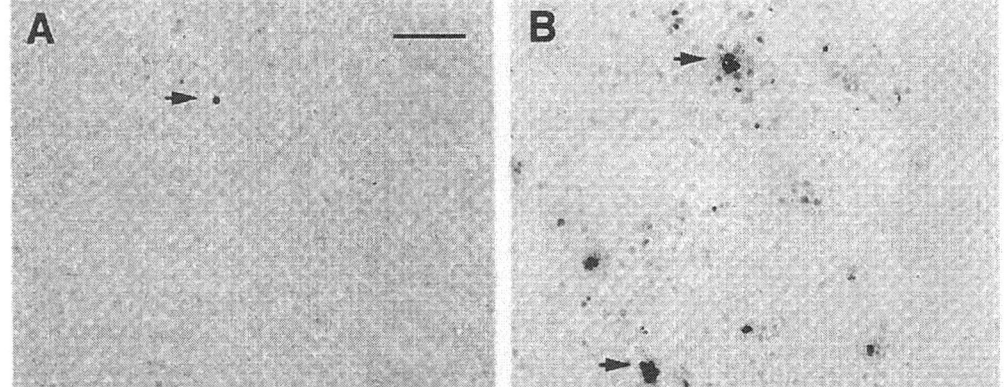

Reprinted from *Neuroscience*, volume 125, issue 1, Husson et al, page 165, copyright (2004) with permission from Elsevier.

Striatum of Mouse Brain: Cell Death as Demonstrated by TUNEL Labeling

These photos demonstrate lethal changes in the *striatum* of a mouse pup. The first image (left) shows limited cell death after a single injection of saline. In contrast, ***exposure to a single injection of methylphenidate (5 mg/kg) on postnatal day 5 resulted in a substantial (five-fold) and statistically significant increase in cell death***.

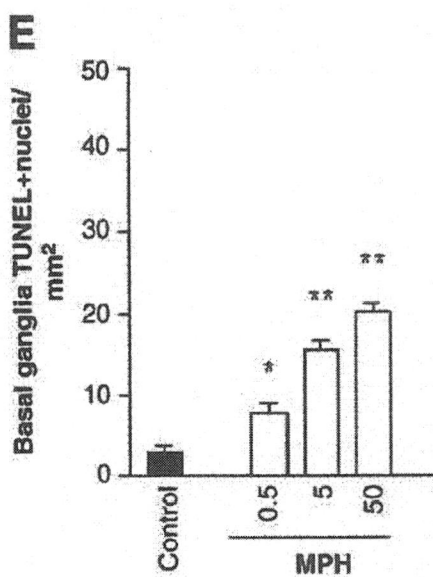

Reprinted from *Neuroscience*, volume 125, issue 1, Husson et al, page 165, copyright (2004) with permission from Elsevier.

This bar graph depicts the dose-related effects of methylphenidate upon cell death in the basal ganglia of the mouse pup. Most remarkably, ***even a single injection of 0.5 mg/kg of the stimulant drug induced a large (2.7-fold) increase in neural death***.

Stimulants

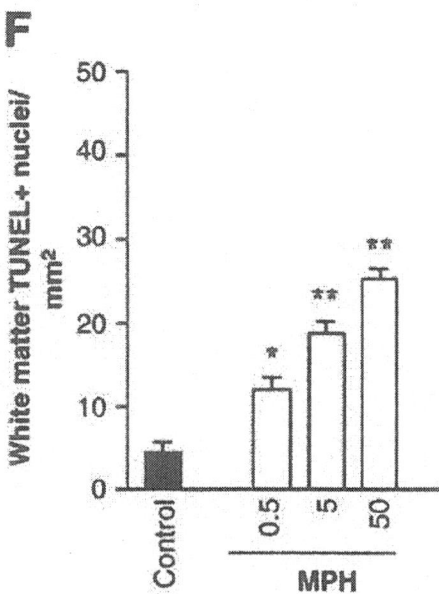

Reprinted from *Neuroscience*, volume 125, issue 1, Husson et al, page 165, copyright (2004) with permission from Elsevier.

This bar graph demonstrates cell death (via TUNEL labeling) in the *periventricular white matter* of the mouse pup's brain. Here again, early exposure to methylphenidate resulted in a dose-related pattern of neurotoxicity.

In Sum:

Mouse pups were exposed to single injections of methylphenidate at various doses, two of which (0.5 mg/kg, 5 mg/kg) were clinically relevant for humans.

All three doses of methylphenidate in this experiment induced significant ***increases in cell death in the basal ganglia (striatum) and white matter of the immature brain***.

This experiment involved the use of mouse pups at the age of postnatal day 5. According to research scientists, the maturation of the mouse brain at this stage of development corresponds to the third trimester of human gestation. Thus, the results of this experiment strongly suggest that methylphenidate is a brain teratogen.

Scientists have only recently begun to contemplate the potential of medications to induce harmful effects *throughout* critical periods of neural development. To the extent that the toxicity of methylphenidate impacts the fundamental processes of neurogenesis, myelinogenesis, and/or synaptogenesis, its hazards would be expected to extend well past gestation and into infancy, toddlerhood, and beyond.

Study #3   Lagace et al (2006)   *Methylphenidate Damages the Hippocampus*

In another neurodevelopmental experiment, researchers associated with universities in Florida and Texas investigated the immediate and delayed effects of methylphenidate with respect to cell viability within the temporal lobe.[16]

Experimental Procedure

The experiment began with the selection of male Sprague-Dawley rats. Starting on **postnatal day 20**, all of these juvenile animals were subjected to a 16-day period of chemical treatment. This consisted of twice-a-day, intraperitoneal injections with saline (1 mg/kg) or methylphenidate (2 mg/kg). *The goal of this study was to explore age-related differences in the proliferation and survival of neural progenitors (early cells) in the hippocampus.* Following the two-week period of exposure, the animals were divided into subgroups (5 to 6 per group) and observed for variable periods of time (cell proliferation vs. cell survival protocols).

Cell Proliferation Protocol
In the proliferation protocol, the pre-medicated animals at three different life stages (*PND 46, PND 77, PND 90) were administered a single injection of BrdU (150 mg/kg). This procedure was employed in an effort to detect neurogenesis (new neurons). Two hours later, the animals were euthanized. Brains were removed, processed, and immediately examined using light microscopy. Using a select number of brain slices from each animal, the researchers performed a count of cells which had incorporated BrdU. [Recall, from chapter four, that BrdU signifies DNA synthesis – *not* the division or migration of new neurons]. In a related protocol, the researchers employed additional techniques (immunofluorescent cell labeling) and instrumentation (confocal microscopy) in an effort to identify the nature of the allegedly "new" cells (glia or neurons).

Cell Survival Protocol
One group of animals was subjected to the aforementioned treatment protocol in the juvenile period (PND 20 through PND 35). BrdU was injected on day 85. Following a four-week period of further observation, these animals were euthanized and examined as described above. The purpose of this protocol was to evaluate the longevity of ***"new cells" born in the hippocampus of adult rats*** (i.e., neurons born on PND 85) following juvenile exposure to saline or methylphenidate.

PND = postnatal day
PND 20-35 = approximate rat brain development of human child/adolescent
PND 46/77/90 = approximate rat brain development of human late teens/30s/middle age

Findings

Following a sixteen-day period of exposure to methylphenidate during early development (PND 20 through 35), the appearance of the brain was examined at three different periods of maturation: **late adolescence** (PND 46), **early adulthood** (PND 77), and **late adulthood** (PND 90).

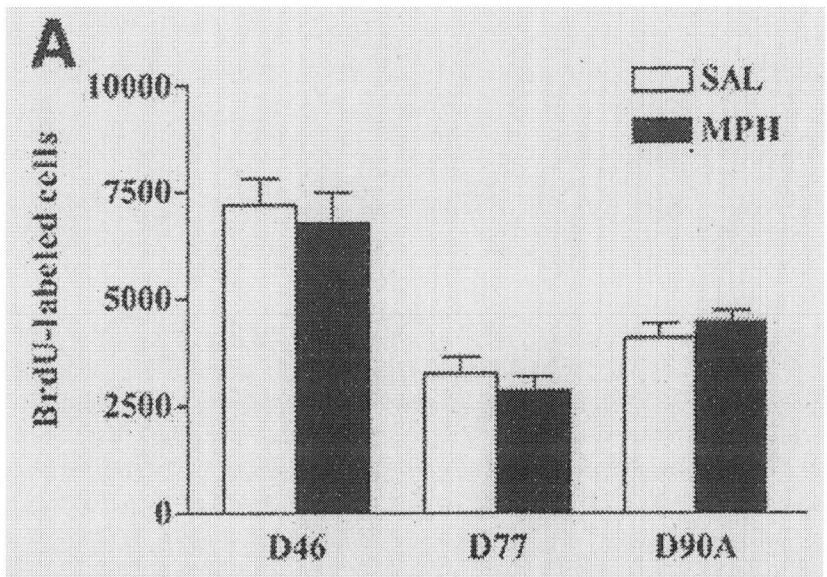

Reprinted from *Biological Psychiatry*, volume 60, D.C. Lagace, J.K. Yee, C.A. Bolanos, and A.J. Eisch, "Juvenile Administration of Methylphenidate Attenuates Adult Hippocampal Neurogenesis," page 1124, copyright (2006), with permission from Elsevier.

Above: According to **hippocampal cell counts** (number of cells which stained positive for BrdU), there were *no significant differences in "cell proliferation"* when scientists compared the brains of formerly medicated animals and drug-free controls at three different stages of maturity.

However, a further analysis of *discrete regions of the hippocampus* revealed a more complicated set of developments.

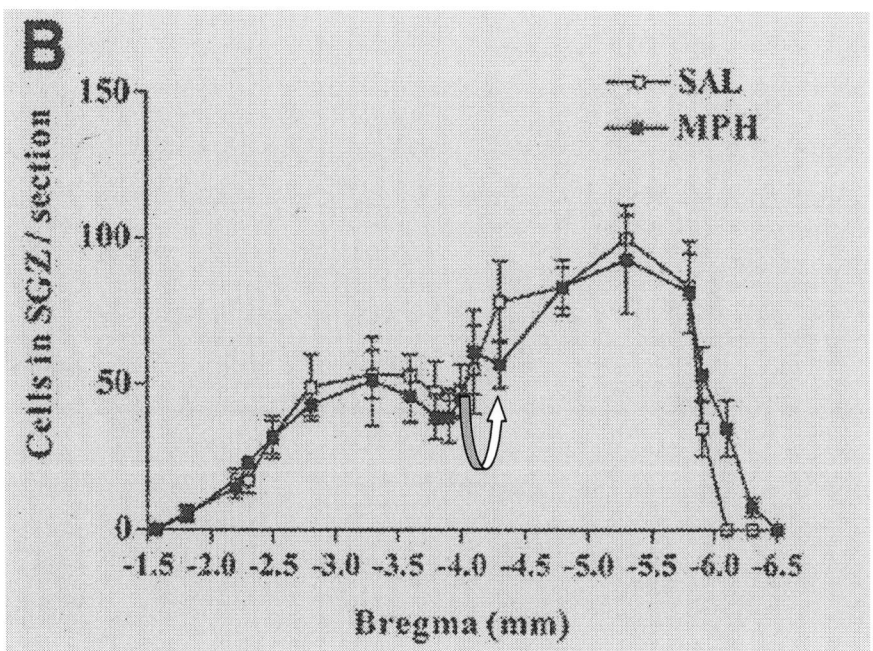

Reprinted from *Biological Psychiatry*, volume 60, D.C. Lagace, J.K. Yee, C.A. Bolanos, and A.J. Eisch, "Juvenile Administration of Methylphenidate Attenuates Adult Hippocampal Neurogenesis," page 1124, copyright (2006), with permission from Elsevier. (arrow added by author)

BrdU Labeled Cells (Cell Count Per Section) at PND 46 - Adolescent Rats

This graph demonstrates changes in DNA synthesis (BrdU uptake) within the hippocampus (subgranular zone or SGZ) of **adolescent rats** following early life exposure to saline or methylphenidate.

The **bregma** (x-axis) is an anatomical reference point on the rat skull. Using this as a map of neuroanatomy (from front to back of the brain = -1.5 mm towards -6.5 mm), this graph provides evidence of a ***30% regional decrease in BrdU uptake in the brains of the formerly medicated animals (arrow)***.

Although the research team did not call attention to this result, their graph strongly implied the existence of at least one potentially important **focal change** in physiology in the temporal lobe (-4.1 through -4.7 mm).

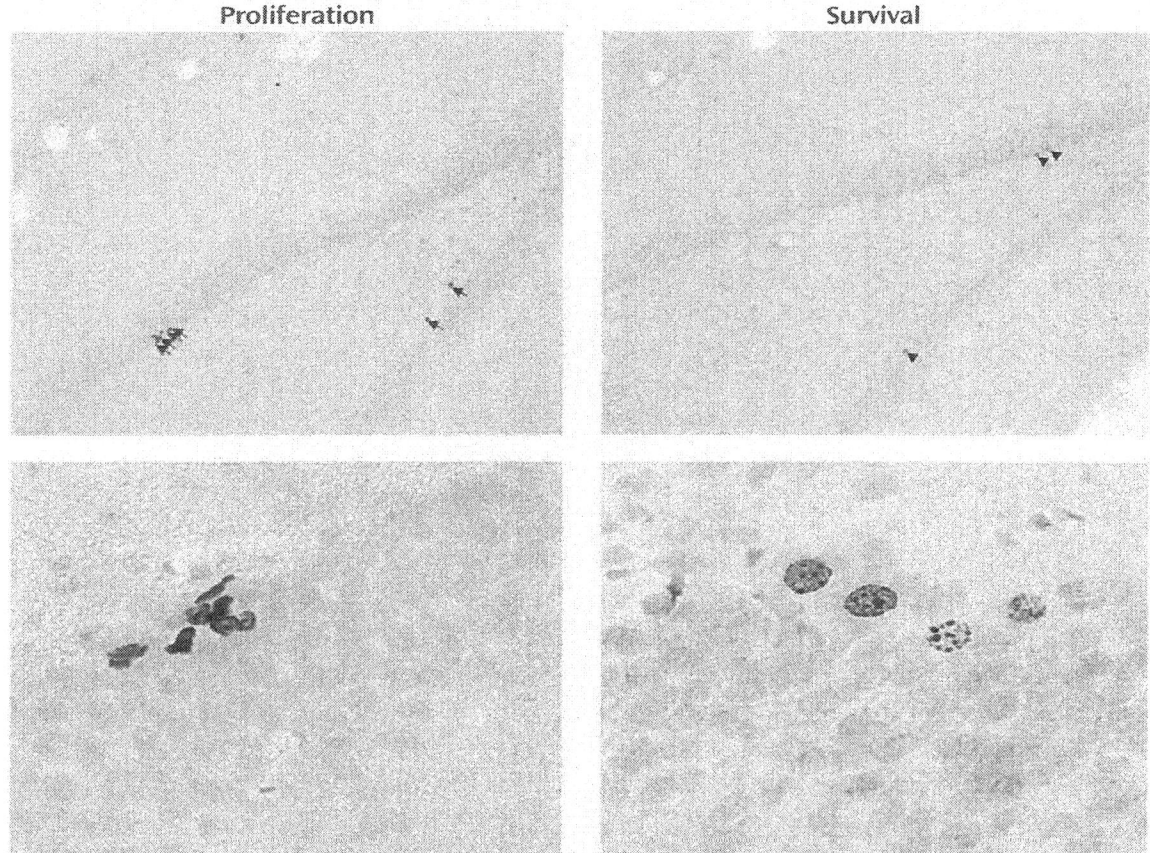

Source: D.C. Lagace, M.A. Noonan, and A.J. Eisch, "Hippocampal Neurogenesis: A Matter of Survival," volume 164, issue 2, page 205. Reprinted with permission from the *American Journal of Psychiatry*, (Copyright 2007). American Psychiatric Association.

Hippocampal Cells in the Adult Rat: BrdU Staining in "New" Neurons

In the **Cell Survival** protocol, Lagace et al. explored the effects of early exposure to methylphenidate upon the longevity (viability) of "new cells" within the adult brain.

The slides on the left demonstrate the appearance of brain tissue two hours after the injection of BrdU (i.e., cells undergoing DNA synthesis). The photos on the right depict the survival of BrdU-labeled cells (mature neurons) in the hippocampus of an adult rat, four weeks after the addition of BrdU.

In the methylphenidate experiments performed by Lagace et al, *__juvenile exposure to the stimulant resulted in a 30% decrease in the survival of neurons within the subgranular zone of the hippocampus__*.

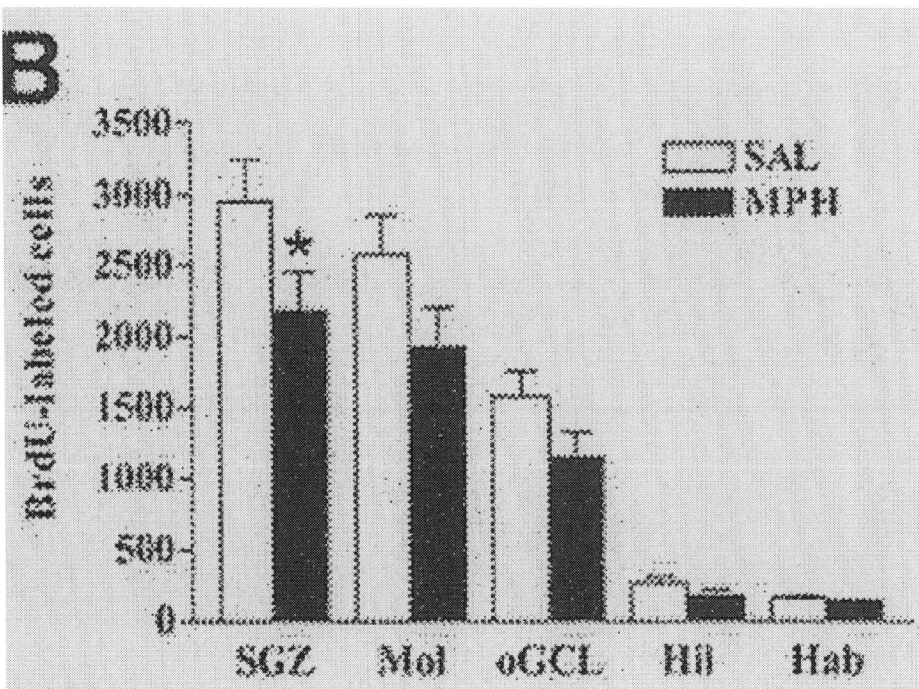

Reprinted from *Biological Psychiatry*, volume 60, D.C. Lagace, J.K. Yee, C.A. Bolanos, and A.J. Eisch, "Juvenile Administration of Methylphenidate Attenuates Adult Hippocampal Neurogenesis," page 1125, copyright (2006), with permission from Elsevier.

Cell Survival in the Adult Rat (PND 112): Effects of Early Methylphenidate

This graph demonstrates differences in **cell survival** in various regions of the hippocampus (SGZ = subgranular zone, Mol = molecular layer, oGCL = outer granule cell layer, Hil = hilus) and in one additional region of the brain (Hab = habenula). Two key findings in this graph should be appreciated.

First, animals exposed to methylphenidate (MPH) during youth grew up to experience a significant *reduction in cell survival*, relative to unmedicated (saline) controls. *Whether one interprets the BrdU-labeled cells to be "new neurons" or healthy mature neurons with extra DNA (aneuploidy), early exposure to methylphenidate reduced this population by 30% in older adults (day 112).*

Second, the *detrimental effects of early exposure to methylphenidate appeared to be extensive* (the decreased survival of BrdU-labeled cells was apparent in every region of the hippocampus).

In Sum:

Researchers exposed juvenile rats (PND 20 to 35, analogous to human childhood) to sixteen days of a clinically relevant dose of methylphenidate. When scientists examined brain tissue as these animals matured, they discovered that ***early treatment with the stimulant drug impaired the proliferation and/or survival of cells within the hippocampus***.

Short-term effects of early exposure to methylphenidate included the ***focal suppression of BrdU uptake in specific regions of the temporal lobe***. This change was possibly indicative of decreased neurogenesis or regeneration.

Early exposure to methylphenidate also exerted a ***delayed toxic effect*** upon the brain. When the fate of neurons was analyzed in the hippocampus of older rats (PND 85 through 112), ***early stimulant therapy was linked to a 30% reduction in cell survival***.

Amphetamine

Study #4   Ricaurte et al (2005)   *Amphetamine Damages the Basal Ganglia*

In a study that was unique for its use of non-human primates and for its use of therapeutically relevant doses, a team of scientists affiliated with Johns Hopkins University explored the anatomical consequences of amphetamine upon the **basal ganglia** (the subcortical brain structures which mediate the involuntary control of movement, as well as motivation, cognition, and mood).[17]   Two facts sparked this investigation. First, the researchers were aware of the large body of evidence which had described how recreational amphetamines (street drugs) damage nerve endings in the basal ganglia. Second, they were well acquainted with the link between recreational stimulants and neurodegenerative conditions, such as Parkinson's disease.

Experimental Procedure

The goal of this experiment was to understand the potential hazards of *therapeutic doses* of an amphetamine mixture identical to that contained in a popular medicinal drug (Adderall).

Therapeutic Doses of Amphetamine in Rodents and Primates [18-21]

species	dose	plasma drug level
*rats	0.1-3.0 mg/kg	NA
baboon	0.7-1.0 mg/kg p.o. b.i.d.	143-193 ng/mL
squirrel monkey	0.68 mg/kg p.o. b.i.d.	111-157 ng/mL
**humans	0.1-1.0 mg/kg	
d-amphetamine	1 mg/kg p.o. b.i.d.	120-160 ng/mL
Adderall XR	30 mg p.o. q.d.	117 ng/mL

p.o. = by mouth   NA = not available   b.i.d. = twice a day   q.d. = daily

*   Rat doses are based upon drug-induced changes in behavior. Doses above 3 mg/kg are considered to be toxic based upon the induction of intense stereotypic behaviors.

** Human values are based upon measurements in children following 3 to 5 weeks of continuous drug therapy. Steady-state blood levels for adults have not been published.

In the Johns Hopkins experiment, adult animals were trained to administer the stimulant drug (a 3:1 mixture of dextro-amphetamine and levo-amphetamine) orally, via a fruit-flavored drink. The experiment consisted of two protocols.

Protocol #1

Adult baboons of both genders (10 to 15 years old) were exposed to increasing oral doses of mixed amphetamine on a twice-a-day schedule for four weeks. Using a set of different baboons who received similar doses, the corresponding plasma levels of amphetamine were measured on a weekly basis:

Baboon Group #1 – oral doses of mixed amphetamines (mg/kg, twice per day)				
	day 1	days 2-5	days 6-13	days 14-27
baboon #1	0.12	0.24	0.48	0.95
baboon #2	0.13	0.25	0.50	1.00
baboon #3	0.13	0.25	0.50	1.00
Baboon Group #2 – oral doses of mixed amphetamines (mg/kg, twice per day)				
	days 1-7	days 8-14	days 15-21	days 22-28
baboon #4	0.17	0.33	0.50	0.67
baboon #5	0.17	0.33	0.50	0.67
baboon #6	0.25	0.50	0.75	1.00
amphetamine blood levels after four weeks:				143-193 ng/mL

Protocol #2

Adult male squirrel monkeys (exact age, unknown) were exposed to the same amphetamine mixture via orogastric lavage (tube feeding). Plasma drug concentrations were measured on a weekly basis, and drug doses were adjusted in order to reach and maintain a blood level of 100 to 150 ng/mL for the remainder of the four-week study.

Squirrel Monkeys – oral doses of mixed amphetamines (mg/kg, twice per day)				
	days 1-7	days 8-14	days 15-21	days 22-28
monkey #1	0.30	0.64	0.68	0.65
monkey #2	0.28	0.63	0.65	0.63
monkey #3	0.26	0.58	0.65	0.64
monkey #4	0.30	0.64	0.68	0.68
amphetamine blood levels after four weeks:				111-139 ng/mL

Two to four weeks after the end of treatment, animals were euthanized. Brains were removed and processed using a variety of biochemical techniques. These lab procedures were designed to detect changes in the neurons which make and release dopamine in the striatum.

Background Information

Within the brain, dopamine-containing cells are classified according to location and function. Anatomists have historically referred to **four dopamine pathways** based upon the positions of the cell bodies which produce neurochemical precursors (e.g., enzymes, such as tyrosine hydroxylase) and the dendrites and axons which complete dopamine synthesis and release it. These four networks are identified below:

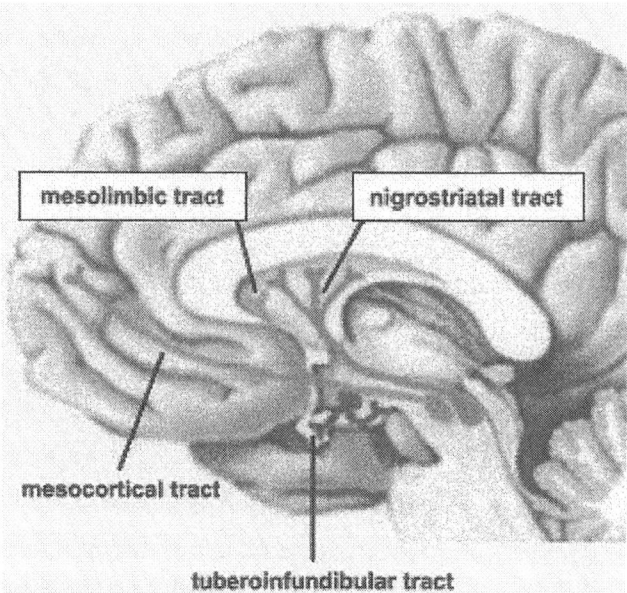

Source: McGill University, The Brain from Top to Bottom

the **mesolimbic tract**: a network of dopamine neurons whose axons travel from the midbrain (meso = middle) to the limbic system of the brain. The latter system includes a variety of brain regions – the amygdala, hippocampus, thalamus, hypothalamus, and the septal nuclei – which regulate mood and emotion.

the **mesocortical tract**: a network of dopamine neurons whose axons travel from the midbrain to the cortex

the **tuberoinfundibular tract**: a network of dopamine neurons whose axons travel from the hypothalamus to the pituitary gland

the **nigrostriatal tract**: a network of dopamine neurons whose axons travel from the substantia nigra (midbrain) to the **striatum** (caudate and putamen)

In the Johns Hopkins amphetamine study, the scientists explored drug effects upon the **nerve terminals** of dopamine neurons within the **striatum**. Although they did not employ advanced microscopic techniques to directly confirm degenerative changes in the dendrites, axons, or nerve terminals, they employed several *indirect methods to identify markers of cell damage*:

Indirect Markers of Damage to Dopamine Neurons in the Brain

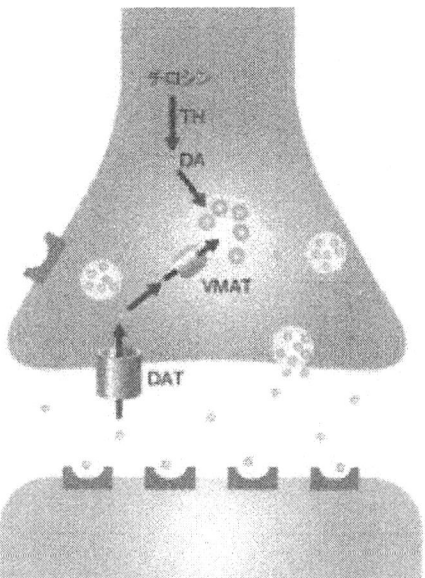

Source: Dr. Ichiro Sora, http://sendaibrain.org/eng/members/sora.html

Indirect Marker #1: reduction in the number of dopamine transporters (DATs)
The dopamine transporter (aka, dopamine reuptake pump) is located on the dendrites and axons of dopamine neurons. When these cell parts are damaged by chemicals or metabolic activities, DAT density is reduced.

Indirect Marker #2: reduction of vesicular monoamine transporters (VMATs)
The vesicular monoamine transporter, or VMAT, is located on the membrane of intracellular storage sacs (vesicles). When dopamine neurons are damaged or destroyed, the internal population of VMATs is reduced.

Indirect Marker #3: reduction of the dopamine content in the brain
Dopamine is made within the axons and dendrites of the four pathways mentioned above. Although damage to the cell bodies in any region – such as the midbrain or the hypothalamus – can harm the machinery which is used to make dopamine, damage to **nerve terminals** (shown above) can result in the loss of dopamine after it has been made and stored.

Findings

Based upon changes in several markers of cell integrity (reduced DAT, reduced VMAT, and reduced dopamine content), researchers confirmed that a ***one-month exposure to clinically relevant doses of mixed amphetamines resulted in damage to striatal dopamine neurons in all three groups of experimental animals***:

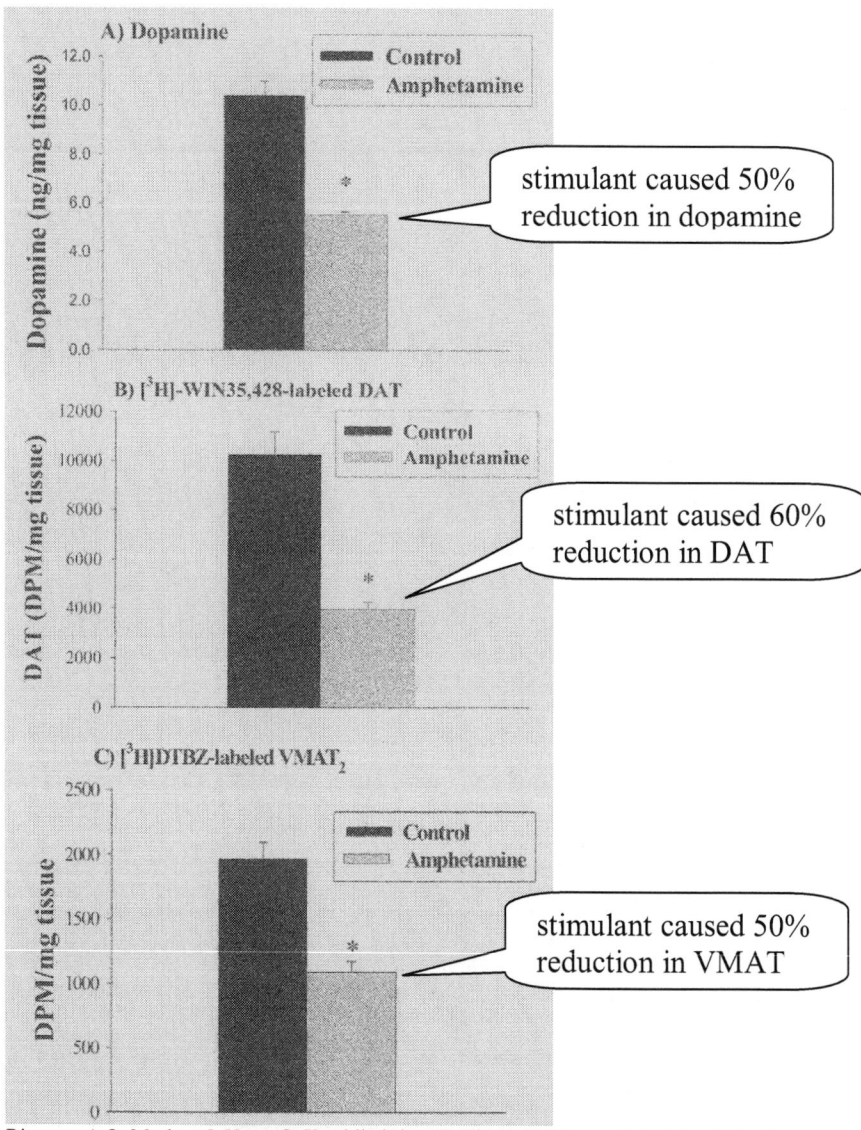

Reprinted with permission, G.A. Ricaurte, A.O. Mechan, J. Yuan, G. Hatzidimitriou, T. Xie, A.H. Mayne, et al, "Amphetamine Treatment Similar to That Used in the Treatment of Adult Attention Deficit/ Hyperactivity Disorder Damages Dopaminergic Nerve Endings in the Striatum of Adult Nonhuman Primates," *Journal of Pharmacology and Experimental Therapeutics* 315:1 (2005): 96.

Baboon Group #1: These bar graphs reflect dysfunction in the dopamine neurons of the striatum. Reductions in the content of dopamine and in the density of DAT and VMAT strongly implied neuronal damage or death following treatment with mixed amphetamines.

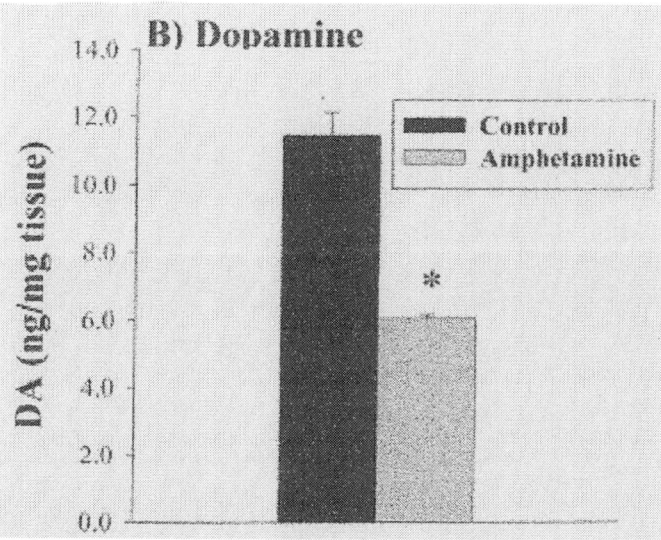

Reprinted with permission, G.A. Ricaurte, A.O. Mechan, J. Yuan, G. Hatzidimitriou, T. Xie, A.H. Mayne, et al, "Amphetamine Treatment Similar to That Used in the Treatment of Adult Attention Deficit/ Hyperactivity Disorder Damages Dopaminergic Nerve Endings in the Striatum of Adult Nonhuman Primates," *Journal of Pharmacology and Experimental Therapeutics* 315:1 (2005): 96.

Baboon Group #2: This bar graph demonstrates a **50% reduction in the dopamine content of the striatum** of animals exposed to mixed amphetamines, relative to drug-free controls. (Similar reductions were found for the other two markers: 40% reduction in DAT density, and 40% reduction in VMAT).

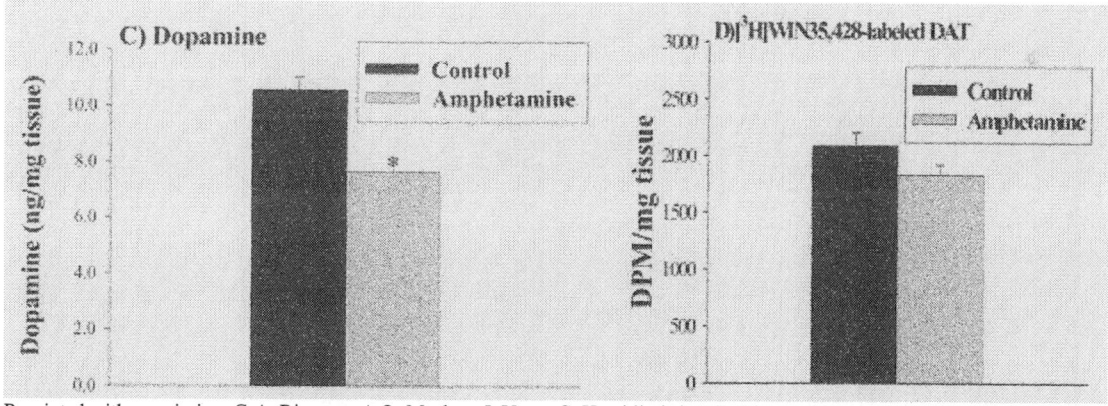

Reprinted with permission, G.A. Ricaurte, A.O. Mechan, J. Yuan, G. Hatzidimitriou, T. Xie, A.H. Mayne, et al, "Amphetamine Treatment Similar to That Used in the Treatment of Adult Attention Deficit/ Hyperactivity Disorder Damages Dopaminergic Nerve Endings in the Striatum of Adult Nonhuman Primates," *Journal of Pharmacology and Experimental Therapeutics* 315:1 (2005): 94.

Squirrel Monkeys: Qualitatively similar effects were observed in the striatum of monkeys exposed to mixed amphetamines. These histograms reveal an approximate **30% reduction in the dopamine content of the striatum** following treatment with amphetamines (left) and an approximate **10% reduction in DAT density** (right). A comparable reduction in VMAT density was also observed (graph not shown).

In Sum:

Adult baboons and adult squirrel monkeys were exposed to a four-week regimen of mixed amphetamines, identical to a popular prescription stimulant (Adderall). ***Doses were likely relevant for humans, based upon plasma drug concentrations*** which have been measured in patients receiving treatments for ADHD.

Following the period of active drug treatment, animals were withdrawn from medication for a period of two to four weeks. Brain tissue was then examined for signs of injury to dopamine neurons in the **striatum** (a brain region associated with the control of movement, cognition, and mood).

In all three groups of animals, treatment with mixed amphetamines reduced brain levels of dopamine and reduced the density of specific cell membrane components (DAT, VMAT). ***These changes were consistent with drug-induced degeneration of dopamine cells within the basal ganglia.***

The findings have important implications for patients and physicians. The abnormalities observed here were identical to those which have been documented in human neurological conditions, such as Parkinson's disease and Lewy body dementia.

Stimulants

Study #5 Armstrong et al (2004)   *d-Amphetamine Induces Subcortical Gliosis*

Recognizing the link between amphetamine and brain damage, on the one hand; and appreciating the connection between brain damage and gliosis, on the other, researchers in California designed a study to confirm several markers of drug-related toxicity.[22] The procedure combined methods for detecting harm to dopamine cells (i.e., **reduction in DAT density**) with methods designed to identify the brain's reaction to injury (**increase in GFAP**).

Background Information

GFAP Refresher

Glial **F**ibrillary **A**cidic **P**rotein or **GFAP** is a skeletal component which contributes to the structural integrity and motility of astrocytes. GFAP levels are often monitored in neuroscience research as an *indirect signal of brain injury*. As astrocytes respond to chemical insults, they generally increase in number and size. This results in a higher level of astrocytic proteins, such as GFAP.

Experimental Procedure

Scientists exposed adult male Sprague-Dawley rats to clinically relevant amounts of d-amphetamine or to equal volumes of saline – the latter, as a non-drug control. Daily intraperitoneal injections of the drug, in the dose of 2 mg/kg, were administered for seven days. Animals were then subjected to a 10-day period of drug withdrawal. On the 11th day, they received a single injection of the medication (same dose), followed by a brief period of observation. All animals were then euthanized. Brain tissue was dissected and prepared chemically in order to permit evaluation by light microscopy.

After identifying neurons and astrocytes by the presence of DAT or GFAP, one researcher performed a manual count of these cell populations in samples of tissue obtained from several brain regions: the striatum (caudate and putamen), the nucleus accumbens, and the prefrontal cortex.

Findings

Consistent with the findings of other investigators (e.g., Ricaurte et al), *the seven-day exposure to d-amphetamine resulted in significant disruption in the striatum of adult rats (34 to 45% increase in GFAP)*. In addition to gliosis, treatment with d-amphetamine was associated with a *13% reduction in the number of dopamine transporters* in the ventral striatum:

of Cells Staining Positive for DAT and GFAP
(based upon samples of brain tissue)

	d-amphetamine n = 10	saline controls n = 10
GFAP		
dorsal striatum	**100.3**	74.8
ventral striatum	**96.6**	66.8
DAT		
dorsal striatum	85.4	78.4
ventral striatum	**78.0**	89.7

Although DAT density was not *universally* reduced, and although the changes in GFAP *might* have reflected adaptive rather than degenerative changes, it is equally likely that d-amphetamine damaged neurons in the striatum. Moreover, since this was a study which investigated brain changes **after a 10-day period of drug withdrawal** (with limited re-exposure), the results could also be interpreted as providing evidence for *persistent* drug-related injury.

Study #6   Frey et al (2006)   *Further Evidence of Stimulant-Induced Gliosis*

Similar to the team of Armstrong et al, a team of Brazilian scientists was curious to know more about the biological effects of *low doses of prescription stimulants*.[23]

Experimental Procedure

Adult male Wistar rats were divided into two experimental groups. Group one received a single intraperitoneal injection of saline or d-amphetamine (1 mg/kg, 2 mg/kg, or 4 mg/kg). Group two received daily injections (one per day) of the same treatments for a period of seven days. Following these brief exposures, the researchers observed the animals for changes in motor activity. Animals were then sacrificed. Brains were removed, and slices from two regions – the **cerebral cortex and the hippocampus** – were processed for microscopic analysis. The explicit goal of this experiment was the *identification of neurotoxicity*. Specifically, researchers employed a sophisticated lab technique (known as **Enzyme Linked ImmunoSorbent Assay**, or **ELISA**) to identify tissue levels of **GFAP** (glial fibrillary acidic protein).

Findings

With respect to motor movement, both acute and chronic exposures to d-amphetamine resulted in an increased level of rodent activity. (The researchers hypothesized that this might be the basis of drug-related mania in humans.) With respect to toxicity, a single injection of the stimulant did not provoke an increase in GFAP. However, **continued treatment for seven days, *irrespective of dose*, resulted in the proliferation of glial cells within the hippocampus**:

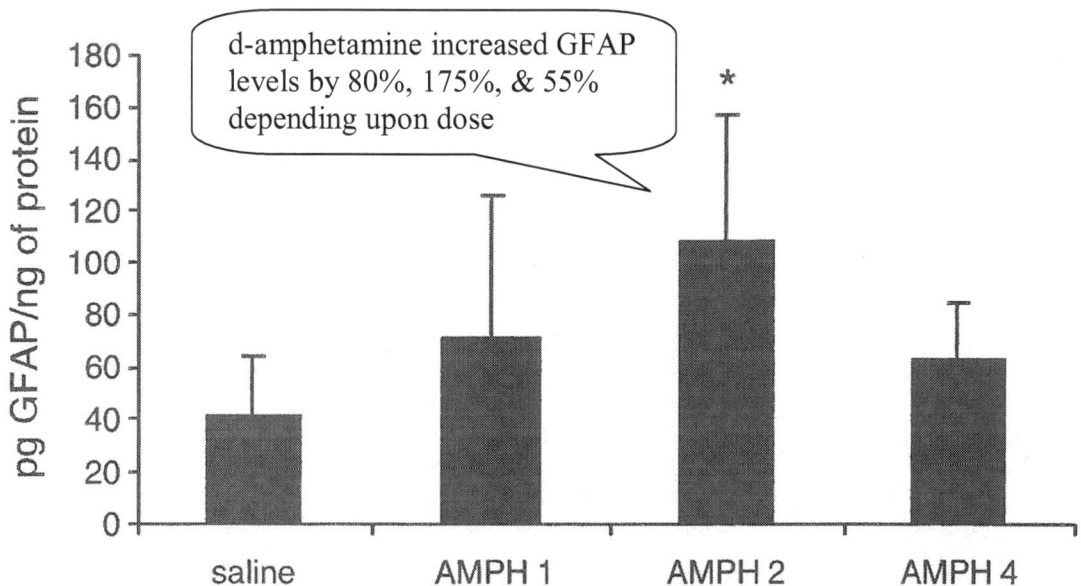

Reprinted from *Progress in Neuro-Psychopharmacology & Biological Psychiatry*, volume 30, issue 7, B.N. Frey, A.C. Andreazza, K.M. Cereser, M.R. Martins, F.C. Petronilho, D.F. de Souza, et al, "Evidence of astrogliosis in rat hippocampus after d-amphetamine exposure," page 1233, copyright (2006), with permission from Elsevier.

Hippocampal Levels of GFAP Following Chronic Treatment

This bar graph depicts brain levels of GFAP, a presumptive marker of brain injury. Each histogram depicts hippocampal changes following a 7-day period of treatment with saline or d-amphetamine (average changes observed in 10 animals, per group).

```
AMPH 1 =    1 mg/kg intraperitoneal injection each day
AMPH 2 =    2 mg/kg intraperitoneal injection each day
AMPH 4 =    4 mg/kg intraperitoneal injection each day
```

In Sum:

Using intricate lab methods to detect indirect markers of brain injury (decreased DAT density, increased GFAP levels), two teams of researchers have confirmed worrisome changes in the **striatum** and the **hippocampus** following short-term treatment with d-amphetamine.

The doses of d-amphetamine used in these experiments were low by the standards of animal experimentation, but they were clinically relevant doses for humans who consume the drug for medicinal (rather than recreational) purposes.

Both studies revealed drug-related increases in GFAP. These findings strongly implied the chemical induction of brain injury, resulting in the compensatory proliferation of astrocytes.

Based upon autopsy studies in humans in whom increased levels of gliosis correlate with the symptoms and pathologies of dementia, the detection of gliosis in these experiments was both important and ominous.

Stimulants

Neuroimaging Studies of Humans

Background Information on Growth

The medical literature features numerous neuroimaging studies in which researchers have reported the anatomical consequences of *illicit drug use*. Without exception, publications on this theme have revealed various stages and locations of brain anomalies following exposure to synthetic stimulants. Common to these reports has been the unambiguous declaration that the chemicals *were the source of brain pathology*.[24-27] The consistency of these declarations stands in stark contrast to the prevailing consensus about psychiatric drugs.

For example, whenever *prescription stimulants* have been found to produce young brains which remain small or grow smaller, the medical community's response has been to blame the patient: either bad genes or pre-existing conditions have been held responsible. An apologist for this perspective might defend it on the grounds of ambiguous research. To be fair, no scientists have yet conducted neuroimaging exams in children using *optimal* study designs, and none so far have explored the structural brain effects of *prescription* stimulants in *adults*. On the other hand, the critical observer cannot help but wonder what it means when an entire health care industry *assumes* safety where none has been appropriately demonstrated – and worse yet – disregards important evidence of iatrogenic harm. A case in point involves the drug-induced suppression of growth.

For more than fifty years, pharmaceutical companies and many clinicians denied that prescription stimulants could disturb bone metabolism. In fact, according to a Harvard doctor who proposed the "maturational delay theory of ADHD," slow skeletal development in hyperactive and poorly focused children was presumed to be the result of an inherent brain defect. Tragically, despite the existence of many case reports, case-control, and animal studies which had long suggested otherwise, it was not until the publication of two large governmental studies that the so-called KOLs (key opinion leaders) in American medicine weakly conceded that there was a problem.

The first of these federally sponsored investigations, known as the Multimodal Treatment of ADHD (or MTA) Study, involved 579 children between the ages of 7 and 10.[28-29] Although the project was initiated in 1991 as a one-year study, and follow-up examinations were subsequently performed for *an undisclosed number of years, the first published report of *growth effects* did not appear until 2004.[30]

---

* undisclosed: The MTA Study has generated numerous publications. Despite the fact that the initial paper (1999) described the study as a 14-month comparison of different therapies, the researchers have sporadically disclosed 2-, 3-, 6- and 8-year outcomes. At this point in time, there is no way of knowing how many follow-up examinations have actually been performed on the MTA cohort, nor is there any way of knowing why reports of negative outcomes and drug toxicities have been repeatedly delayed.

At that time, investigators divulged that the ADHD children had *exceeded the population norms for height and weight at the start of the project*. **In other words, none of these children had suffered from a pre-existing growth delay**. Under the influence of prescription stimulants, however, these same children developed significant reductions in *growth velocity* which persisted throughout the entire treatment period:

Growth Rate Reduction on Stimulants - MTA Study		
Year 1	Year 2	Estimated Growth Suppression Under the Influence of Stimulant Drugs
-0.90 cm per year	-1.04 cm per year	~ 1 cm (1/2 inch) per year
-2.55 kg per year	-1.22 kg per year	~ 1.25 kg (3 lb) per year

In 2007, growth outcomes for the MTA children were updated when the research team announced three-year findings in terms of z-scores (standard deviations).[31] Raw data for growth reductions per year (as reported previously) were not disclosed. Despite this obfuscation, however, the MTA team confirmed that three-year outcomes were identical to the 14-month and 24-month research results. ***Once again, they observed a persistent reduction in the growth of the chronically medicated subjects.***

Similar effects have been detected in even younger children. In a second federally sponsored experiment (the **Preschool ADHD Treatment Study**, or **PATS**), researchers from the National Institute of Mental Health and outside academic centers examined the effects of a specific stimulant (methylphenidate) upon 140 previously unmedicated children under the age of five.[32] Like the MTA study before, measurements at baseline showed that this experimental cohort *exceeded* the population norms for height and weight. ***This negated the possibility that these children had been suffering from a pre-existing maturational delay***.

Following a ten-month period of treatment with methylphenidate, preschoolers experienced a ***20% reduction in height gain*** (based upon an expected gain of 6.79 cm/year) and a ***55% reduction in weight gain*** (based upon an expected gain of 2.39 kg/year):

Stimulants

Growth Rate Reduction on Stimulants – PAT Study

	growth rate deficit	change in *percentile points	change in **z-units
height	- 1.38 cm/year (0.54")	- 7.53 pts per year	- 0.26 per year
weight	- 1.32 kg/year (2.9 lbs)	-13.18 pts per year	- 0.49 per year

* percentile points – Percentile points refer to changes in growth patterns as they appear on a standard growth chart. Healthy growth is associated with very little variation in percentile points, as a child will generally find a steady line of growth and follow it. Children growing along the 50th percentile are taller or shorter than 50% of their peers. Those growing along the 10th percentile are taller than 10%, but shorter than 90%, etc.

** z-units – a statistical method of reporting data in terms of the distribution around the mean (average score); a z-unit of 1 refers to a measurement which is exactly 1 standard deviation greater than the mean. Z-scores are used in research in order to permit comparisons between different populations in various studies.

The importance of these results can only be fully grasped in the context of normal childhood growth. **Between the ages of 4 and 10, healthy children grow at the rate of 2 to 2.4 inches (5 to 6 cm) per year.** Prior to puberty, there is a slight decrease in this rate, followed by an acceleration known as the *pubertal growth spurt*. This results in height increases of 3 to 4 inches per year (3.33 inches or 8.5 cm per year for females, 3.75 inches or 9.5 cm per year for males).[33-34] All of these changes can be appreciated on a **standard growth velocity chart**:

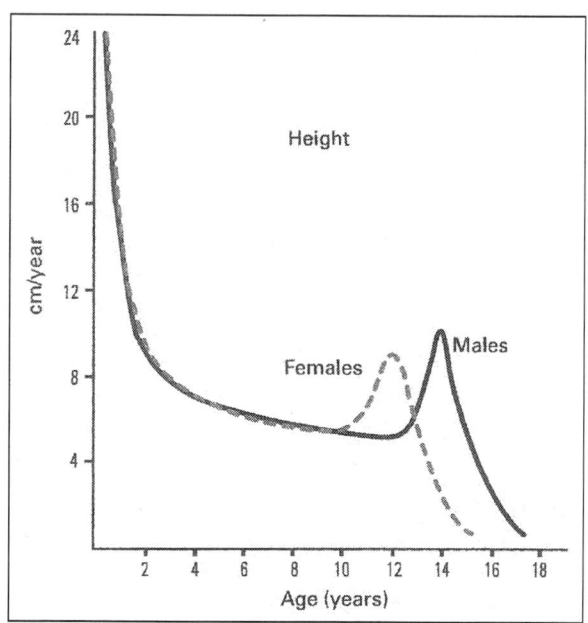

© 2009 *Australian Family Physician*. Reproduced with permission from The Royal Australian College of General Practitioners. Text and images copyright of *Australian Family Physician*. Permission to reproduce must be sought from the publisher, The Royal Australian College of General Practitioners.

Changes in Growth Velocity by Age			
birth to 12 months:	18 to 25 cm	= 7 to 10 inches	per year
12 to 24 months:	10 to 13 cm	= 4 to 5 inches	per year
24 to 36 months:	7.5 to 10 cm	= 3 to 4 inches	per year
3 yrs to puberty:	5 to 6 cm per	= 2 to 2.4 inches	per year
puberty:	8.5 to 9.5 cm	= 3.4 to 3.7 inches	per year

Ironically, physicians are trained to aggressively seek out and correct *growth changes which are inappropriate for a child's age*. The technical name for this condition is ***growth failure***.[35-36] While the authors of the MTA and PAT Studies revealed that medicated children experienced *reductions in growth velocity*, relative to their peers, an undisclosed number of these children may have also satisfied the threshold for frank *growth failure* (i.e., change in height < 2 inches per year).

However, even if one assumes that all of these children had been lucky enough to avoid this level of growth suppression, both of the government-sponsored studies still provided several legitimate reasons for concern:

> ➢ the findings invalidated the theories of several leaders in academic psychiatry, who had long held that growth retardation during stimulant therapy was the result of a pre-existing growth delay

> ➢ the findings prompted the investigators to recommend a revision of clinical practice parameters regarding the use of stimulants in children

> ➢ the findings confirmed the role of catecholamines (and catecholamine-disrupting agents) in the development of the skeletal system.

Nevertheless, after concluding that stimulants were indeed the cause of diminished somatic development, ***not a single member of the MTA or PAT research teams expressed curiosity about the possibility of bidirectional communications between the bones and the brain***. It is difficult to imagine that scientists from the National Institute of Mental Health had been unaware of advances in the field of neuroskeletal biology, but if this were true, it might account for their failure to consider the broader implications of drugs which inhibit bone growth.

Especially with respect to the drugging of children, it is significant that advances in biological research have shown that various neurochemicals – including serotonin, dopamine, and small proteins known as neuropeptides – assume critical roles in skeletal development.[37-44] Similarly, it is significant that investigators have identified substances in the skeleton (such as bone morphogenetic proteins or BMPs) which play essential roles in the development of dopamine pathways within the brain.[45-46] On this account, it is highly pertinent that scientists at the National Institute of Drug Abuse have confirmed a role for methamphetamine in suppressing one such protein (BMP7) – a finding which may explain why and how that particular stimulant suppresses recovery after brain injury.[47-48]

When one considers the totality of these developments, the revelations from neuroskeletal research underscore the urgent need for physicians to ask the following question:

Is it possible for drugs to impair the growth and remodeling of bones without simultaneously impairing the growth and remodeling of the brain?

A preliminary answer may be found in the neuroimaging studies of children who have been treated with prescription stimulants for the symptoms of ADHD.

Neuroimaging Studies of Humans

Similar to the design of epidemiological investigations, neuroimaging studies exist in two varieties. ***Cross-sectional studies*** refer to comparisons between subjects based upon *scans obtained at one point in time*. ***Longitudinal studies*** refer to comparisons within and between subjects based upon *repeated brain scans over some interval of time*.

Background Information: Normal Brain Development

By employing a variety of magnetic resonance imaging (MRI) techniques, brain mappers have typically measured the size of various brain regions in subjects across the lifespan. Using this raw data, they have then applied mathematical formulas in order to generate curves which correspond to various patterns of neurodevelopment.[49]

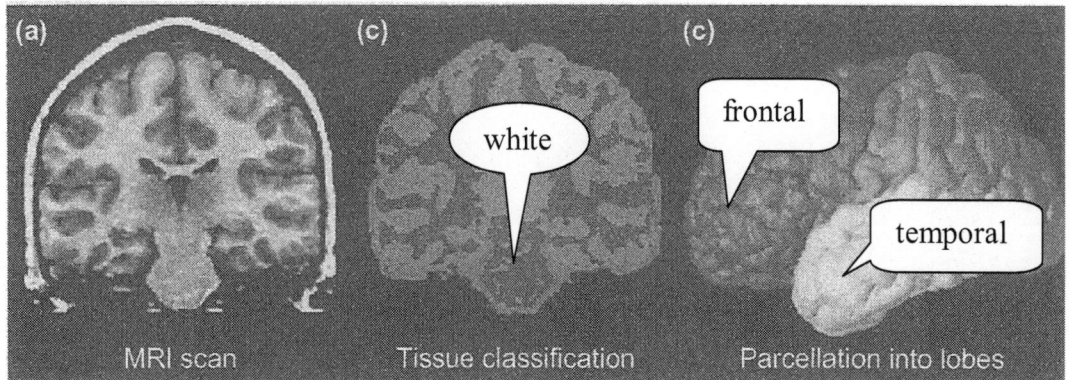

Reprinted from *Trends in Neuroscience*, volume 29, issue 3, A.W. Toga, P.M. Thompson, and E.R. Sowell, "Mapping brain maturation," page 149, copyright (2006), with permission from Elsevier.

How Brain Mappers Make Pictures of the Brain

Far Left: ***MRI image of the human brain (coronal section):***
mappers begin with an ordinary brain scan (black/white/gray)

Center: ***tissue classification:*** researchers employ various methods to identify brain tissue according to the intensity of its appearance on the MRI. Computer software can be used to assign different colors, such as green for *gray matter* (neurons, astrocytes, microglia); blue for *white matter* (myelinated axons, oligodendrocytes).

Far Right: ***lobar parcellation:*** Using various computer techniques, researchers can also assign colors to different lobes or regions of the brain: blue (frontal), green (parietal), yellow (temporal), and red (occipital).

Based upon maps of various brain sizes and features as they change according to age, research teams have been able to determine several properties which are common to healthy or so-called "normal" brain development. These can be summarized succinctly by the following rules and pictures.

Rule #1: Different Brain Regions Mature at Different Rates

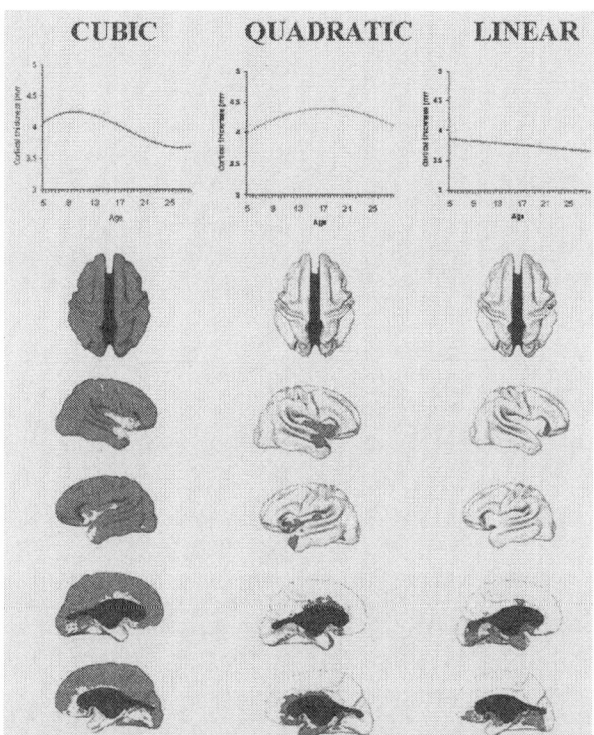

Reprinted from *The Journal of Neuroscience*, volume 28, issue 14, P. Shaw, N.J. Kabani, J.P. Lerch, K. Eckstrand, R. Lenroot, N. Gogtay, et al, "Neurodevelopmental Trajectories of the Human Cerebral Cortex," page 3588, copyright (2008), with permission of the Society for Neuroscience.

According to researchers from the National Institute of Mental Health (NIMH), the development of the cerebral cortex assumes three different curves or trajectories (depicted above).[50] The shape of these curves is believed to correspond to the complexity of the underlying tissue. Pertinent to the studies of children who have been diagnosed with ADHD, investigators have found that the cortex expands dramatically in size (breadth and depth) from infancy to early childhood. The emergence of *peak thickness*, and the *timing* of its resolution, varies according to brain region. **In other words, each brain zone has its own unique schedule of development**. For example, Giedd et al (1999) observed the following variations in the timing of *peak gray matter volume* according to location:[51]

	frontal lobe	parietal lobe	temporal lobe	occipital lobe
peak gray matter	age 12	age 12	age 16	after age 20

Rule #2: Gray Matter and White Matter Mature at Different Rates

According to researchers at UCLA, **brain volume** expands between infancy and puberty (~ age 11).[52] In general, **gray matter** proliferates rapidly until the age of 7, declines slightly through adolescence (as neuronal connections are eliminated via a process called synaptic pruning), and finally plateaus until the seventh decade of life. **White matter** (so-named for the appearance of the fatty layer of insulation which speeds transmission along the axons of the brain) increases from birth until well into middle age (mid-40s). **Cerebrospinal fluid** increases in a fairly linear fashion across the lifespan:

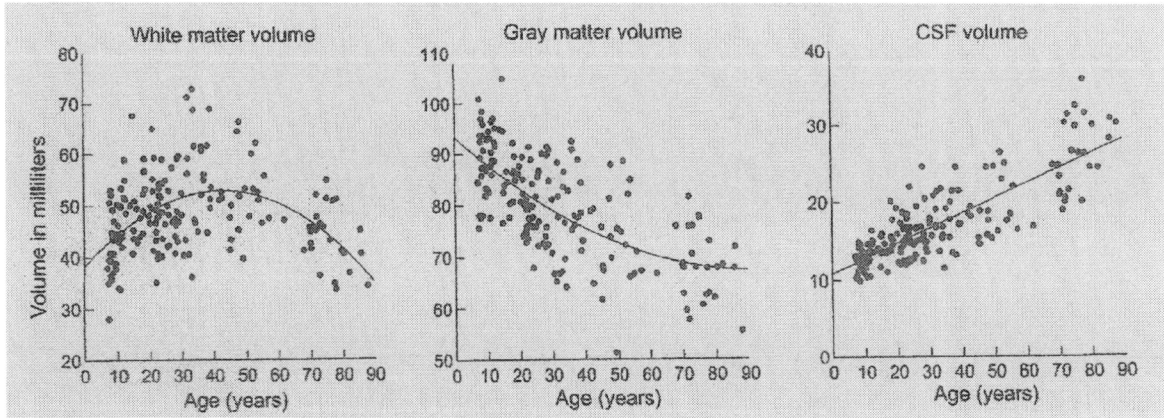

Reprinted by permission from Macmillan Publishers Ltd: *Nature Neuroscience*, E.R. Sowell, B.S. Peterson, P.M. Thompson, S.E.Welcome, A.L. Henkenius, and A.W. Toga, "Mapping cortical change across the human life span," volume 6, issue 3, page 314, copyright (2003).

Rule #3: Progressive and Regressive Processes Occur Simultaneously

Between the ages of 7 and 30, progressive and regressive changes occur simultaneously within the brain.[53-55] Cortical thickness (outer layer on the MRI) expands at the rate of 1 mm per year until puberty. However, following this period of growth, complex changes occur in the *composition of the underlying tissue*. As neuronal connections are eliminated (pruned) during adolescence, white matter is *simultaneously* added to the cells. This results in a *declining ratio of gray matter to white matter* in the brain. On an MRI, this changing ratio appears as the progressive *thinning of the cortical layer*. The rate and timing of **cortical thinning** varies according to brain region. Researchers at UCLA have documented the appearance of brain maturation between the ages of 5 and 11. In their investigation of 45 children, maximal gray matter loss occurred in the dorsolateral prefrontal, parieto-occipital, and inferior temporal regions at the rate of 0.1 to 0.3 mm per year. At the same time, maximal gray matter gain (thickening) occurred in the left inferior frontal and left posterior temporal regions (0.05 mm to 0.2 mm per year).

Stimulants

Rule #4: Intellectual Ability May Correspond to the Pattern of Brain Development

Although its design and findings have not yet been replicated, a recent investigation by scientists at NIMH explored changes in cortical development and intellectual ability.[56] The study involved the neuropsychological testing and neuroimaging of 220 children, ranging in age from 7 to 16. On the basis of standardized "IQ" testing, the subjects were stratified into three groups according to the Wechsler intelligence scale:

83-108	= average intelligence
109-120	= high intelligence
121-149	= superior intelligence

All subjects were scanned by MRI on one or more occasions (58% received at least two scans) with repeat imaging conducted at an average interval of two years. Growth curves were then constructed in order to depict age-related changes in cortical thickness and thinning. Three major findings were observed: 1) the **magnitude of peak thickness** *was greatest in children with high and superior intelligence*; 2) the **timing of peak thickness** *occurred later in children with high and superior intelligence*; and 3) the **speed of cortical thinning** (shown below) *was accelerated in children with superior intelligence*.

	superior intelligence	average intelligence
peak thickness	4.45 to 4.8 mm	4.35 to 4.75 mm
age at peak thickness	11 to 12	prior to age 6
rate of cortical thinning	> 0.05 mm/yr (age 14-16) > 0.08 mm/yr (age 16-19)	0 to 0.05 mm/yr

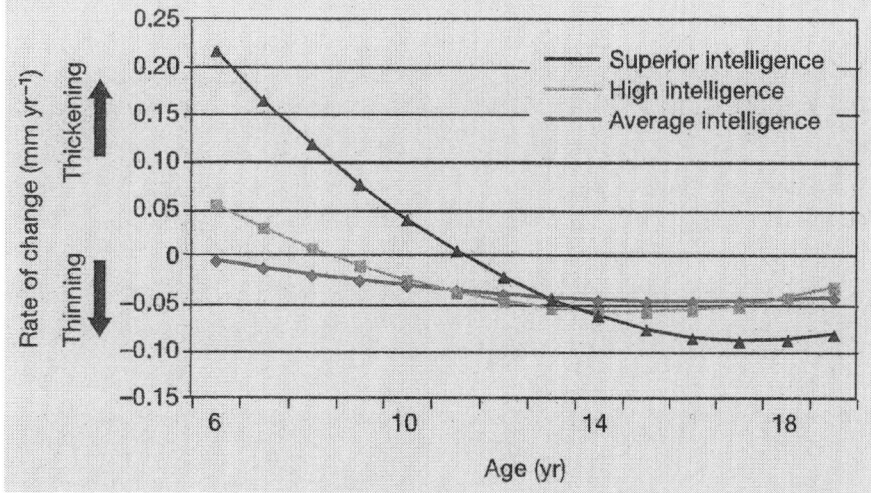

Reprinted with permission from Macmillan Publishers Ltd: *Nature*, P. Shaw, D. Greenstein, J. Lerch, R. Lenroot, N. Gogtay, A. Evans, et al, "Intellectual ability and cortical development in children and adolescents," volume 440, page 677, copyright (2006).

Cross-Sectional Neuroimaging Studies of Children Diagnosed with ADHD

Despite the fact that stimulants have been used to manage the behavior of children since the 1960s, there have been very few studies comparing the brains of drug naïve (never medicated) subjects relative to medicated peers and non-ADHD controls. Without these types of investigations, clinicians and patients cannot clearly know the kinds of changes which are caused by pharmaceuticals, rather than by other biological and environmental factors.

Cross-sectional imaging studies are limited by the fact that they provide snapshots of brain features at only one point in time. Because they do not compare views of the brain from the same person "before and after" drug use, they cannot *definitively* confirm the anatomical effects of medication. However, cross-sectional studies may at least permit an answer to the following question:

> *Does exposure to a given chemical enhance or hinder age-appropriate development of the brain?*

With respect to the use of stimulants in children diagnosed with ADHD, dozens of cross-sectional neuroimaging investigations have been performed and reported over the past 30 years. As far as this writer is aware, **none of these studies has detected a restorative or growth-enhancing effect of medication**:

Cross-Sectional Studies – A Sampling [57-62]

Study Group	Study Design	Key Findings in ADHD
Ashtari et al (2005)	*diffusion tensor imaging (MRI)*	
18 ADHD vs. mean age: 8y11m **72% on stimulants in past** average duration: 1 ½ years **67% on stimulants at time of scan** height: 138.4 cm weight: 30.65 kg	18 controls 9y2m 139.95 cm 38.39 kg	reduced white matter in right premotor cortex, right internal capsule, right and left pons, anterior lobe of cerebellum, and left parieto-occipital region
Carmona et al (2005)	*optimization voxel based morphometry (MRI)*	
25 ADHD vs. ages 6 to 16 21 males, 4 females **100% medicated at time of scan**	25 controls ages 6 to 16 21 males, 4 female	5% smaller brain volume, 5% reduction in gray matter, 6% reduction in white matter; predominant decreases noted in frontal lobes

Stimulants

Cross-Sectional Studies (continued)		
Study Group	**Study Design**	**Key Findings in ADHD**
Durston et al (2004)	*volumetric analysis (MRI)*	
30 ADHD vs. 30 controls mean age: 12y1m 10y7m **90% on stimulants at time of scan** Tanner stage: 1.6 Tanner stage: 1.8		transient motor tics (27%); *sexual development delayed* (Tanner stage); all brain region volumes and gray matter reduced in ADHD children by 1-4%, ***despite being older by 1 ½ years***
Mostofsky et al (2002)	*volumetric analysis (MRI)*	
12 ADHD vs. 12 controls 8y1m-13y10m (8y4m-13y7m) mean: 10y1m 10y2m **83% on drugs at time of scan**		7-11% reduction in total brain volume, total gray matter, and total white matter
Pliska et al (2006)	*volumetric analysis (MRI)*	
14 drug-naive ADHD (age 12y6m) vs. 21 controls 16 medicated ADHD (age 12y9m) (age 13y2m) duration of drug use: 2 to 8 years		ADHD children had ↓ caudate volumes, 16% smaller than controls (drug therapy failed to normalize size of caudate)
Sowell et al (2003)	*brain surface morphometry (MRI)*	
27 ADHD vs. 46 controls 11y7m-12y10m 11y10m-12y2m **56% on stimulants at time of scan** high lifetime history of drug treatment		smaller brain surfaces; delayed maturation of posterior temporal lobes, reduced total white matter

The collective body of evidence from these and other cross-sectional studies refutes the premise that stimulants either prevent or reverse abnormal brain development. In fact, any one of the aforementioned findings – such as the alarming rate of motor disturbances (motor tics), delayed sexual maturation, and impaired myelination – would have been reason enough for the writers and journal editors to expound upon the potential hazards of drug therapy. Curiously, researchers have not only failed to question the safety of drugging children and teenagers, but they have also repeatedly emphasized why stimulants could *not* have contributed to the observed anatomic anomalies.

For example, in their 2003 report which appeared in the British medical journal, *The Lancet*, the California team of Sowell et al made the following remarks:

> "The effects of stimulant drugs could have confounded our findings of abnormal brain morphology in children with attention-deficit hyperactivity disorder, because 15 of the 27 patients with attention-deficit hyperactivity were taking a stimulant at the time of imaging…Results of a comprehensive report, however, suggest that stimulant drugs do not contribute to group differences in brain morphology in patients with attention-deficit hyperactivity disorder." [63]

Influenced by the same "comprehensive report," the Spanish team of Carmona et al conveyed similar beliefs in their 2005 article published in *Neuroscience Letters*:

> "It is improbable that our findings could be attributed to the psychostimulant treatment. In other studies, ADHD GM [gray matter] abnormalities did not differ significantly between medicated and unmedicated patients." [64]

However, a close inspection of the favored report (Castellanos et al 2002) supports a far different conclusion.

Longitudinal Neuroimaging Studies of Children Diagnosed with ADHD

Notwithstanding the fact that numerous cross-sectional investigations have failed to demonstrate a "curative" effect of stimulant medications, with respect to delayed or atypical brain development, academic psychiatrists have consistently stated the opposite. In fact, in the federal government's press release announcing the publication of its first major *longitudinal* ADHD brain-scan study (October 8, 2002), the leader made the following statement:

> "'There is no evidence that medication harms the brain,' said Castellanos, who conducted the study at NIMH before joining New York University. 'It's possible that medication may promote brain maturation.'" [65]

Other members of the NIMH research team supported this view:

> "'Fundamental developmental processes active during late childhood and adolescence are essentially healthy in ADHD.'" [66]

Subsequently, the content of the governmental press release was uncritically regurgitated by members of the national news media (*New York Times*, *Detroit Free Press*) who apparently never grasped the methodological flaws and full findings of the study. Had medical reporters performed their jobs responsibly and competently, the public might have received a far different message than that which was conveyed by the title of the NIMH Press Release: **"Brain Shrinkage in ADHD Not Caused by Medications."**

... *What did the results of the Castellanos Study really show?*

Study #1   Castellanos et al (2002)   *Evidence of Prolonged Growth Delay*

Initiated in 1991 and continued until 2001, the Castellanos Study enrolled 152 children with the diagnosis of ADHD and 139 non-psychiatric controls.[67] Originally planned to provide longitudinal information about the development of "the ADHD brain," based upon the expectation of obtaining serial brain scans, the investigators were able to procure repeat imaging in only 60% of the sample. This resulted in the mixing of cross-sectional and longitudinal data via elaborate statistical techniques.

Widely hailed by the news media and the federal government as the largest study to ever compare unmedicated and medicated children, the integrity of the project was severely marred by four major limitations: 1) a failure to perform appropriate age matching; 2) a failure to compare patients according to *current* versus *past* stimulant use; 3) a failure to provide information about long-term treatment status; and 4) a failure to include in the final statistical analyses data from all of the participants.

Failure to Perform Appropriate Age Matching

The investigators at NIMH failed to enroll equal numbers of children matched appropriately, according to age:

	mean age of subjects in each subgroup
ADHD, prior drug treatment	10 years 11 months
ADHD, no prior drug treatment	8 years 4 months
non-psychiatric control	10 years 6 months

The importance of this error hinged upon the fact that the NIMH team repeatedly compared the features of older, medicated children with unmedicated patients who were 2 ½ years younger (and hence, 2 ½ years behind them in brain development). This was one source of the claim that "Brain Shrinkage in ADHD is Not Caused by Medication." The other source of the claim was a retrospective massaging of *partial datasets*.

To compensate retroactively for these age disparities, the NIMH researchers performed a special subgroup analysis in which they examined the brain sizes of children matched precisely for age. ***Unfortunately, this subgroup analysis excluded 25 of the 49 unmedicated subjects***. In other words, what might have been the "largest ever study of unmedicated children" failed because of poor study design. The subgroup analysis was also complicated by a failure to match children according to *age **and** gender*. The integrity of the subgroup analysis was further hindered by the investigators' failure to consider (or to report) other critical influences upon brain size, including height, skull size, and pubertal status.

Castellanos Study: *Covert Subgroup Analysis			
	ADHD children never medicated	ADHD children **medicated	controls
total # of children in each group	49	103	139
# included in special subgroup analysis	24	50	54
% of children excluded	**51%**	**51%**	**40%**

* *covert* = NIMH has not published the raw data from this analysis

** **medicated** = in a footnote which appeared below a table in the *JAMA* publication, the NIMH researchers divulged that this category included children with ***current and/or prior*** treatment with stimulant medications

Failure to Compare Children According to Current vs. Past Use of Stimulants
Although the subgroup analysis was performed in an effort to match children appropriately (by age), the researchers failed to segregate subjects according to the *timing of drug treatment*. When they created the categories of medicated vs. drug-naïve patients, the former subgroup included all of the ADHD children with ***current and/or past stimulant use***. This necessarily confounded the brain imaging results, as the medicated subgroup must have included findings from formerly drugged children whose brains were experiencing rebound growth. In other words, by mixing data from those who had stopped taking stimulants, with data from children who remained medicated, the NIMH researchers decreased the ability of the study to produce a valid assessment of chemical effects upon the anatomy of the immature brain.

Failure to Provide Data About Long-Term Medication Use
When the researchers moved on to the longitudinal (rather than baseline) differences between the ADHD subjects and controls, they failed to provide information about longitudinal exposures to drug therapy. At no point in the published report which appeared in *JAMA (Journal of the American Medical Association)* did the authors divulge how many ADHD patients remained on stimulants or stopped stimulants in between the first and subsequent brain scans.

Failure to Include Data From All Subjects
The brain-growth curves (see below) which have become the focus of drug advocates were, in fact, a reflection of a highly selective dataset. Few commentators seem to appreciate the fact that the ***graphs which were reproduced in the JAMA article omitted results from more than 10% of the available scans***. How this occurred requires some explanation.

An unavoidable limitation of radiological studies is the vulnerability of the technology to movement artifacts. The excessive movement of subjects in the Castellanos Study resulted in the ***exclusion of images from 11% of the ADHD children***, and from *6% of the normal controls*. Next, the researchers focused upon the construction of graphs to depict the trajectory of brain development. For the sake of producing a quadratic curve (inverted U), the team members ***omitted brain scan data from the youngest and oldest subjects in each subgroup, in order to achieve the best "fit" between their results and the curve-generating, mat hematical formula***:

> "....graphs of developmental curves are restricted to the central 90% of each sample's age distribution because fitted polynomial curves may be heavily influenced by outliers at the age range extremes..." [68]

Findings

Despite the methodological confounders reviewed above – all of which served to *minimize* differences between medicated and drug-free children, and all of which undermined the study's interpretability– several findings remained disquieting.

While it was true that the baseline (starting) brain size of medicated children was generally larger than the drug-naïve subjects, it is also true that this was **not** a universal finding throughout the brain. For example, never-medicated children had more total gray matter (by volume), larger caudates, and arguably more appropriate cerebellar development [never-medicated children were 2 years younger than controls and had 6% smaller cerebella; medicated children were 5 months *older* than controls but still had 4% smaller cerebella].

Interestingly, when compared to the ADHD subjects exposed to stimulants, the drug-naïve children demonstrated superior linguistic ability based upon a standardized test of intelligence (Wechsler Intelligence Scale, vocabulary subtest):

	ADHD drug naive n = 49	ADHD stimulant-exposed n = 103	Controls n = 139
# of females	22 (45%)	41 (40%)	56 (40%)
# of males	27 (55%)	62 (60%)	83 (60%)
average age	8 yrs 4 months	10 yrs 11 months	10 yrs 6 mo
vocabulary	12.5	11.5	12.55
total gray matter (volume)	704.2 cc	699.3 cc	727.9 cc
caudate (volume)	10.50 cc	10.29 cc	10.75 cc
cerebellum (volume)	121.8 cc	125.1 cc	129.8 cc

Stimulants

The highlight of the Castellanos Study was supposed to have been the objective analysis of longitudinal changes in the brains of ADHD children, relative to non-psychiatric controls.

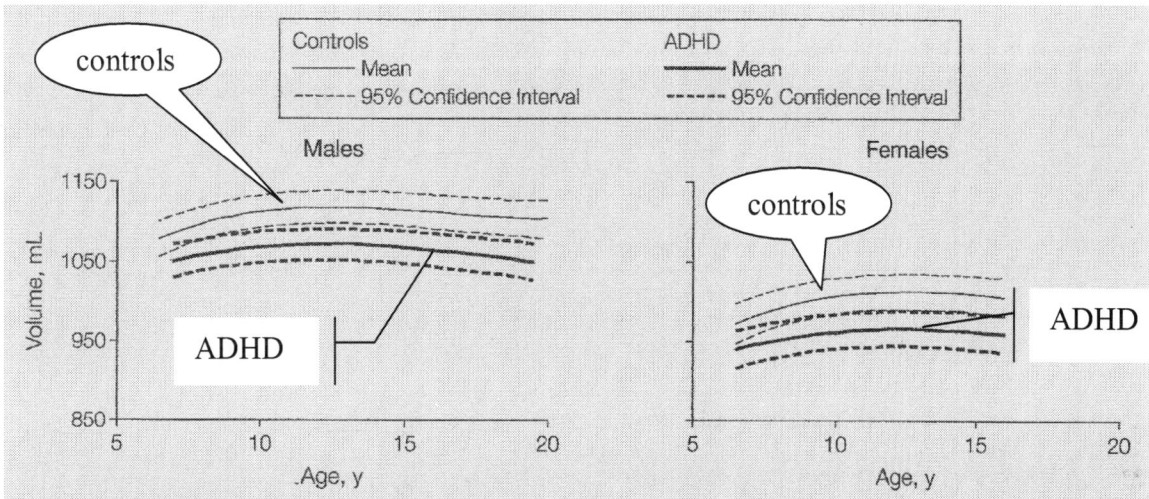

Reprinted with permission, F.X. Castellanos, P.P. Lee, W. Sharp, N.O. Jeffries, D.K. Greenstein, L.S. Clasen, et al, "Developmental Trajectories of Brain Volume Abnormalities in Children and Adolescents With Attention-Deficit/Hyperactivity Disorder," *JAMA*, volume 288, issue 14, page 1746, October 9, copyright © (2002) American Medical Association. All rights reserved.

Castellanos Study – Changes in Brain Volume According to Age

These images demonstrate the larger brain volumes of normal controls (top line) versus the smaller brain volumes (bottom line) of the children with ADHD.

> male age range: 5 years 1 month to 18 yrs 5 months
> female age range: 5 years 4 months to 16 years

Based upon the finding that the trajectory of *total brain volumes* in the ADHD patients remained *parallel* with normal controls, the NIMH researchers proposed that "fundamental developmental processes" had remained unchanged. **Their conclusion was unjustified for the following reasons.**

A parallel growth curve does not invalidate the existence of an underlying pathological process. Within the field of endocrinology, medical professionals use the term *canalization* to refer to the tendency of the human body to maintain a narrow and predictable track of growth. This can occur even while the organism is developing in a restricted fashion in the context of illness (respiratory, renal, gastrointestinal), malnutrition, and/or endocrine disease. Moreover, the fact that the NIMH investigators mixed data from pre-pubertal, pubertal, and post-pubertal children doubtlessly impaired their ability to detect growth-rate divergence between drug-naïve and medicated children. Data from medicated children who were undergoing a pubertal growth spurt, combined with data from previously medicated children who were experiencing rebound ("catch-up") growth, had to have overwhelmed the data from younger subjects who remained under the heaviest influence of drug-related endocrine dysregulation. (In other words, by failing to match children according to pubertal status, the researchers introduced yet another confound which masked the full neuroskeletal toxicity of stimulants.)

Although the NIMH press release and interviews emphasized normal development among children exposed to stimulants, the government possessed data which showed otherwise. In a separate analysis of long-term changes in the size of the cerebellum, the growth curves of the ADHD children (68% exposed to stimulants at baseline) clearly diverged from controls after puberty:

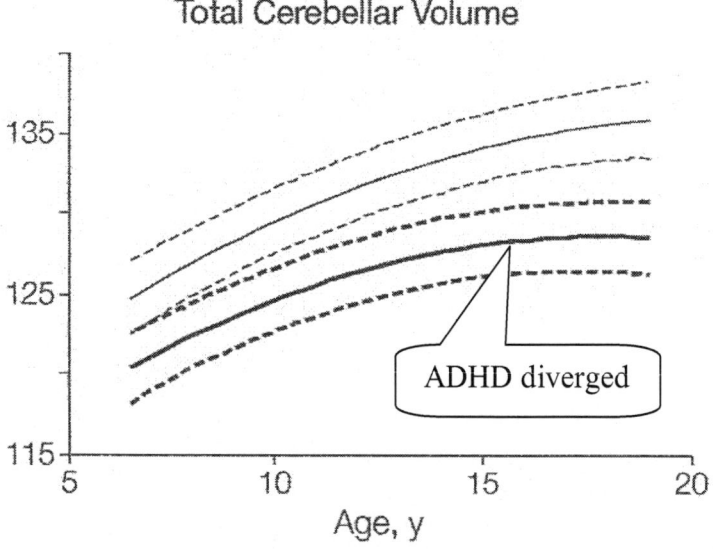

Reprinted with permission, F.X. Castellanos, P.P. Lee, W. Sharp, N.O. Jeffries, D.K. Greenstein, L.S. Clasen, et al, "Developmental Trajectories of Brain Volume Abnormalities in Children and Adolescents With Attention-Deficit/Hyperactivity Disorder," *JAMA*, volume 288, issue 14, page 1746, October 9, copyright © (2002) American Medical Association. All rights reserved.

Changes in Cerebellar Volume According to Age

top solid line = controls   bottom solid line = ADHD patients

In Sum:

The Castellanos Study has been repeatedly cited by the news media and by health care professionals as the largest longitudinal comparison of unmedicated and medicated ADHD children. However, the term medicated was applied ambiguously, as it included patients with *past and/or current drug exposures*.

Baseline measurements of brain features (total volumes, gray matter, white matter) were difficult to interpret due to a variety of flaws in the study design: a failure to perform adequate age-matching, a failure to segregate medicated subjects according to past or current drug use, and a failure to compare subjects according to other critical variables (height, skull size, pubertal status).

A separate subgroup analysis compared 24 of the 49 drug-naïve ADHD children against unequal numbers of ADHD subjects with past or continuing drug use, and without simultaneously matching for height, gender, and pubertal status. Reporting only that this subgroup analysis produced results which were "essentially" unchanged from non-age-matched comparisons, this became the basis of the government's claim that "medications do not harm the brain." Of concern, the researchers did not divulge the raw data from this assessment.

Despite the numerous problems with methodology, the study still documented toxic drug effects. At baseline, **never-medicated ADHD subjects had more gray matter, larger caudates, and superior linguistic ability** than children with past or continuing exposures to stimulants.

The longitudinal analysis of brain development was also marred by poor study design. Specifically, the researchers failed to report the treatment status of patients over time (how many children stopped or continued stimulants) and failed to produce curves which compared the growth of ADHD subjects according to this variable. Furthermore, in order to fit data to a mathematical formula which generated a quadratic (inverted U) growth curve, the investigators omitted brain scan results from a substantial number of subjects.

Even with the numerous confounders which undermined the integrity of the growth curve analysis, however, **parallel trajectories demonstrated the continuing suppression of brain development under the influence of past or current treatment with stimulants**.

The cerebellum was especially vulnerable to medication. After puberty, the trajectory of cerebellar growth in the ADHD subjects (68% of whom had past exposure to stimulants) diverged from controls, suggesting a long-lasting, dystrophic drug effect.

Study #2   Mackie et al (2007)   *Atrophy of the Cerebellum*

Presumably reacting to their earlier discoveries of impaired cerebellar development in children with ADHD, the NIMH team undertook a strictly longitudinal study of 36 psychiatric subjects and 36 controls.[69] Unique to this investigation was the fact that every participant underwent at least three MRIs. These were performed at 3-year intervals between the ages of 10 ½ and 16 ½ years. Also unique was a specific focus upon changes in discrete regions of the cerebellum:

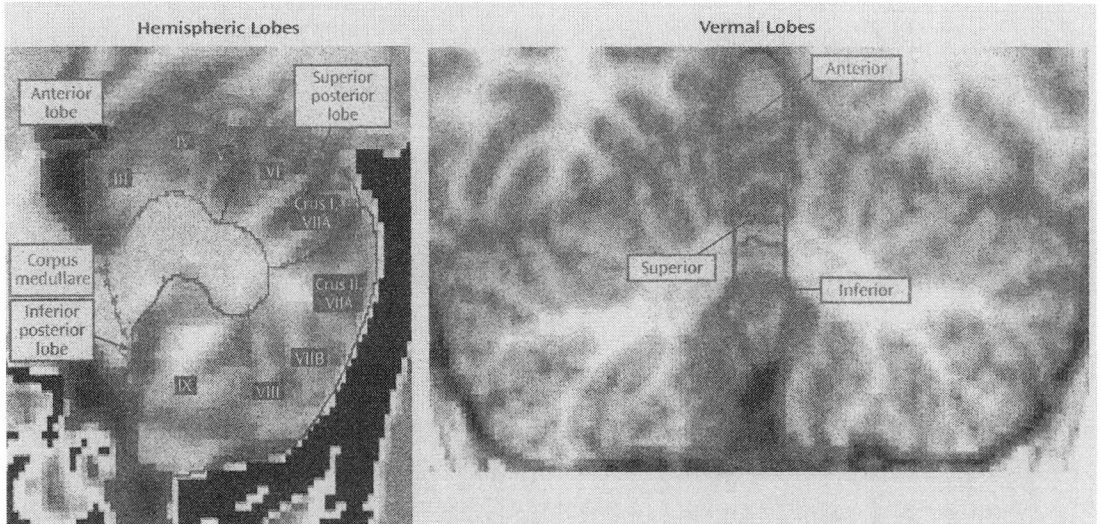

Reprinted with permission from the American Journal of Psychiatry, (Copyright 2007). American Psychiatric Association. S. Mackie, P. Shaw, R. Lenroot, R. Pierson, D.K. Greenstein, T.F. Nugent III, et al, "Cerebellar Development and Clinical Outcome in Attention Deficit Hyperactivity Disorder," *American Journal of Psychiatry* 164 (2007): 649.

Magnetic Resonance Imaging: Regions of the Cerebellum

Left: hemispheric lobes of the cerebellum (anterior, superior posterior, inferior posterior)

Right: lobes of the cerebellar vermis (anterior, superior, inferior)

Limitations

The NIMH researchers failed to provide data about the height, skull size, or pubertal status of the participants. Seventy per cent of the ADHD subjects had histories of prior exposure to stimulants at the time of the first brain scan, and ***ninety-four per cent of the ADHD patients were consuming stimulants at the time of last scan***. Unfortunately, no data were provided for baseline brain features, and no subgroup analyses were reported for the children who avoided stimulants completely.

Findings

Based upon the "average" volume of brain regions measured repeatedly over six years, the ADHD children showed diminished features (smaller total gray matter, smaller total white matter) when compared to controls. Most importantly, the ADHD children exhibited **consistent reductions in all but one region of the cerebellum** (right anterior hemisphere).

Demographics and *Average Brain Features		
	ADHD n = 36	controls n = 36
mean age at first scan	10 yrs 6 mo	10 yrs 4 mo
males/females	38.9%	38.9%
prior exposure to stimulants	70%	none
# with 3 MRI scans	100%	100%
# with 4 MRI scans	39%	50%
gray matter + white matter	**1118 cc**	1144 cc
total cerebellum	**137 cc**	140.5 cc
vermis, whole	**8.72 cc**	9.62 cc
vermis, superior	**2.18 cc**	2.56 cc

* average = average of baseline data + data from serial scans

Using the Children's Global Assessment Scale (CGAS) – a numeric tool employed by clinicians to rank children's social, academic, and behavioral functioning – the researchers assigned patients to two subgroups.[70] Those scoring less than 62 points on the CGAS qualified for the *worse outcome* category. Those scoring 62 or above were assigned to the *better outcome* category. The purpose of this analysis was to investigate possible links between regional changes in brain size and clinical results.

Notably, this element of the study showed that ***poorer outcomes were associated with earlier (and potentially longer) exposures to stimulant medication***. Children with poorer outcomes were also more apt to receive treatment with additional psychiatric drugs:

	ADHD worse outcomes n = 18	ADHD better outcomes n = 18
mean age at first scan	10 yrs	11 yrs 1 mo
time between first stimulant exposure and first brain scan	3 yrs 3 mo	2 yrs 9 mo
prior exposure to stimulants	67%	72%
mean age at start of stimulants	6 yrs 9 mo	8 yrs 4 mo
# taking anti-adrenergic drug	2 (11%)	0
# taking anti-anxiety drug	1 (6%)	0
# taking antidepressant drug	3 (17%)	2 (11%)

As they had done in previous pursuits, the NIMH research team applied a mathematical formula to model the longitudinal trajectory of brain changes. *On this occasion, the investigators excluded 20% of the data (retaining the middle 80%) in order to produce a quadratic curve (inverted U).* Drug-related deficits were still observed:

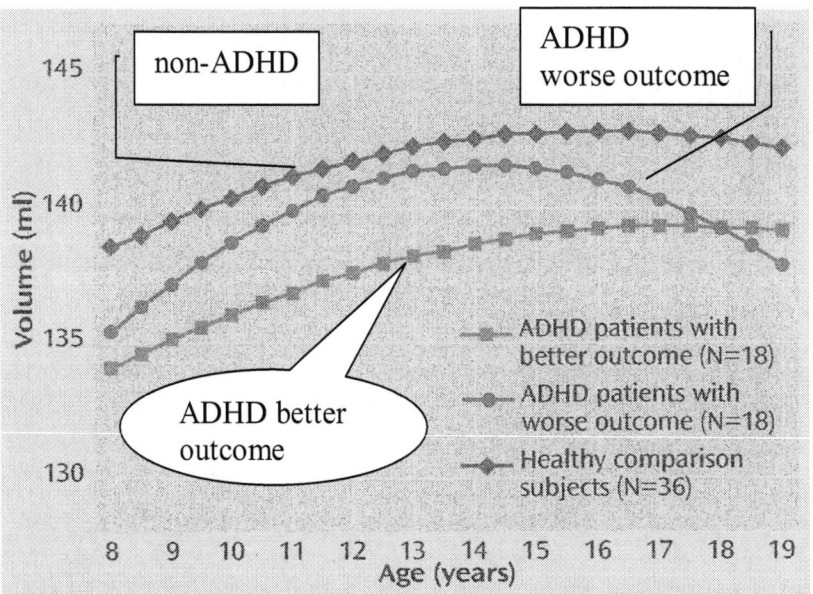

Reprinted with permission from the American Journal of Psychiatry, (Copyright 2007). American Psychiatric Association. S. Mackie, P. Shaw, R. Lenroot, R. Pierson, D.K. Greenstein, T.F. Nugent III, et al, "Cerebellar Development and Clinical Outcome in Attention Deficit Hyperactivity Disorder," *American Journal of Psychiatry* 164 (2007): 652.

Changes in Whole Cerebellar Volume by Age and Condition

All of the ADHD children exhibited **smaller total cerebellar volumes**, relative to controls (70% of the ADHD children began the study with past or continuing use of stimulants; 94% were taking stimulants at the time of the last brain scan).

Stimulants

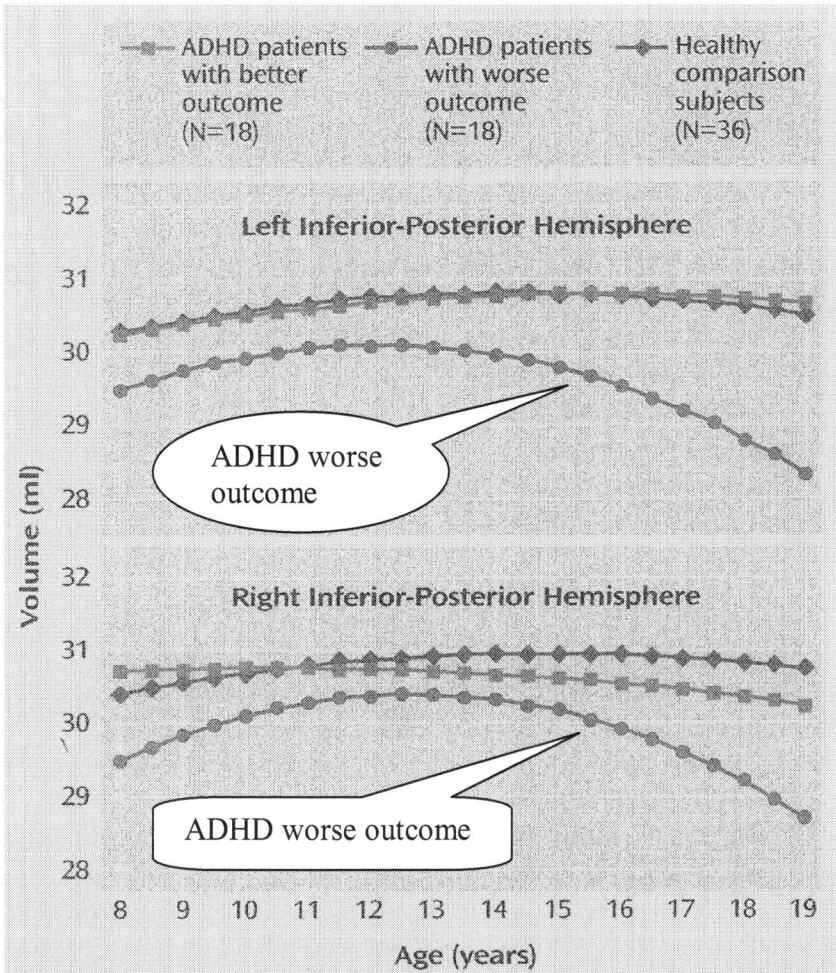

Reprinted with permission from the American Journal of Psychiatry, (Copyright 2007). American Psychiatric Association. S. Mackie, P. Shaw, R. Lenroot, R. Pierson, D.K. Greenstein, T.F. Nugent III, et al, "Cerebellar Development and Clinical Outcome in Attention Deficit Hyperactivity Disorder," *American Journal of Psychiatry* 164 (2007): 653.

Changes in the Inferior-Posterior Hemispheres of the Cerebellum

Under the influence of stimulants, all of the ADHD patients – "better" and "worse" outcomes – experienced the progressive loss of tissue from the right inferior-posterior hemisphere of the cerebellum. These changes appeared to escalate after puberty.

Studies #3-4 Shaw et al (2006, 2007)   *Stimulants Impair Cortical Maturation*

As neuroimaging techniques advanced in the new millennium, permitting the evaluation of changes in *cortical structure*, various teams of researchers became interested in characterizing the **peak thickness** of brain tissue in children with ADHD and non-ADHD controls. One of these studies, conducted by NIMH investigators, enrolled more than 400 participants between the ages of 5 and 20.[71] Twenty-five percent of these children underwent neuroimaging repeatedly at an interval of three years.

Longitudinal Study: Timing of Peak Cortical Thickness (2007)		
	ADHD n= 223	controls n = 223
mean age at first scan	10 yrs 2 mo	10 yrs 7 mo
# of male subjects	63%	63%
prior stimulant exposure:	66%	none

While the researchers did not report *baseline thickness* of the cortex for drug-naïve versus formerly medicated patients, and while they did not track the trajectories of thickness according to drug status (i.e., continuing treatment versus medication withdrawal), they did report one clear and relevant finding. ***Among the ADHD children (66% with past histories of stimulant use), the maturation of the cortex was delayed by approximately three years***:

Longitudinal Study: Timing of Peak Cortical Thickness (2007)		
	ADHD n = 223	controls n = 223
median age for peak thickness:		
cortex (overall)	10 yrs 6 mo	7 yrs 6 mo
prefrontal	10 yrs 5 mo	7 yrs 6 mo
middle/superior temporal cortex	10 yrs 7 mo	6 yrs 10 mo
primary motor cortex	7 yrs	7 yrs 5 mo

By itself, delayed maturation would not necessarily be harmful, provided that "normal" peak thickness could eventually be attained. Although Shaw et al failed to divulge the **magnitude of cortical thickness** in the aforementioned subjects (2007), a smaller study which they had described the previous year (2006) provided cause for alarm.[72] The NIMH researchers again failed to disclose raw data for the drug-naïve versus medicated subjects, and again failed to convey how many children were taking stimulants at the time of follow-up. Nevertheless, the results which they *did* report suggested that the past or continuing use of stimulants contributed to persistent suppression of cortical growth (peak thickness).

Longitudinal Study of Cortical Thickness (2006)

	ADHD n = 163	controls n = 166
mean age at first scan	10 yrs 1 mo	10 yrs 5 mo
prior stimulant use	**66%**	none
follow-up stimulant use	not revealed	none
# with at least 2 scans	60%	56%
interval between first 2 scans	2 yrs 10 mo	2 yrs 4 mo
statistically significant differences in mean thickness:		
cortical thickness (total)	**4.06 mm**	4.15 mm
right superior/medial prefrontal cortex	**4.15 mm**	4.69 mm
left superior/medical prefrontal cortex	**4.34 mm**	4.51 mm
right anterior/medial temporal cortex	**3.98 mm**	4.20 mm
left precentral cortex	**3.73 mm**	3.85 mm

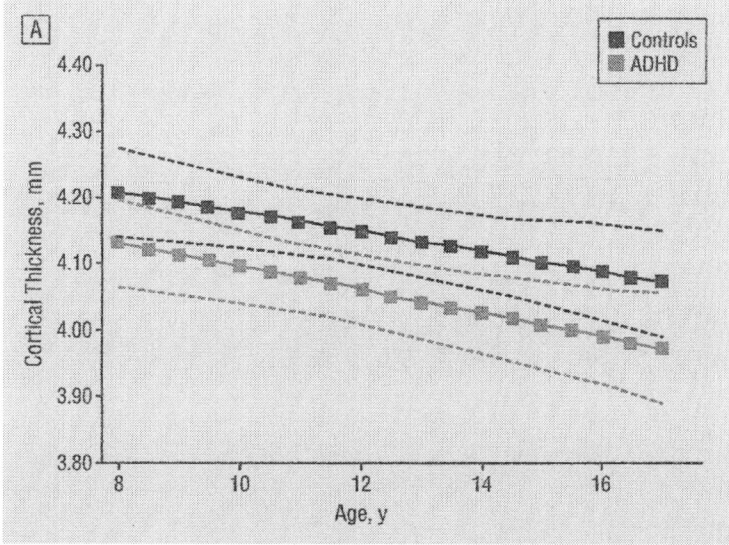

Reproduced with permission, P. Shaw, J. Lerch, D. Greenstein, W. Sharp, L. Clasen, A. Evans, et al, "Longitudinal Mapping of Cortical Thickness and Clinical Outcome in Children and Adolescents With Attention-Deficit/Hyperactivity Disorder," *Archives of General Psychiatry,* volume 63, page 546, copyright © (2006) American Medical Association. All rights reserved.

Changes in Overall Cortical Thickness According to Age

This graph shows the continuing suppression of cortical development (peak thickness) under the influence of past or continuing stimulants.

[Dashed lines refer to 95% confidence intervals for each group.]

Study #5 Shaw et al (2008)   *Stimulants Impair Cortical Thinning*

In the first study of its kind, researchers from NIMH conducted an experiment which explored the ***cortical effects of continuous medication versus drug withdrawal***.[73] This time, the investigators reported the results of serial brain scans (MRIs) from 43 children between the ages of 12 and 17 years. Over a four-year period of follow-up, comparisons were made between 19 children who stopped taking stimulants and 24 children who remained continuously drugged.

Remarkably, the cessation of drug therapy was associated with several superior results. Children withdrawn from stimulants were more successful in achieving full remission of ADHD symptoms (26% vs. 13% of the chronically medicated children); more successful in avoiding additional drug therapy (16% versus 30% of the chronically medicated children); and more successful in achieving robust rates of cortical thinning (based upon changes in two regions of the frontal lobes and the right parieto-occipital cortex):

	ADHD stopping stimulants n = 19	**ADHD chronically medicated** n = 24
mean age at first scan	13 yrs	12 yrs 1 mo
mean age at last scan	16 yrs 8 mo	16 yrs 1 mo
# taking stimulants at baseline (scan 1)	90%	96%
% taking non-stimulant drugs between scans	16%	30%
# achieving full remission of ADHD symptoms	26%	13%
average rate of cortical thinning	0.15 mm per yr	0.03 mm per yr

Stimulants

Despite the fact that the drug-stoppers, as a group, demonstrated multiple improvements, their aggregate rate of cortical thinning (0.15 mm per year) was interpreted by Shaw et al as an *abnormality which represented **excessive** change.* Furthermore, the governmental investigators used this observation to justify a recommendation for uninterrupted drug therapy. While this was certainly one possible interpretation of the structural changes, it was not the only one. Even more incredibly, it was the members of this same research team who had previously performed the study confirming elevated rates of cortical thinning (*above 0.08 mm per year*) in conjunction with *advancing age* and *superior intelligence*.

Ultimately, the significance of these findings depends upon the functional consequences of rapid gray matter pruning and myelination. Given the developmental toxicities of stimulants which have been demonstrated in animal experiments, the detection of accelerated thinning in the immediate aftermath of drug withdrawal could be reasonably interpreted as a sign of *rebound growth*. While this rate may *seem* excessive in comparison to healthy, unmedicated controls, the experiences of the latter subjects may be misleading here. In other words, in the context of *detoxification* from neuroskeletal inhibitors, the emergence of robust cortical thinning quite likely represents an effect of *healthy* (rather than defective) maturational processes.

To review:	Longitudinal Neuroimaging Studies
Castellanos et al (2002)	152 ADHD children (49 drug naïve) vs. 139 controls age range: 5 to 16 (female), 5 to 18 ½ (male) *deeply flawed study design* ***past stimulant exposure was associated with smaller caudates, restricted growth of cerebellum, and inferior linguistic ability***; under the influence of past, new, or continuous drug therapy, ***brain development was persistently suppressed; cerebellar atrophy emerged after puberty***
Mackie et al (2007)	MRI study of 36 ADHD subjects and 36 controls ≥ 3 brain scans per subject, obtained at 3-year intervals age of subjects: 10 ½ years at start; 16 ½ years at end 94% of patients were taking stimulants at time of last scan children exposed to stimulants had poorest social, academic, and behavioral functioning; ***under the influence of stimulants***, children experienced ***persistent deficits in vermal volume and post-pubertal atrophy of the cerebellum***
Shaw et al (2007)	longitudinal MRI study (timing of brain maturation) 223 ADHD subjects and 223 controls ages: 5 through 20 25% underwent repeat scanning at 3-year intervals 66% of ADHD patients had prior stimulant exposure ***maturation of the cortex was delayed by ~ 3 years*** (maturation = age at onset of peak thickness)
Shaw et al (2006)	longitudinal + cross-sectional MRI study (peak thickness) 163 ADHD subjects and 166 controls ages: 10 to 10 ½ years >> 16 ½ to 18 ½ 60% of subjects were scanned more than once 66% of ADHD subjects had prior stimulant use ***under influence of past, new, or continuing medication, cortical thickness never reached level of controls***
Shaw et al (2008)	longitudinal MRI study 43 ADHD subjects, aged 12 to 17 years 19 children stopped taking stimulants, 24 continued drugs brain scans were performed over a 4-year interval ***children who stopped taking stimulants were:*** more successful in achieving full remission of ADHD more successful in avoiding other drug therapies ***more successful in achieving accelerated cortical thinning***

Epilogue

For more than 60 years, advocates of psychopharmaceuticals have encouraged or coerced the perpetration of a perfect crime: *the medical murder of brain cells*. This book has described the causes and consequences of this development by reviewing the epidemiological and scientific evidence for psychiatric drugs as agents of dementia (aka, drug-induced dementia).

The words of the famous Christian theologian, C.S. Lewis, have never been so pertinent:

> "Of all tyrannies a tyranny sincerely exercised for the
> good of its victims may be the most oppressive."[1]

A medical tyranny has been created by the courts and pharmaceutical companies, using the services of academicians, "key opinion leaders," ghostwriters, PR firms, journalists, and government officials. As a result, the standard therapies – which some observers erroneously construe to be a capital "S" Standard of Care – have given rise to numerous Standards of Harm. [As an aside, physicians who oppose this tyranny must now fear for their licensure, their hospital privileges, and their livelihoods, every bit as much as patients must fear for their lives. What has transpired in the past six decades in many nations is the birth of "pharmacratic totalitarianism," as writers from Solzhenitsyn to Szasz have forewarned.]

Ironically, at the very moment that researchers within neurology are celebrating the use of imaging technologies to monitor the progression of dementia, and at the very moment that they express alarm about the rates of dementia-related atrophy which occur in Alzheimer's disease (2 to 10% per year), they remain woefully silent about the research evidence which has confirmed equal or greater brain atrophy among the consumers of psychiatric drugs.[2-4]

Where does this leave clinicians and patients who aspire to integrity in medicine, and who desire the right to avoid neurotoxic drugs? I believe it leaves them with two essential tasks:

    1) **the task of identifying the problem clearly**, and
    2) **the task of responding to the problem on personal and systemic levels**.

Epilogue

Task #1 - Identifying the Problem of Drug-Induced Dementia

Do psychiatric drugs induce or enhance dementia? Before one can answer this question, one must employ a logical method for deciding disease causation. In the introduction of this book, the Hill criteria were introduced as an example of one such system.

Although the Hill criteria have several crucial limitations which demand a cautious and thoughtful approach to their use (see **Appendix A**), it is important for clinicians to appreciate the fact that *the dementia-related effects of psychopharmaceuticals satisfy every single one of the Hill criteria* – particularly, those variables which are most relevant for mind-altering drugs:

strength of association	psychiatric drugs have consistently been found to *double* the risk of dementia *(often raising the risk even more)*
consistency of association	numerous research teams, using numerous research methods (postmortem studies, neuroimaging studies, epidemiological investigations) have confirmed the relationship between psychiatric drugs and dementia
timing of association	dementia has predictably emerged or worsened following exposure to psychiatric drugs
the biological likelihood (plausibility) of the association	psychiatric drugs have been shown to exert numerous toxicities in the soma and brain, most of which are linked to dementia
the coherence of the association	the neurotoxic effects of psychiatric drugs have been consistent with biological, toxicological, and neuroscience research in dementia
experimentation	dementia-inducing effects have been verified in research animals using clinically relevant doses of medication

Drawing upon the evidence which has been presented in this book, the inherently dementing properties of psychiatric medications can be summarized in the following way:

Epilogue

Drug-Induced Dementia – Evidence Reviewed in This Book

	AD	AP	AA	MS	STIM
Strength:					
≥ 2-fold risk (epidemiology)	✓	✓	✓	✓	
Timing:	✓	✓	✓	✓	✓
Consistency:					
epidemiology	✓	✓	✓	✓	✓
postmortem	✓	✓		✓	
neuroimaging	✓	✓	✓	✓	✓
Biology/Coherence:					
autoimmune		✓		✓	
↓ seizure threshold	✓	✓			✓
REM suppression	✓	✓	✓	✓	✓
cardiac dysrhythmia	✓	✓		✓	✓
hypertension	✓	✓		✓	✓
BBB	✓	✓		✓	✓
immunosuppression	✓	✓	✓	✓	
TBI risk	✓	✓	✓	✓	✓
endocrine disruption	✓	✓	✓	✓	✓
mitochondrial toxicity/ oxidative stress	✓	✓	✓	✓	✓
nutritional defect		✓		✓	
storage disorder	✓	✓			✓
cancer/anti-cancer	✓	✓	✓	✓	✓
vascular disease	✓	✓	✓	✓	✓
Experimentation:					
animal research	✓	✓	✓	✓	✓

BBB = blood-brain barrier disruption
AD = antidepressant   AP = antipsychotic (neuroleptic)   AA = anti-anxiety
MS = mood stabilizer   STIM = stimulants

Epilogue

Task #2 – Responding to the Problem of Drug-Induced Dementia

Physicians and patients who respect the validity of the foregoing evidence may reasonably decide to avoid, limit, and remediate the neurotoxic effects of psychiatric drugs. The response to drug-induced dementia must involve individual (personal) awareness and action. Ideally, one would hope for changes in the legal, medical, and politico-economic *systems*, as well. (Ten possible systemic reforms are presented in **Appendix C**.)

What Individuals Can Do

On an individual level, one approach which patients should demand and physicians must prioritize is the methodical consideration of four variables whenever drug therapy is involved: 1) pharmacodynamic effects; 2) pharmacokinetic effects; 3) unique biology; and 4) target organ toxicity.

Pharmacodynamic Effects: What Does Each Drug Do to the Body?

Pharmacodynamic effects refer to the impact of drugs upon the body. Before beginning, adjusting, or stopping any medication, physicians are obligated to consider how medications interact with target and non-target organs. Unfortunately, training programs and standard textbooks currently emphasize only immediate and short-term drug effects. As a result, doctors are accustomed to using medications and predicting responses based upon time-limited experiments in animals, and based upon clinical trials in human patients which have typically spanned only six to eight weeks.

For the safety of patients, the entire paradigm of allopathic medicine must expand to include considerations of **allostatic load**: the ***maladaptive responses that emerge under the influence of chronic therapy***. Training curricula must be revised to incorporate the research evidence which describes changes in pharmacodynamic properties – such as changes in receptor density, changes in receptor sensitivity, and changes in the cellular processes which control neurochemical synthesis and release. These revisions must include the problem of **genomic imprinting**. Long ignored by neurologists and psychiatrists, this refers to the phenomenon through which chemical therapies *alter gene expression* and *re-wire brain circuits* in ways that can result in *delayed or persistent harm*.

Pharmacokinetic Effects: What Does the Body Do With Each Drug?

Pharmacokinetic effects pertain to the impact of the body upon each drug. When a prescriber considers the pharmacokinetics of specific therapies, he or she must evaluate how the body absorbs, distributes, metabolizes (breaks down), and eliminates (excretes) each substance.

Epilogue

Two metabolic processes—known as oxidation and glucuronidation – affect the length of time that it takes the body to eliminate a given chemical. With respect to psychiatric drug treatments, many agents are substrates ("targets") of a metabolic machine known as the *cytochrome P450 enzyme system* (a component of Phase I metabolism). This system refers to a group of proteins which promote chemical reactions in the mitochondria and the endoplasmic reticula of cells throughout the soma and brain. A similar system of enzymes, known as *UGTs* (short for uridine-5'-diphosphate glucuronosyltransferases) helps in a second step of metabolism and detoxification (Phase II metabolism, primarily within the endoplasmic reticula of the liver and gut.)

In order to avoid potentially lethal interactions between two or more drugs, physicians must know and consider the kinds of collisions which may occur via these enzyme pathways. These are called *drug-drug interactions or DDIs.* Undetected or ignored DDIs can be fatal because of the fact that P450 and UGT conflicts can result in toxic levels of medications or drug metabolites. Particularly helpful resources for determining potential DDIs include:

1) the handbooks authored by Drs. Cozza, Armstrong, and Oesterheld
   (e.g., *Drug Interaction Principles for Medical Practice, 2nd Edition*)

2) the Internet website of Dr. Sheldon Preskorn:
   http://www.preskorn.com/

3) the P450 Drug Interaction Table of Dr. David A. Flockhart:
   http://www.medicine.iupui.edu/Flockhart/table.htm

4) the product labels of specific drugs, many of which can be searched at the FDA's website:
   http://www.accessdata.fda.gov/Scripts/cder/DrugsatFDA/   and

5) the National Library of Medicine's search engine known as PubMed
   (e.g., one can perform a search of specific P450 enzymes, such as 2D6 or 3A4). This last resource is especially valuable as it permits reviews of the latest findings from neuroscientific, toxicological, and pharmacological research.
   http://www.ncbi.nlm.nih.gov/pubmed/

As important as it for clinicians to consider the P450 and UGT enzyme systems, it is equally important for them to consider another unique aspect of psychiatric drugs. Some of these agents pass through the plasma membranes of brain cells and enter vesicles where they are stored. Subsequently, it can take longer for these chemicals to leave or clear the brain tissue than it does to wash out of the peripheral bloodstream (a property known as "blood-brain dissociation"). Thus, when physicians consider the full extent of possible DDIs, they must be aware of the fact that *biologically relevant levels of previous treatments may still be present in the brains of some patients, even when blood concentrations of the same drugs are undetectable or very low.*

Epilogue

Unique Biology

The third step in evaluating drug effects consists of the consideration of unique biology. Here, it is essential for the clinician to review ***unique characteristics of each patient*** which may affect the absorption, distribution, metabolism, and excretion of specific drugs. For example, disease processes which affect the liver, the thyroid, or the kidney may reduce the body's ability to detoxify medications and their metabolites, and extend the time that it takes to eliminate those substances from the body.

Although the science of pharmacogenetics remains in its infancy, some clinicians and researchers strongly advocate that patients be tested for pertinent metabolic attributes (P450 genotypes) which may influence responses to medication. For example, it is well known that approximately 7 to 10% of Caucasians possess combinations of genes which impair the clearance of drugs via the 2D6 (P450) pathway. This genetic predisposition can result in higher blood and brain levels of commonly prescribed chemicals – such as fluoxetine, paroxetine, haloperidol, and risperidone – thereby increasing the likelihood and severity of certain adverse effects.

Target Organ Toxicity

Finally, it is essential for clinicians to consider the long-term consequences of past and continuing illnesses, injuries, and drug therapies with respect to target organ toxicity. For example, one would not predict clinically significant benefits when giving L-dopa to patients with advanced stages of Parkinson's disease, because the neurons which convert L-dopa to dopamine (the target of the therapy) would already be gone. Similarly, one would not predict clinically significant benefits from the use of cholinesterase inhibitors (anti-Alzheimer's drugs) in patients who had already lost significant numbers of cholinergic neurons.

Although it is not widely known by the public, and although it is valued neither by insurance companies, nor by health care systems, there is abundant information available to individuals who desire thoughtful, methodical reviews of drug effects with the goal of avoiding or responding to the problem of drug-induced dementia. An example of this **method of drug review (*pharmacodynamics* + *pharmacokinetics* + *unique biology* + *target organ toxicity*)** can be found in **Appendix B**.

Hopefully, this book will succeed in supplying evidence which is needed to protect physicians and patients from the most violent physical assault which can befall any member of our species: the destruction of the human brain. Whether this assault upon the locus of creativity, volition, and personality occurs via involuntary mechanisms, or via manipulated and misadvised consent to voluntary care, its destructiveness remains the same. Unless and until health care consumers and providers demand the end of this destruction, the epidemic of drug-induced dementia will continue to flourish as one of society's most perfect crimes.

Appendix A

Systematic Errors in the Evaluation of Drug Safety

Reason #1 – Strength of Association

Epidemiologists have increasingly interpreted this variable to mean that causation depends upon the relative degree of disease risk (*typically, a relative risk ≥ 2*). In other words, the research question becomes: "by how much does a given drug increase the risk of a certain negative event?" However, this approach can be a red herring in the search for causation. First, the relative risk is irrelevant to the question of whether or not a specific agent induces biological changes which are known to produce disease. Second, there are occasions when a comparison of relative risk might result in illogical or erroneous conclusions about causation. A hypothetical example should illustrate why this is so. Suppose that 5% of a group of methylphenidate (Ritalin) injectors experience strokes. Suppose that the same percentage of methamphetamine injectors experience strokes. The relative risk of stroke for methylphenidate compared to methamphetamine would be equal to one (5% divided by 5%). This would be interpreted as evidence of *non-causation*, because the strength of association would be regarded as weak (actually, it would be regarded as non-existent).

Nevertheless, even when drug regulators interpret **strength of association** with reference to *absolute risk* or *risk difference* (how would a 5% rate of stroke among Ritalin injectors compare to the overall rate of stroke among non-injectors the same age?), the sources of data which are generally used by the FDA are an unreliable means to achieve this end. For example, most **strength-of-association** calculations are drawn from randomized, placebo-controlled drug trials – otherwise known as **RCT**s. Regrettably, these are short-term studies where patients are monitored for six to eight weeks or less. Furthermore, the patients selected for these studies bear little resemblance to individuals who are exposed to the same products in the "real" clinical world.[1-3] For example, most RCTs in psychiatry exclude people who are pregnant, too young, too old, too medically ill, too mentally healthy, too mentally disabled, or too medicated with psychiatric and/or other drugs.

There is yet another important reason why the FDA's strength-of-association calculations have been severely flawed. In interpreting the findings of psychiatric RCTs, the FDA has routinely ignored a common design feature which should have been used to negate, rather than to validate, many new drug applications (NDAs). Specifically, it has become standard practice for drug companies to conduct studies which incorporate a ***placebo run-in*** phase. However, this a fraudulent research design which rigs trials in favor of the drug companies. It works in the following way.

Appendix A

First, all of the study participants are informed that they must immediately stop taking their current psychiatric drug or drugs. For a period of four to ten days, they are all given capsules which contain an inert substance (such as sugar). Next, half of the participants are randomly assigned to continue the study in the *placebo control* group. These individuals continue to receive the same inert compound in the same capsule that was used in the run-in phase. The remaining subjects are assigned to the group which receives the "active" (experimental) drug that has been cleverly disguised in a capsule identical to the previous placebo.

Because of the fact that the placebo run-in phase begins with an abrupt period of chemical cessation (for those individuals who were recently consuming other drugs), psychiatric RCTs are essentially analyses of what happens to different groups of people when they experience *different kinds of drug withdrawal*: one group (placebo) enduring the effects of prolonged, uninterrupted withdrawal; the other group (active drug) enduring the effects of compensated withdrawal. This is hardly the basis of authentic or reliable estimates of *treatment efficacy*, let alone a valid basis for determining the quality or frequency of adverse effects.

It is precisely for this reason that the RCT database should either be rejected completely, or failing that, used only with utmost caution by serious researchers who are appropriately sensitive to these and other confounders. Ideally, the results of well designed, clinically relevant, and ethically executed investigations (such as case reports, observational studies, and experimental research in basic science) should be used *instead of RCTs* for evaluating the absolute and relative risks of drug toxicity.

Reason #2 - Consistency of Association

In the post-Daubert era of medical practice and drug regulation, ***attorneys, scientists, and physicians who represent the interests of the pharmaceutical industry have constructed a mythological hierarchy of evidence for disease causation***. According to these commercial spokesmen and spokeswomen (and the federal rules which they have been allowed to craft), it is *only certain kinds of evidence* which should be used by the courts or the FDA to determine whether or not a medication is causally related to any given harmful event.[4-6]

*This prevarication has made a mockery of the Hill criterion of **consistency of association**, according to which physicians were formerly encouraged to consider an entire spectrum of research results.* Hill appreciated the fact that the findings of disease causation became increasingly relevant, and increasingly clear, as evidence came to be corroborated by different groups of researchers, using different methods of study, and employing diverse populations of patients – even if those populations were small. Hill never believed and never conveyed the notion that RCTs (randomized, placebo-controlled clinical trials) or large population studies (non-RCT epidemiological investigations) were the only or even best sources of information for establishing disease causation.[7]

Appendix A

In fact, the history of medicine is replete with examples where evidence from case reports, case series, and/or biological research was analyzed by astute clinicians using **inductive logic** (*reasoning from specific case >> to general theory*).[8] On these occasions, insightful doctors and scientists were able to protect patients and save lives, despite the fact that data from RCTs or large observational studies were either contradictory or lacking. For example, it was a *series of cases* which spurred Drs. William McBride and Dr. Widukind Lenz to recognize the link between thalidomide and birth defects in 1961.[9-10] Similarly, it was a *series of cases* which spurred physicians at Massachusetts General Hospital to recognize the link between DES and vaginal cancer in 1971.[11-13] More recently, it was a *series of case reports combined with biological research* which provided clear and sufficient evidence of the causal link between cisapride (a gastrointestinal drug) and a potentially lethal disturbance in heart rhythm known as QT_c prolongation.[14-15] Without these sources of "real world," commercially unfiltered, and scientifically authenticated data – and without the application of inductive or abductive reasoning (especially, when applied to events that occur rarely in the general population), the quality of medical care would be perpetually compromised.

Reason #3 – Specificity of the Association

Hill's specificity variable was a holdover from the early days of microbiology research, when scientists believed that *specific* microorganisms could be demonstrated to cause *specific* diseases (e.g., the tuberculosis bacterium >> tuberculosis). Unfortunately, it did not take long to prove the limitations of this variable, and it is now quite clear that most microorganisms (viruses, bacteria, parasites) have the capacity to induce a wide variety of illnesses (e.g., *Escherichia coli* >> gastroenteritis, urinary tract infections, pneumonia, meningitis). Furthermore, it has become equally clear that most illnesses have more than one possible cause (e.g., pneumonia >> viral, bacterial, protozoal, fungal, autoimmune, toxic, etc.).

Particularly with respect to the human brain, though, the expectation of "specific" drug effects is a complete misnomer. This is because of the enormous complexity of the target organ (10 billion to 1 trillion neurons; 100 billion to 10 trillion glia; more than 100 different neurotransmitters) and the inseparability of different brain regions which modulate complex behaviors and functions.[16-18]

Reason #4 – Temporality

Hill's temporality variable was introduced in an effort to ensure that exposure to a causal agent occurred *before*, rather than *after*, the onset of disease. However, FDA safety reviewers have come to apply this criterion in a restricted and medically inappropriate fashion. Unless all patients experience a given side effect within the same time frame (a phenomenon which the FDA refers to as *temporal clustering*), the FDA denies that the event is drug-related. This denial has no basis in real science and, instead, raises questions about the competence of FDA safety reviewers when it comes to appreciating the complexity of human responses to medication – especially with regard to the human brain.

Why is it unreasonable to expect temporal clustering with respect to psychiatric drug effects?

First, there is the influence of fraudulent research design. As described previously, (*Strength of Association* section, above) psychiatric drug RCTs commonly employ the ***placebo run-in*** methodology. What the FDA safety reviewers habitually neglect or ignore is the fact that many psychiatric medications enter brain cells, undergo a prolonged period of storage within those sites, and pass out of the brain tissue more slowly than they clear the bloodstream.[19-21] This means that the four- to ten-day placebo run-in periods in most RCTs are not long enough to remove certain chemicals from the target organ (the brain). By extension, this means that different people in an RCT will be experiencing adverse events at different times, depending upon the properties of previously used and/or new medications.

Second, there is a tremendous amount of variability in individual biology. This dissimilarity includes differences in liver and kidney function; differences in genetic structure; and differences in lifestyle (diet, habits), all of which influence drug metabolism and drug effects. Perhaps most critically, though, patients in psychiatric RCTs differ in terms of their *total lifetime exposures to psychiatric drugs*. These differences in *cumulative exposure* modify the timing and intensity of the onset of certain adverse events, such as the symptoms of Parkinson's disease and tardive phenomena.

Third, the diverse nature of psychiatric drug toxicity invalidates the relevance of temporal clustering. Research has repeatedly demonstrated that the onset of drug-induced lupus, drug-induced anemia, drug-induced encephalopathy, drug-induced heart disease, and/or drug-induced endocrinopathies (diabetes) ***can occur within weeks, months, or even years of treatment*** depending upon differences in patient susceptibility, differences in treatment history, and differences in pathophysiology. For example, when the FDA reviewing officer denied that olanzapine was a cause of heart disease prior to that drug's approval in 1996, he did this on the basis of "no temporal clustering." Unfortunately, this ignored the epidemiological and basic science research which had shown that other analogous neuroleptics could damage the heart muscle in a variety of ways and *with variable timing*: via inflammation (myocarditis), via endothelial damage and vasoconstriction (myocardial ischemia and infarction), via mitochondrial toxicity and direct harm to myocytes (cardiomyopathy and conduction defects), via ion channel dysfunction (cardiac dysrhythmias), and via abnormalities in hemostasis (blood clots and emboli).

Appendix A

Fourth, the expectation of temporal clustering ignores the problem of allostatic load – the phenomenon through which drugs, themselves, become a source of physiological stress for the body.[22] ***Allostatic load includes acute and delayed processes.*** Within the brain, this stress involves changes in cell physiology (such as changes in the number or sensitivity of neuronal receptors). It also involves changes in gene expression (chemical or epigenetic imprinting), changes in protein synthesis, and cell-to-cell re-wiring of the brain.[23-25] Finally, it includes the long-term toxicities which occur under the continuing influence of psychopharmaceuticals due to the loss of cell parts and eventual cell death. ***In other words, the FDA's continued reliance upon temporal clustering ignores fundamental facts from basic science.*** The biology of allostatic load has repeatedly shown that drug toxicities appear at different times in the course of exposure, due to the varying time course of adaptations and degenerative events.

Reason #5 – Dose-Response Relationship

Hill's criterion for a biological gradient was based upon the example of environmental hazards, such as exposures to chimney soot, tobacco (cigarettes), or dust. Intriguingly, research in the field of environmental toxicity has since revealed remarkable diversity in dose-response effects. The history of modern toxicology has substantiated the complexity of living organisms, demonstrating that adverse outcomes are not always commensurate with the intensity of acute or chronic exposures.

For example, research in the field of endocrine toxicology has negated the famous adage of Paracelsus, according to which "the dose makes the poison." ***Non-linear dose-response relationships have been discovered for numerous chemicals***, including phthalate (DHEP), bisphenol A (BPA), and the pesticide hexachlorobenzene (HCB).[26] In these cases, low doses *far below regulatory health standards* have been found to induce profound disturbances in animal immunology and metabolism. These and other studies have shown that existing safety requirements have miscalculated the dangers of many hormonally active compounds, because they have not actively investigated the effects of *extremely low levels of exposure*.

Within psychiatry, there exist numerous examples of drug-induced toxicity which also violate Paracelsus' rule. Nevertheless, the lack of a dose-response relationship does not invalidate the reality of these hazards. For example, the association between valproate (Depakene) and encephalopathy frequently bears no direct correspondence to the blood level of the anticonvulsant, nor to the blood level of ammonia. Another example of the same phenomenon occurs in the case of lithium toxicity, where patients experience tremor, confusion, slurred speech, lethargy, and/or unconsciousness despite the consumption of regular doses, and despite the persistence of so-called "therapeutic" drug levels when samples are obtained from the peripheral bloodstream.

Appendix A

Reasons #6-9 – Biological Plausibility/Coherence/Experimentation/Analogy

The criteria of disease causation with *reference to science* include the observation of plausible mechanisms of disease; coherence with accepted theories of disease; confirmation via experimental manipulation; and/or analogy to the properties or effects of other agents which are known to produce disease. These criteria are only useful to the extent that the existing state of knowledge is correct, and to the extent that the person who applies these criteria is qualified or competent to do so.

Unfortunately, history has repeatedly shown that the FDA's leadership, consultants, and reviewing officers have failed to apply these four Hill criteria proficiently. This has been reflected by the federal government's consistent inability (or unwillingness) to recognize or respond to any of the following psychiatric drug effects and associated toxicities:

1) mitochondrial dysfunction (suppression of cell respiration and oxidative stress)

2) metabolic dysregulation (depletion of nutrients; suppression of essential cellular processes – such as beta-oxidation, Krebs cycle, etc.)

3) endocrine disruption (hormonal imbalances, neurotransmitter depletion)

4) anti-proliferative / cytotoxic effects (chemotherapy properties)

5) allostatic load (chronic adaptations and genomic/epigenetic imprinting)

Unless and until drug regulators acknowledge the fundamental pathways through which psychopharmaceuticals damage all cells in the body – but especially, the cells of the brain – their opinions pertaining to treatment hazards will continue to reflect the chaotic approach which has resulted in the disability and death of many patients.

Appendix B

Avoiding Drug-Induced Dementia

There are four major variables to consider when evaluating the potential dangers and interactions of drug therapies: 1) pharmacodynamic effects; 2) pharmacokinetic effects; 3) unique biology; and 4) target organ toxicity. To illustrate how a patient or clinician might perform a methodical drug review, the following example refers to the first case report described in the introduction of this book.

Case #1   **44-year-old female with bipolar disorder**

recent treatments: fluphenazine, quetiapine, risperidone, carbamazepine
new treatments: risperidone + valproate

pharmacodynamics:

    risperidone   potent blockade of D_2 receptors (upregulated by past treatments – blockade typically results in depression, cognitive impairment, Parkinsonian features, high prolactin); potent blockade of H_1 receptors (sedation, low blood pressure upon standing, weight gain); potent alpha$_1$-adrenergic blockade (low blood pressure upon standing; possible sedation, hypersalivation, urinary incontinence)

    valproate   blockade of sodium and calcium channels; possible induction of epileptiform activity (status epilepticus) via "paradoxical" stimulation of glutamate (glutamate transmission may have been activated by previous exposures to carbamazepine, fluphenazine, and quetiapine)

pharmacokinetics:

Possible P450 and Phase II Interactions[1-2]			
	drug is metabolized by (substrate of)	drug inhibits	drug induces
recent drugs			
carbamazepine	1A2, 3A4, 2C8, 2C9, UGTs	--	1A2, 3A4, 2C19
fluphenazine	2D6, 1A2	1A2, 2D6	--
quetiapine	3A4 (also Phase II)	--	--
current drugs			
risperidone	3A4, 2D6	2D6 (mild)	--
valproate	2A6, 2C9, 2C19 (also Phase II)	2C9, 2D6, UGTs	--

Appendix B

unique biology: a) COPD, unstable high blood pressure, congestive heart failure

b) presumptive damage to cortical and subcortical brain regions (due to chronic treatment with mood stabilizers and neuroleptics)

c) possible misdiagnosis of COPD (could be respiratory dyskinesia)

d) possible hypothyroidism (not investigated by clinicians)

e) possible cerebellar damage from SILENT (?) (would need to know more about past medical treatments)

f) possible lifestyle effects: unclear if patent smokes (if so, this would accelerate the clearance of certain drugs; conversely, hospital restrictions on smoking could contribute to higher levels of some medications in the blood and brain)

Any Illnesses Affecting Drug Clearance ?

heart disease	congestive heart failure (↓ perfusion of liver and kidneys)
liver disease	investigated and ruled out
renal disease	investigated and ruled out
thyroid disease	not investigated
metabolic defect	investigated via patient interview and collateral history and denied (i.e., no one in family had MELAS or other form of inherited metabolic disease)

Appendix B

Target Organ Toxicity

Dementia-Inducing Mechanisms of Current Drug Therapies

	risperidone	valproate
autoimmune/lupus	✓	✓
↓ seizure threshold	✓	(paradoxical effect)
REM suppression	✓	✓
cardiac dysrhythmia	✓	✓
hypertension [with metabolic syndrome]	✓	✓
blood-brain barrier disruption	✓	✓
immunosuppression	✓	✓
TBI risk	✓	✓
endocrine disruption	✓	✓
mitochondrial toxicity/oxidative stress	✓	✓
nutritional defect		✓
storage disorder		
tumor/cancer[3-5]	✓	✓
anti-cancer		✓
vascular	✓	✓

Possible Pathologies Associated with Current Drug Therapies		
	risperidone	valproate
Alzheimer's disease	✓	
CJD		
frontotemporal dementia	✓	✓
hydrocephalus		✓
Lewy body/Parkinson's disease	✓	✓
other (cerebellar degeneration)		✓

Appendix B

Interpretation:

In any patient who consumes psychiatric drugs, one must consider the combined impact of past and current illnesses and treatments. This case involved a patient who had experienced more than 30 years of exposure to psychopharmaceuticals, including combinations of potent neurotoxins (neuroleptics + mood stabilizers).

Under the influence of recent medications, the 2D6 cytochrome system had been chronically suppressed (fluphenazine + risperidone). This inhibition continued under the influence of risperidone and valproate, more than likely delaying the breakdown of the neuroleptic. The recent discontinuation of carbamazepine (an inducer of 2C19) presumably contributed to *declining activity* in the 2C19 pathway. As the patient progressed through the stages of cytochrome adaptation, but as she remained continuously exposed to a drug whose metabolism depended upon the 2C19 system (valproate), she was vulnerable to the crescendo-like accumulation of valproate and ammonia. The suppressive effects of valproate upon the 2C9 and UGT systems would have further contributed to increased blood and brain levels of the anticonvulsant via the process of autoinhibition.

This patient's lifelong history of exposure to multiple chemical agents, combined with respiratory and cardiac disease, made her particularly vulnerable to the problems of drug resistance (tachyphylaxis), drug accumulation, drug-related seizures, and drug-induced neurobehavioral decline. The treatment team's decision to continue risperidone – along with a new, unidentified anticonvulsant – increased the likelihood of progressive neurodegeneration, worsening disability, and premature death.

Appendix C

Preventing and Responding to Drug-Induced Dementia – What Can Systems Do?

Beyond the level of the individual, concerned observers in positions of authority and leadership might respond to the epidemic of drug-induced dementia by introducing a number of systemic reforms. The following is a list of ten possible measures which might facilitate this goal.

#1 Recognition of Target Organ Toxicity (TOT)
A priority must be placed upon the identification, avoidance, and mitigation of Target Organ Toxicity due to drug therapies and other medical interventions.

For governmental regulators, this precedence should involve the revision of product labels to include **explicit warnings about TOT**. For example, TOT could easily be incorporated in existing or new Black Box Warning sections. Patient Information Guides (Warning Sheets) could be prepared for pharmacies. Legislatures could mandate the distribution of these materials with every new prescription and every refill.

For federally financed researchers (such as scientists associated with the National Toxicology Research Center, the National Institute of Drug Abuse, and the National Institute of Mental Health), a priority should be placed upon the characterization of TOT in humans and animals, using clinically relevant doses; using multiple species; and using broad ranges of doses and durations of exposure – including those which permit the determination of non-monotonic dose-response curves (*see Appendix A*).

For academicians and educators, textbooks and curricula must be expanded and revised in order to promote the explicit recognition of TOT as a likely and predictable effect of allopathic treatments. ***Within each and every specialty of medicine***, clinicians must prioritize the identification of TOT which occurs under the influence of standard therapies in order to avoid, limit, or correct the consequences of iatrogenic injury.

#2 Non-Reimbursement of Medical Neglect
The current system of medical care within the United States emphasizes 10- to 15-minute "med checks," during which clinicians are encouraged to review only superficial changes in symptoms. Discussions of adverse drug effects, to the extent they occur at all, must be truncated. Careful inspections of DDIs seldom occur. Within psychiatry, these problems are intensified by the bad habits of some clinicians who completely omit the consideration of physical illnesses and non-psychiatric medications (a behavior best exemplified by the concise entry on Axis III: "*please see medical chart*"). The insurance programs and treatment facilities which promote this form of care collude in medical neglect. For the safety of patients, these systems must be dismantled.

Appendix C

#3 Compensation of High Quality Care
The antithesis of a for-profit medical scheme (which rewards doctors for processing high volumes of patients as though they were pill-consuming widgets) is a system which values (rather than punishes) high quality, competent care. How might this be achieved?

First, physicians must be allowed to perform **appropriate evaluations** of each patient, based upon the unique needs of the individual and his or her clinical condition. This demands flexibility in scheduling in order to accommodate the varying complexity of some encounters. There are many situations where algorithm-based or "cookbook" medicine cannot, does not, and should not apply.

Second, physicians must be permitted to perform the kinds of intellectual work – such as **medical chart reviews** – which are necessary to detect the effects of past treatments and illnesses, and which are necessary to perform proficient evaluations of DDIs (pharmacodynamics + pharmacokinetics + unique biology + target organ toxicity). This can only be done when adequate time for these activities is protected within the daily schedule of each provider.

Third, physicians who engage in quality care must receive **protection from job termination, decredentialing actions, and other forms of professional retribution**, as currently happens in systems which reward and prioritize efficiency, productivity, and the high-volume processing of patients.

Fourth, **compensation must be provided** for the competent and thoughtful review of medical records, DDIs, and adverse responses to drug therapies. Just as attorneys are allowed to bill their clients for the time which is spent in research, writing, and preparation for litigation, so, too, must physicians be able to bill insurance companies and patients for the time which they spend in researching, writing, and preparing for the safe and effective delivery of treatment.

#4 Restoration of Medical Ethics (End of Corporate Tyranny)
As practiced for many centuries, medicine was regarded as a *calling*. Those who entered the healing professions were motivated by a genuine desire to alleviate suffering and to serve mankind. Over the past century, however, the combined effects of "corporatization" and the pursuit of wealth have undermined this legacy. Certainly within the United States, the purpose of medicine has been diverted away from service to profit by three essential pieces of legislation. However unrealistic it may be to expect the repeal of these laws, their reversal would go a long way towards restoring integrity in the American system of health care.

Appendix C

Repeal #1 - The Bayh-Dole Act
Under the terms of the Bayh-Dole Act of 1980, American universities, small businesses, and non-profit organizations which receive federal funds were granted patent protection and intellectual control of their inventions and discoveries.[1] Prior to this legislation, the presumption of title rested with the source of federal research funds: the U.S. taxpayer. Critics believe that the Bayh-Dole Act has corrupted basic science and medical research,
largely by converting academic and governmental institutions into research and marketing arms of commercial industries; by stifling academic freedom of expression; and by privatizing knowledge.

Repeal #2 – PDUFA
The 1992 passage and subsequent renewals of the Prescription Drug User Fee Act (PDUFA) transferred money to the Food and Drug Administration from the very industry which that agency was created to regulate. Under the terms of this legislation, fees from the drug companies have been used to fund everything from the review of new drug applications, to the post-marketing surveillance of drug safety. Critics believe that PDUFA has engendered political and economic conflicts of interest which preclude legitimate oversight and imperil the public's safety.

Repeal #3 - DTCA
In 1997, Congress relaxed historical restrictions on drug manufacturers with respect to commercial advertising. By loosening federal policies related to Direct to Consumer Advertising or DTCA, the government permitted the promotion of pharmaceuticals over the airwaves with only limited descriptions of product efficacy or risk.[2-3] The marketing of illnesses and drug therapies proliferated. Critics claim that DTCA has fueled excessive and hazardous prescribing practices by manipulating public and professional demand for the newest treatments.

#5 Avoidance of Type III Medical Malpractice
In his books, *The Last Well Person* and *Worried Sick*, Dr. Nortin Hadler has defined two different forms of malpractice.[4-5] Type I malpractice refers to "medical or surgical performance that is unacceptable." Type II malpractice involves "doing something to patients very well that was not needed in the first place." Although Hadler himself has not described it, his line of reasoning clearly suggests a Type III malpractice which is especially relevant for psychiatry: ***doing something to patients that is not needed, and doing it very dangerously.***

How has Type III malpractice flourished in the era of *biological psychiatry?

---

* biological psychiatry – this term refers to the hegemony of a philosophy of care which begins by interpreting existential problems, odd actions, or disturbing thoughts as the essential byproducts of brain disease; it ends by promoting or coercing the use of drugs magnets, surgery, and/or electroshock. Accordingly, moral agency and volition are denigrated in deference to the view that the brain (not the Self) determines behavior.

First, as all physicians who train in American medical schools and American psychiatry residencies can attest, a fundamental rule of practice involves ***medical clearance***. Before any psychiatric label can be applied to a patient's symptoms, doctors must first confirm that the symptoms are not explained by an organic, objective disease in the soma or brain. It is ***only after the confirmation of physical normalcy*** (normal blood tests, normal urine tests, and normal brain scans) that psychiatric drugs are initiated for purely psychiatric conditions (e.g., major depressive disorder, bipolar disorder, schizophrenia).

Second, the use of psychiatric drugs in the physically normal is mutually justified by physicians and patients who agree to "assume" or "pretend" that the subject's brain is dysfunctional or diseased.

Third, the fact that the profession of psychiatry lacks any objective diagnostic test for any mental illness renders the use of psychiatric labels entirely speculative and subjective.

These three features – the performance of medical clearance, the "pretend" nature of brain dysfunction and corrective prescribing, and the application of purely subjective labels – constitute the first dimension of Type III Medical Malpractice: the "doing something to patients" that is ***never medically needed*** in the first place. The second dimension of Type III Medical Malpractice – doing something dangerously – arises from the use of drug therapies and other measures which are dementing and deadly.

#6 Protection of Whistle Blowers

When ethical clinicians inform their patients about the limited utility or true dangers of prevailing treatments, and when they decide to deliver services which are more effective and safe, they are currently afforded no systemic protections from retribution. This reprisal assumes the form of **Sham Peer Review** and **Psychiatric Profiling** (e.g., Disruptive Physician Proceedings).

Sham Peer Review refers to the unfounded persecution of clinicians who oppose Group Think. Currently, American physicians who refuse to comply with unacceptably harmful or bogus practices can be easily driven out of medicine via fraudulent competency hearings and fraudulent decredentialing actions.

Psychiatric Profiling occurs when critically reflective practitioners are diagnosed implicitly or explicitly in retaliation for their concerns. Under the veil of Consensus Medicine, the career of any health care professional in America can be quickly destroyed by fellow practitioners, merely on the basis of allegations that he or
she is impaired or disruptive: impaired, by virtue of wrong thought; disruptive, by virtue of challenging the status quo. In fact, State Medical Boards and health care facilities are actively coached by attorneys, and actively empowered by law, to terminate the careers of competent clinicians who question illegitimate practices and corporate fraud.

Appendix C

What Federal and State Governments Can Do
Only the creation of nation-wide Whistle Blower protections will end the scourges of Sham Peer Review, Disruptive Physician Proceedings, and physician persecution by State Medical Boards.

One step that should be taken immediately is the dissolution of the National Practitioner Data Bank or NPDB. Created under the auspices of the Health Care Quality Improvement Act of 1986, it is the NPDB which serves as a ***permanent repository*** of the names of suspect providers, along with a list of attacks on licensure or clinical privileges *for any reason.* Even when attacks have subsequently been proven unfounded (Sham Peer Review or Psychiatric Profiling by State Medical Boards), innocent physicians have not been allowed to correct reports by disputing the merits or bases of erroneous information in the Data Bank.[6]

Because of the fact that the NPDB is used by employers and insurance companies to determine physician qualifications, and to deny the livelihoods of professionals who have been falsely accused, its continued existence undermines the quality of health care in the U.S.A.

#7 Prosecution of Corporations for Negligence, Manslaughter, and Ecocide
Although the makers of pharmaceutical products are *theoretically* responsible for revealing drug hazards and drug limitations to regulatory authorities and to investors, the sad reality which has transpired in America is a system which rewards corporate medical fraud. How has this occurred?

By allowing corporations to control access to raw data from animal and human drug tests; by allowing corporations to manipulate the methodological designs and statistical results of pre- and post-marketing research; by allowing corporations to coerce the opinions and silence of scientists and clinicians (via the imposition of confidentiality and non-disclosure agreements); by allowing corporations to monopolize the content of journal articles, training programs, treatment algorithms, and other authoritative sources of medical and legal information; and by allowing corporations to intimidate judges, juries, academicians, public watchdogs, federal overseers, and publishers (e.g., via the manipulation of jury settlements and public elections, and via the threat of litigation) the legislatures and the courts have enabled the pharmaceutical equivalent of **corporate homicide** and **corporate ecocide** (the latter, referring to the destruction of the environment).[7-8] Each of these loopholes must be closed in order to protect the health and survival of the *entire biosphere.*

Appendix C

#8 Compensation of the Iatrogenically Injured

Given the fact that most patients who have been harmed by psychopharmaceuticals have experienced these injuries without the benefit of adequate or properly informed consent; and given the fact that many more individuals have been victimized by involuntary exposure to these chemicals as wards of the State (in prisons, in foster care, and in State mental institutions; or, more recently, as outpatients subjected to Community Treatment Orders), there is an urgent need to *compensate these victims for damages*.

In the same way that the U.S. government has historically authorized payments to individuals (or their families) as compensation for injuries caused by nuclear tests and unauthorized medical experimentation (radiation, syphilis), a trust fund should be prepared and distributed to those who have been harmed by psychiatric treatments through the negligence of the State.

One aspect of compensation which could occur immediately is the transformation of the health care system to recognize drug-induced dementia; to prioritize the avoidance and withdrawal of the drugs which cause it; and to compensate victims via the delivery of rehabilitation services (see below).

Another aspect of compensation should involve the initiation of a national clean-up and environmental protection plan. Communities must be compensated with civil engineering and environmental health programs which are dedicated to eliminating and preventing pharmaceutical contaminants in the air, in the water supply, and in the food chain.

#9 Rehabilitation of the Iatrogenically Injured

Rehabilitation efforts can be optimally achieved only if they include the removal of existing toxins. Whereas current treatment programs encourage or force mental health providers and patients to comply with new pills for each and every new symptom, a system committed to rehabilitation must shift the focus of care towards the reduction and removal of brain-offending agents.

Whenever possible, drug-withdrawal protocols should be carefully planned, closely monitored, and – in most cases – slowly implemented, respecting the fact that discontinuation signs and symptoms should be expected to emerge in three phases:

1) an acute stage, reflecting the elimination of medication from the bloodstream and brain;

2) an intermediate stage, reflecting the reversal of drug-induced neural adaptations (i.e., the resetting of cell receptor numbers, cell receptor sensitivity, and intracellular metabolic processes); and

3) a delayed and potentially long-lasting stage, reflecting the reversal of epigenetic imprinting (i.e., changes in gene expression and the re-wiring of brain circuits) and the mitigation or repair of drug-related damage.

Although it may not be possible to reverse the effects of all psychiatric medications, and although it may not be possible for all patients to tolerate even small adjustments in substances that have induced physiological and psychological dependence, it is always ethically and medically appropriate to attempt changes in treatment regimens which are associated with reduced longevity and neurobehavioral decline.

#10 Defense of Patient and Physician Privacy (Freedom from Bodily Assault)

For patients and physicians equally, there must be a right to avoid unwarranted chemical assault upon the brain.

Patient Privacy

Barring extraordinary acts of litigation and medical advocacy, anyone in America who is deemed incompetent for medical decision-making can be subjected to coercive treatment. However, because the State empowers doctors to deliver psychiatric drugs involuntarily, and because doctors erroneously conform to Consensus Medicine (corporate medical fraud), the drugs which are now administered in the so-called "best interests" of many patients are, regrettably, inflicting tremendous physical and psychological harm.   Given the improbability that that these practices will change, there remain relatively few protective strategies.

Some observers have recommended the creation of **Psychiatric Living Wills** or **Psychiatric Advanced Directives** while a person remains competent to record his or her treatment preferences. (In fact, this would be the only time that a patient in America could object to specific types of therapies, without the objection *itself* being interpreted as proof of mental disease.)   Unfortunately, there is no guarantee that the State or medical facilities will actually pay attention to these instruments. Nevertheless, it is possible that the existence of legal directives could at least form the basis of a future claim for compensation, in the all-too-likely event that instructions of this kind were ubiquitously ignored.

On a systemic scale, the best thing that could happen in any democratic society would be the eradication of unwarranted, coercive therapies. Only within psychiatry is the fundamental right of patient autonomy, and the fundamental duty of physician non-maleficence, routinely trampled. Preferably, law schools and medical schools would prioritize the creation of an entire network of professionals (such as Jim Gottstein, Esq., and the Law Project for Psychiatric Rights) who would be dedicated to the task of preventing psychiatric assault.

Appendix C

As a corollary to each patient's right to medical privacy, the residents of any free society should be safeguarded from several other forms of potential psychiatric abuse. This right should include protection from pre-employment or pre-admission brain scans (designed to detect "wrong thoughts" in workers, immigrants, or travelers); protection from genetic profiling (designed to predict future manifestations of "mental deviancy"), and protection from the unauthorized collection and dissemination of health data (i.e., there must be a right to opt out of electronic medical records systems, pharmaceutical registries, and other computerized databases).

Physician Privacy (Freedom from Perpetrating Bodily Assault)
At the present time, clinicians who desire the freedom to avoid the use of dementing drugs are accused of incompetence or negligence. Their right to privacy must begin with the freedom to challenge Medical Consensus (Group Think) for the good of their patients. Legislative and policy measures to protect this right might include:

1) **the termination of Pay-for-Performance programs**
   Pay-for-Performance refers to insurance programs and health-care payment policies which provide incentives and punishments to doctors, based upon their conformity or non-compliance with Consensus Medicine.[9-10]

2) **the termination of corporate preference in Continuing Medical Education (CME) programs**
   Currently, physicians are required by State Medical Boards to accumulate specific numbers of **Category 1 CME credits** each year. These units – almost always the most costly – are usually associated with commercially prepared and commercially biased activities. A simple solution to this problem would be the elimination of CME requirements completely, or the elimination of the current hierarchy which awards privileged credits for indoctrination by the drug industry.

3) **the protection of physicians who base their treatment decisions upon basic science and non-censored epidemiological research**
   Critical thinkers who appreciate the biases and limitations of Consensus Medicine, and who refuse to prioritize therapies which have limited benefits and unacceptable risks, are now persecuted for their convictions. One measure that would protect insightful physicians would be the rejection, by clinics and hospitals, of flowcharts, guidelines, and algorithms which reflect corporate medical fraud. Another measure would be a renewed emphasis in medicine upon basic science, rather than clinical drug trials. By returning the focus of practice to the modification of *essential causes* of illness, rather than symptom suppression, one might hope to end the present crisis of performativity: a crisis which has resulted in the development of dubious consensus statements and copy-cat drugs, in order to secure an interminable supply of research funds, job security, and power.

Appendix D

Permissions for Images and Tables

Cover photo
Kindergehirne.jpg. Reproduced with permission of Raum und Zeit, Wolfratshausen Deutschland. Source: H. Kremer, "Ritalin – Tatort Gehirn: Wissen die Therapeuten, was sie Kindern antun?," *Raum und Zeit* 115:1 (2002).

Prologue

Image #1: Retrocollis
Reproduced with permission of Mr. Justin G. Aquines, Executive Director. National Spasmodic Torticollis Association. 9920 Talbert Avenue Fountain Valley, CA 92708 Tel: 1-800-HURTFUL (487-8385) Email: NSTAmail@aol.com
Website: www.torticollis.org

Chapter 2 – Mechanisms

Image #1: Tootsie Pops. Reproduced with permission of Tootsie Roll Industries. Permission obtained from Mr. Barry Bowen. Image accessed from:
http://www.cybercandy.co/uk/acatalog/190.jpg

Image #2: Brain human sagittal section.svg – Wikimedia Commons, public domain.

Image #3: Brain cells. Source: R.S. Snell. *Clinical Neuroanatomy for Medical Students, 4th Ed.* (Philadelphia: Lippincott-Raven Publishers, 1997), 78. Reproduced with permission of Wolters Kluwer and LWW.

Image #4: Pituitary Gland – Reproduced with permission of Mayo Foundation for Medical Education and Research.

Image #5: Hypothalamus Pituitary Adrenal Axis. Reproduced with permission of Dr. Richard Bertram and Dr. Eugene Izhikevich. Source: R. Bertram, J. Tabak, and N. Toporikova (2006), "Models of Hypothalamus." *Scholarpedia* 1:12, 1330.

Image #6: Human Cell. Artist: Dan Lietha. Reproduced with permission of Tiffany Winkler at Answers in Genesis. Source: http://www.answersingenesis.org

Image #7: Chromosome/DNA – public domain, NIH – National Human Genome Research Institute
www.genome.gov/Pages/Hyperion/DIR/VIP/Glossary/Illustration/Pdf/chromosome.pdf

Image #8: Mitochondrial DNA, copyright Russell Knightley Media. Reproduced with permission. Source: rkm.com.au (Australia).

Appendix D

Image #9: Image of cell. Reproduced via Creative Commons License.
https://eapbiofield.wikispaces.com/Chapter+6+A+Tour+of+the+Cell+HNE

Image #10: Electron Transport Chain. Source:
http://en.wikipedia.org/wiki/Image:Etc2.png

Image #11: ETC inhibitors. P.C. Champe, R.A. Harvey, and D. R. Ferrier. *Lippincott's Illustrated Reviews: Biochemistry, 3rd Ed.* (Philadelphia: Lippincott Williams and Wilkins, 2005), 76. Reproduced with permission of LWW.

Image #12: Meth toxicity - T. Kita, G.C. Wagner, and T. Nakashima, "Current Research on Methamphetamine-Induced Neurotoxicity: Animal Models of Monoamine Disruption," *Journal of Pharmacological Sciences* 92 (2003): 187. Reproduced with permission of the Japanese Society of Neuropsychopharmacology.

Image #13: One carbon transfer cycle. Source: R. Diaz-Arrastia, "Homocysteine and Neurologic Disease," *Archives of Neurology* 57 (2000): 1423. Reproduced with permission. Copyright 2000, American Medical Association. All rights reserved.

Image #14: Myelin whorls after fluoxetine. Reproduced with permission of the Japanese Society of Toxicologic Pathology from Nonoyama T. et al. Drug-induced Phospholipidosis – Pathological aspects and its prediction. *Journal of Toxicology and Pathology* 21: 9-24, 2008.

Chapter 4 – Antidepressants

Image #1: Brain structures in human and rat brains. Image in public domain. F.T. Crews and K. Nixon (July 2004). "Alcohol, Neural Stem Cells, and Adult Neurogenesis." Accessed at:
http://pubs.niaaa.nih.gov/publications/arh27-2/197-204.htm

Image #2: Rat brain. Image in public domain. C.D. Pozniak and S.J. Pleasure, "Genetic control of hippocampal neurogenesis," Genome Biology 7 (2006): 207.2.

Image #3: Neuron and boutons. Source: http://en.wikipedia.org/wiki/File: Neuron-no_labels.png

Image #4: SSRI mechanism of action. CNS forum, free access. Original source: H.P. Rang, M.M. Dale, and J.M. Ritter. *Pharmacology, 4th Ed.* (Edinburgh: Harcourt Publishers, Ltd, 2001).

Appendix D

Images #5-7: Fluoxetine and sertraline nerve damage. Reprinted from *Brain Research* 858, M. Kalia, J.P. O'Callaghan, D.B. Miller, and M. Kramer, "Comparative study of fluoxetine, sibutramine, sertraline and dexfenfluramine on the morphology of serotonergic nerve terminals using serotonin immunohistochemistry," pages 102, 97, 101 (2000), with permission from Elsevier.

Images #8-10: Serotonin fiber damage in Parkinson's disease. Reprinted from *Brain Research* 1217, E.C. Azmitia and R. Nixon, "Dystrophic serotonergic axons in neurodegenerative diseases," page 188 (2008), with permission of Elsevier.

Image #11: Tree shrew. public domain, accessed on August 6, 2008 at: http://farm3.static.flickr.com/2178/2281788533_27cd6e7e79.jpg?v=0

Images #12 and 13: Astroglial study in tree shrews. Reprinted with permission of Macmillan Publishers Ltd. *Neuropsychopharmacology*, B. Czeh, M. Simon, B. Schmelting, C. Hiemke, and E. Fuchs, "Astroglial Plasticity in the Hippocampus is Affected by Chronic Psychosocial Stress and Concomitant Fluoxetine Treatment," volume 31, p. 1617, copyright 2006.

Image #14: Hippocampus. Public domain. Accessed on March 23, 2009 at: http://commons.wikimedia.org/wiki/File:Gray739-emphasizing-hippocampus.png

Images #15 and 16: Astroglial study in tree shrews. Reprinted with permission of Macmillan Publishers Ltd. *Neuropsychopharmacology*, B. Czeh, M. Simon, B. Schmelting, C. Hiemke, and E. Fuchs, "Astroglial Plasticity in the Hippocampus is Affected by Chronic Psychosocial Stress and Concomitant Fluoxetine Treatment," volume 31, p. 1622, copyright 2006.

Image #17: Neurogenesis with AD treatment. Reprinted from *Biological Psychiatry*, vol. 46, R.S. Duman, J. Malberg, and J. Thome, "Neural plasticity to stress and antidepressant treatment, p. 1182, copyright (1999), with permission from Elsevier.

Image #18: Cell cycle. Reprinted with permission of W. Sombatworapat (March 4, 2008). "Cancer Treatment by Conventional and Herbal Medicine." Accessed on October 25, 2008 at: http://herb4cancer.wordpress.com/2007/11

Images #19-21: Antidepressant effects on hippocampus. Source: *Journal of Neuroscience*, M. Sairenen, P. Ernfors, M. Castren, and E. Castren, vol 25, "Brain Derived Neurotrophic Factor and Antidepressant Drugs Have Different But Coordinated effects on Neuronal Turnover, Proliferation, and Survival in the Adult Dentate Gyrus," p. 1090, copyright (2005), reprinted with permission of the Society for Neuroscience.

Appendix D

Images #22 and 23: Hippocampus effects, loss of circuits. Reprinted from *Current Opinion in Pharmacology*, volume 4, E. Castren, "Neurotrophic effects of antidepressant drugs," p. 60, copyright (2004), with permission from Elsevier.

Image #24: Dentate gyrus. Reprinted and modified with permission of *Arquivos de Neuro-Psiquiatria*, volume 65:4B, M.J. Sa, C. Ruela, and M.D. Madeira, "Dendritic Right/Left Asymmetries in the Neurons of the Human Hippocampal Formation," p. 1106, (2007).

Images #25-26: Dying neurons in hippocampus. Reprinted from *American Journal of Pathology*, vol 158, P.J. Lucassen et al, "Hippocampal Apoptosis in Major Depression is a Minor Event and Absent from Subareas at Risk for Glucocorticoid Overexposure," pages 459 and 460, copyright (2001), with permission from the American Society for Investigative Pathology.

Chapter 5 – Antipsychotics

Image #1: Caudate. W.J. Hendelman, *Atlas of Functional Neuroanatomy* (Washington, D.C.: CRC Press, 2000), p. 55. Reprinted with permission of Taylor & Francis Group LLC books.

Images #2-4: Gliosis in basal ganglia. K. Jellinger, "Neuropathologic Findings after Neuroleptic Long-Term Therapy," in L. Roizin, H. Shiraki, and N. Grcevic, Ed. *Neurotoxicology*. (New York: Raven Press, 1977), 34-35, with permission of Wolters Kluwer / Lippincott Williams & Wilkins.

Images #5-7: Neuronophagia and neuronal loss. L. Roizin, C. True, and M. Knight, "Structural effects of tranquilizers," *Research Publications – Association for Research in Nervous and Mental Disease* 37: (1959), 306, 305, 309.

Image #8: Neuron. McGill University. The Brain from Top to Bottom.
In public domain via Copyleft.
http://thebrain.mcgill.ca/flash/i/i_01/i_01_cl/i_01_cl_ana/i_01_cl_ana.html#1

Image #9: Neurofibrillary tangle. Reproduced with permission of Dr. William I. Rosenblum from "Neuropathology Mini-Course for Residents," VCU Department of Pathology. Accessed on December 31, 2006 at:
http://www.pathology.vcu.edu/WirSelfInst/neur&proc.html

Image #10: Beta-amyloid plaque. U.S. NIH National Institute on Aging. The Hallmarks of AD.
http://www.nia.nih.gov/Alzheimers/Publications/Unraveling/Part2/hallmarks.htm

Appendix D

Image #11: Granulovacuolar degeneration in Alzheimer's. Reprinted with permission of Dr. Charles Y. Shao, Department of Pathology, SUNY Downstate Medical Center, Brooklyn NY. Accessed March 20, 2009 at:
http://www.downstate.edu/pathology/Dr.CharlesShaoMD.htm

Image #12: Lewy bodies. Reprinted with permission of Dr. William I. Rosenblum. from "Neuropathology for Medical Students: Chapter 11, Dementias," VCU Department of Pathology (May 15, 2007). Accessed on September 18, 2008 at:
http://www.pathology.vcu.edu/WirSelfInst/neuro_medStudents/dementias.html

Image #13: Compact plaques, diffuse plaque in AD. Reprinted from *Neurobiology of Disease*, volume 20, issue 2, P.P. Desai, M.D. Ikonomovic, E.E. Abrahamson, R.L. Hamilton, B.A. Isanski, C.E. Hope, et al, "Apolipoprotein D is a component of compact but not diffuse amyloid-beta plaques in Alzheimer's disease temporal cortex," pages 574-582, copyright (2005), with permission from Elsevier.

Image #14: Activated microglia after haloperidol. Reprinted from *Neuroscience*, volume 109, issue 1, I.J. Mitchell, A.C. Cooper, M.R. Griffiths, and A.J. Cooper, "Acute Administration of Haloperidol Induces Apoptosis of Neurones in the Striatum and Substantia Nigra in the Rat," p. 92, copyright (2002), with permission from Elsevier.

Image #15: Parietal lobe in humans. Wikimedia commons.
http://en.wikipedia.org/wiki/File:Brainlobes.svg

Chapter 6 – Anxiolytics

Image #1: Lateral ventricle. Wikipedia Commons. Image from *Gray's Anatomy*, 20th *Ed.*, now in public domain. Accessed on October 20, 2008 at:
http://en.wikipedia.org/wiki/Image:Gray734.png

Image #2: Coronal/horizontal/sagittal planes of brain. Reprinted with permission of Eric H. Chudler, Ph.D., from his website at
http://faculty.washington.edu/chudler/slice.html

Image #3: C.T. scan showing the VBR changes in benzodiazepine user. C. Schmauss and J.C. Krieg, "Enlargement of cerebrospinal fluid spaces in long-term benzodiazepine abusers," *Psychological Medicine* 17 (1987): 870, reprinted with permission of Cambridge University Press.

Image #4: Ridges and grooves (gyri and sulci). Reprinted with permission of Dr. Jean-Francois Mangin. Source: Brain Visa Anatomist. Accessed on October 8, 2008 at: http://brainvisa.info/doc/brainvisa/images/compute8.jpg

Appendix D

Image #5: Hippocampal Anatomy. R.R. Edelman, J.R. Hesselink, M.B. Zlatkin, and J. Crues, Ed. *Clinical Magnetic Resonance Imaging, 3rd Ed.* (Philadelphia: Elsevier-Saunders, 2006), figure 47-4. Reprinted with permission of Dr. John Hesselink and Elsevier.

Images #6-8: DG neurons after midazolam and dark neurons. Reproduced with permission from *Revista de Medicina de la Universidad de Navarra* and from the Department of Anatomy, University of Navarra, Spain. Source: M. Carassiti, A. Ysasi, and L.M. Gonzalo, "Efectos de la administracion cronica de Midazolam sobre el hipocampo de ratas," *Revista de Medicina de la Universidad de Navarra* 1-3 (1998): 23-24.

Image #9: Cell membrane. Wikipedia Commons. Accessed on October 20, 2008 at: http://en.wikipedia.org/wiki/Image:Cell_membrane_detailed_diagram_4.svg

Image #10: GM$_1$. Reprinted with permission of Ines Niehaus. Source: "Acute Neuronal and Systemic Inflammation caused by Lipopolysaccharides of *Salmonella minnesota* can be stopped by treatment with the Antibiotic Ofloxacin," Poster Presentation at 6th Conference of the International Endotoxin Society. Accessed on October 12, 2008 at: http://www.endotoxin.gmx.home.de/artikel3.html

Image #11: Chart of the different gangliosides. Reprinted by permission from Macmillan Publishers Ltd: *Cell Death and Differentiation*, volume 13, A. d'Azzo, A. Tessitore, and R. Sano, "Gangliosides as apoptotic signals in ER stress response," page 406, copyright (2006).

Image #12: Cerebellum. McGill University. The Brain from Top to Bottom. http://thebrain.mcgill.ca/flash/d/d_06/d_06_cl/d_06_cl_mou/d_06_cl_mou.html

Chapter 7 – Mood Stabilizers

Image #1: Cerebellum. McGill University. The Brain from Top to Bottom. Public domain via copyleft.
http://thebrain.mcgill.ca/flash/d/d_06/d_06_cr/d_06_cr_mou/d_06_cr_mou.html

Image #2: Layers of the cerebellum. Deltagen, Inc. (San Mateo, CA). Reproduced with permission of Dr. Winston Thomas, Chief Operating Officer.
Source: http://www.deltagen.com/target/histology/atlas/atlas_files/nervous/cerebellar_cortex_10X.htm

Image #3: Cerebellum after lithium. Reprinted from *Neuroscience Letters*, volume 224, issue 1, S. Dethy, M. Manto, E. Bastianelli, V. Gangji, M.A. Laute, S. Goldman, et al, "Cerebellar spongiform degeneration induced by acute lithium intoxication in the rat," p. 27, copyright (1997), with permission from Elsevier.

Appendix D

Image #4: Changes in CJD. Reproduced with permission of Professor James W. Ironside. National CJD Surveillance Unit, Edinburgh, UK. Accessed on on December 15, 2008 at: http://www.cjd.ed.ac.uk/path.htm

Image #5: Gliosis chart. Reproduced from J.P. O'Callaghan and K. Sriram, "Glial fibrillary acidic protein and related glial proteins as biomarkers of neurotoxicity," *Expert Opinion on Drug Safety* 4:3 (2005): 436, with permission of Informa Healthcare.

Image #6: Reactive astrocytes. Reproduced with permission of Dr. Kar Ming-Fung. Source: A. Ilyas, K. Petropoulou, Z.F. Cheema, and K.M. Fung (Updated February 28, 2005). "A 27 year-old man with headache, visual field loss and right side weakness." Department of Pathology, University of Oklahoma Health Sciences Center. Accessed on November 8, 2008 at:
http://moon.ouhsc.edu/kfung/JTY1/Com04/Com407-2-Diss.htm

Images #7-9: Gliosis after lithium. Source: E. Rocha, *NeuroReport*, volume 9, issue 17, "Lithium treatment causes gliosis and modifies the morphology of hippocampal astrocytes in rats," pages 3972-3973, (1998), reprinted with permission of Wolters Kluwer / LWW.

Image #10: Diffusion. Reprinted from *Neurochemistry International*, volume 45, issue 4, E. Sykova, "Diffusion properties of the brain in health and disease," page 454, copyright (2004), with permission from Elsevier.

Image #11: Damaged mitochondrion (cerebellum) at 9 months. M.E. Sobaniec-Lotowska, "Ultrastructure of Purkinje cell perikarya and their dendritic processes in the rat cerebellar cortex in experimental encephalopathy induced by chronic application of valproate," *International Journal of Experimental Pathology* 82 (2001): 341, with permission from John Wiley and Sons / Wiley-Blackwell.

Image #12: Degenerating neuron (cerebellum) – 12 months. M.E. Sobaniec-Lotowska, "Ultrastructure of Purkinje cell perikarya and their dendritic processes in the rat cerebellar cortex in experimental encephalopathy induced by chronic application of valproate," *International Journal of Experimental Pathology* 82 (2001): 343, with permission from John Wiley and Sons / Wiley-Blackwell.

Image #13: Hippocampus astrocyte and dark neurons (left). M.E. Sobaniec-Lotowska, "Ultrastructure of astrocytes in the cortex of the hippocampal gyrus and in the neocortex of the temporal lobe in experimental valproate encephalopathy and after valproate withdrawal," *International Journal of Experimental Pathology* 84 (2003): 119, with permission from John Wiley and Sons / Wiley-Blackwell.

Appendix D

Image #14: Hippocampus microglial cell (right). M.E. Sobaniec-Lotowska, "A transmission electron microscopic study of microglia/macrophages in the hippocampal cortex and neocortex following chronic exposure to valproate," *International Journal of Experimental Pathology* 86 (2005): 93, with permission from John Wiley and Sons / Wiley-Blackwell.

Image #15: Niethammer SILENT Case #1. M. Niethammer and B. Ford, "Permanent Lithium-Induced Cerebellar Toxicity: Three Cases and Review of the Literature," *Movement Disorders* 22:4 (2007): 570, with permission of John Wiley & Sons, Ltd.

Image #16: Brazil SILENT case 2 atrophy. A.C. Rodrigues de Cerqueira, M. Costa dos Reis, F.D. Novis, J.M.F. Bezerra, G. Canedo de Magahlaes, et al, "Cerebellar Degeneration Secondary to Acute Lithium Carbonate Intoxication," *Arquivos de Neuropsiquiatria* 66:3-A (2008): 578, reproduced with permission of Dr. Antonio Spina-Franca.

Image #17: Cerebellum. McGill University, The Brain from Top to Bottom. In public domain via copyleft.
http://thebrain.mcgill.ca/flash/i/i_01/i_01_cl/i_01_cl_ana/i_01_cl_ana.html

Image #18: Regions of the cerebellum. Wikipedia. "Human brain – midsagittal view – cerebellum." Licensed by creative commons. Author: John A. Beal, PhD, LSU Health Sciences Center, Shreveport. Accessed on December 17, 2008 at:
http://en.wikipedia.org/wiki/File:Human_brain_midsagittal_view_description.JPG

Image #19: Atrophy of brain. R. Guerrini, A. Belmonte, R. Canapicchi, C. Casalini, and E. Perucca, "Reversible Pseudoatrophy of the Brain and Mental Deterioration Associated with Valproate Treatment," *Epilepsia* 39:1 (1998): 28, with permission of Wiley-Blackwell.

Images #20-21: Atrophy of brain and resolution. O. Papazian, E. Canizales, I. Alfonso, R. Archila, M. Duchowny, and J. Aicardi, "Reversible Dementia and Apparent Brain Atrophy During Valproate Therapy," *Annals of Neurology* 38 (1995): 689, with permission of John Wiley & Sons, Inc.

Image #22: Mammalian mitochondrion. R.R. da Fonseca, W.E. Johnson, S.J. O'Brien, M.J. Ramos, and A. Antunes, "The adaptive evolution of the mammalian mitochondrial genome," *BMC Genomics* 9 (2008):119. Open access license. Accessed on January 12, 2009 at: http://www.biomedcentral.com/1471-2164/9/119

Images #23-24: Degeneration at 10 yrs 1 month and resolution. C.A. Galimberti, M. Diegoli, I. Sartori, C. Uggetti, A. Brega, A. Tartara, et al, "Brain pseudoatrophy and mental regression on valproate and a mitochondrial DNA mutation," *Neurology* 67:9 (2006): 1716, with permission of Lippincott Williams & Wilkins.

Appendix D

Image #25: Spongiform changes in cerebellum. *Annals of Neurology*, volume 36, J.A. Schneider and S.S. Mirra, "Neuropathologic Correlates of Persistent Neurologic Deficit in Lithium Intoxication," page 930, copyright (1994), reproduced with permission of John Wiley & Sons, Inc.

Image #26: *Archives of Neurology*, volume 63, August, E.D. Louis, J.P.G. Vonsattel, L.S. Honig, A. Lawton, C. Moskowitz, B. Ford, and S. Frucht, "Essential Tremor Associated With Pathologic Changes in the Cerebellum," page 1190, copyright (2006) American Medical Association. All rights reserved.

Chapter 8 – Stimulants

Image #1: Catechol nucleus. Pyrocatechol. Wikimedia Commons. Accessed on October 28, 2008 at: http://enwikipedia.org/wiki/Image:Pyrocatechol.svg

Image #2: Amine group. Amine 2D. Wikimedia Commons. Accessed on October 28, 2008 at: http://en.wikipedia.org/wiki/Image:Amine-2D-general.png

Image #3: Catecholamine synthesis. Wikimedia Commons. Accessed on March 21, 2009 at: http://commons.wikimedia.org/wiki/File:Catecholamines_biosynthesis.svg

Images #4-6: NE fibers in PFC; NE fibers in HC; striatum. *Journal of Neuroscience*, volume 27, issue 27, J.D. Gray, M. Punsoni, N.E. Tabori, J.T. Melton, V. Fanslow, M.J. Ward, et al, "Methylphenidate Administration to Juvenile Rats Alters Brain Areas Involved in Cognition, Motivated Behaviors, Appetite, and Stress," pages 7201, 7203, 7202, copyright (2007), reproduced with permission of the Society for Neuroscience.

Images #7-9: Striatum cell death, basal ganglia, white matter death.
Reprinted from *Neuroscience*, volume 125, issue 1, I. Husson, B. Mesles, F. Medja, P. Leroux, B. Kosofsky, and P. Gressens, "Methylphenidate and MK-801, an N-Methyl-D-Aspartate Receptor Antagonist: Shared Biological Properties," page 165, copyright (2004), with permission from Elsevier.

Images #10-11: Proliferation bar graph, bregma separation. Reprinted from *Biological Psychiatry*, volume 60, D.C. Lagace, J.K. Yee, C.A. Bolanos, and A.J. Eisch, "Juvenile Administration of Methylphenidate Attenuates Adult Hippocampal Neurogenesis," page 1124, copyright (2006), with permission from Elsevier.

Image #12: BrdU uptake, neurogenesis vs. survival. D.C. Lagace, M.A. Noonan, and A.J. Eisch, "Hippocampal Neurogenesis: A Matter of Survival," volume 164, issue 2, page 205. Reprinted with permission from the *American Journal of Psychiatry*, (Copyright 2007). American Psychiatric Association.

Appendix D

Image #13: Cell survival in adult brain. Reprinted from *Biological Psychiatry*, volume 60, D.C. Lagace, J.K. Yee, C.A. Bolanos, and A.J. Eisch, "Juvenile Administration of Methylphenidate Attenuates Adult Hippocampal Neurogenesis," page 1125, copyright (2006), with permission from Elsevier.

Image #14: Dopamine pathways. McGill University. "The Brain from Top to Bottom: The Reward Circuit." In public domain via copyleft. Accessed on November 3, 2008 at:
http://thebrain.mcgill.ca/flash/a/a_03/a_03_cl/a_03_cl_que/a_03_cl_que.html

Image #15: Dopamine nerve ending. Permission requested from source: Dr. Ichiro Sora. Tohoku University School of Medicine. "Hyperdopaminergic mice lacking of [sic] dopamine transporter." Accessed on November 3, 2008 at:
http://sendaibrain.org/eng/members/sora.html

Images #16-18: Bar graph of markers from group 1 baboons, dopamine in group 2 baboons, DA and DAT in monkeys. G.A. Ricaurte, A.O. Mechan, J. Yuan, G. Hatzidimitriou, T. Xie, A.H. Mayne, et al, "Amphetamine Treatment Similar to That Used in the Treatment of Adult Attention-Deficit/Hyperactivity Disorder Damages Dopaminergic Nerve Endings in the Striatum of Adult Nonhuman Primates," *Journal of Pharmacology and Experimental Therapeutics* 315:1 (2005): 94, 96, 97. Reprinted with permission of the American Society for Pharmacology and Experimental Therapeutics.

Image #19: GFAP bar graph. B.N. Frey, A.C. Andreazza, K.M. Cereser, M.R. Martins, F.C. Petronilho, D.F. de Souza, et al, "Evidence of astrogliosis in rat hippocampus after d-amphetamine exposure," *Progress in Neuro-Psychopharmacology & Biological Psychiatry* 30:7 (2006): 1233, reprinted with permission from Elsevier.

Image #20: Growth velocity chart. Simm PJ and Werther GA. Child and adolescent growth disorders – an overview. *Aust Fam Physician* 2005; 34: 731-737. Permission to reproduce figure 1 granted by the Royal Australian College of General Practitioners © 2009.

Image #21: Brain mapping – how do they make the pictures. Reprinted from *Trends in Neuroscience*, volume 29, issue 3, A.W. Toga, P.M. Thompson, and E.R. Sowell, "Mapping brain maturation," page 149, copyright (2006), with permission from Elsevier.

Image #22: My rule #1 - graph of changes over life span. Reprinted from *The Journal of Neuroscience*, volume 28, issue 14, P. Shaw, N.J. Kabani, J.P. Lerch, K. Eckstrand, R. Lenroot, N. Gogtay, et al, "Neurodevelopmental Trajectories of the Human Cerebral Cortex," page 3588, copyright (2008), with permission of the Society for Neuroscience.

Appendix D

Image #23: My rule #2 - white matter, gray matter, CSF. Reprinted by permission from Macmillan Publishers Ltd: *Nature Neuroscience*, E.R. Sowell, B.S. Peterson, P.M. Thompson, S.E. Welcome, A.L. Henkenius, and A.W. Toga, "Mapping cortical change across the human life span," volume 6, issue 3, page 314, copyright (2003).

Image #24: My rule #4 – intellectual ability and rates of thinning. Reprinted with permission from Macmillan Publishers Ltd: *Nature*, P. Shaw, D. Greenstein, J. Lerch, R. Lenroot, N. Gogtay, A. Evans, et al, "Intellectual ability and cortical development in children and adolescents," volume 440, page 677, copyright (2006).

Images #25-26: Volume of brain growth curves, cerebellar changes over time. F.X. Castellanos, P.P. Lee, W. Sharp, N.O. Jeffries, D.K. Greenstein, L.S. Clasen, et al, "Developmental Trajectories of Brain Volume Abnormalities in Children and Adolescents With Attention-Deficit/Hyperactivity Disorder," *JAMA*, volume 288, issue 14, page 1746, October 9, copyright © (2002) American Medical Association. All rights reserved.

Images #27-29: Cerebellar regions, changes in whole cerebellar volume, changes in inferior posterior hemisphere. Reprinted with permission from the *American Journal of Psychiatry*, (Copyright 2007). American Psychiatric Association. S. Mackie, P. Shaw, R. Lenroot, R. Pierson, D.K. Greenstein, T.F. Nugent III, et al, "Cerebellar Development and Clinical Outcome in Attention Deficit Hyperactivity Disorder," *American Journal of Psychiatry* 164 (2007): 649, 652, 653.

Image #30: Changes in cortical thickness. Reproduced with permission, P. Shaw, J. Lerch, D. Greenstein, W. Sharp, L. Clasen, A. Evans, et al, "Longitudinal Mapping of Cortical Thickness and Clinical Outcome in Children and Adolescents With Attention-Deficit/Hyperactivity Disorder," *Archives of General Psychiatry*, volume 63, page 546, copyright © (2006), American Medical Association. All rights reserved.

Endnotes

Introduction

1. R.B. Carr and K. Shrewsbury, "Hyperammonemia Due to Valproic Acid in the Psychiatric Setting," *American Journal of Psychiatry* 164:7 (2007): 1020-1027.

2. C.J. Pavlis, E.C. Kutscher, R.M. Carnahan, W.K. Kennedy, S. Van Gerpen, and E. Schlenker, "Rivastigmine-induced dystonia," *American Journal of Health -System Pharmacists* 64 (2007): 2468-2470.

3. S. Shorvon, "What is nonconvulsive status epilepticus, and what are its subtypes?" *Epilepsia* 48:Suppl 8 (2007): 35-38.

4. S. Riggio, "Psychiatric Manifestations of Nonconvulsive Status Epilepticus," *Mount Sinai Journal of Medicine* 73:7 (2006): 960-966.

5. J.M. Murthy, "Nonconvulsive status epilepticus: An under diagnosed and potentially treatable condition," *Neurology India* 51:4 (2003): 453-454.

6. F. Centorrino, B.H. Price, M. Tuttle, W.M. Bahk, J. Henne, M.J. Albert, et al, "EEG Abnormalities During Treatment With Typical and Atypical Antipsychotics," *American Journal of Psychiatry* 159 (2002): 109-115.

7. A. Yoshino, M. Watanabe, K. Shimizu, T. Goto, N. Ichinowatari, H. Yoshimasu, et al, "Nonconvulsive status epilepticus during antidepressant treatment [abstract]," *Neuropsychobiology* 35:2 (1997): 91-94.

8. C. Marini, L. Parmeggiani, G. Masi, G. D'Arcaneglo, and R. Guerrini, "Nonconvulsive status epilepticus precipitated by carbamazepine presenting as dissociative and affective disorders in adolescents [abstract]," *Journal of Child Neurology* 20:8 (2005): 693-696.

9. D.W. Hedges and K.G. Jeppson, "New-onset seizures associated with quetiapine and olanzapine [abstract]," *Annals of Pharmacotherapy* 36:3 (2002): 437-439.

10. O. Dogu, S. Sevim, and H.S. Kaleagasi, "Seizures associated with quetiapine treatment [abstract]," *Annals of Pharmacotherapy* 37:9 (2003): 1224-1227.

11. J. Kruk, P. Sachdev, and S. Singh, "Neuroleptic-induced respiratory dyskinesia," *Journal of Neuropsychiatry and Clinical Neuroscience* 7:2 (1995): 223-229.

12. M.W. Rich and S.M. Radwany, "Respiratory dyskinesia. An under recognized phenomenon," *Chest* 105:6 (1994): 1826-1832.

Endnotes

13. C. Leung, D.W. Chung, I.W. Kam, and K.H. Wat, "Multiple rib fractures secondary to severe tardive dystonia and respiratory dyskinesia [abstract]," *Journal of Clinical Psychiatry* 61:3 (2000): 215-216.

14. H.F.K. Chiu, S. Lee, and C.H.S. Chan, "Misdiagnosis of respiratory dyskinesia," *Acta Psychiatrica Scandinavica* 83 (1991): 494-495.

15. P. Sachdev, *Akathisia and Restless Legs* (New York: Cambridge University Press, 1995).

16. T.R. Barnes and W.M. Braude, "Akathisia variants and tardive dyskinesia [abstract]," *Archives of General Psychiatry* 42:9 (1985): 874-878.

17. S.K. Mattoo, G. Singh, and A. Vikas, "Akathisia – diagnostic dilemma and behavioral treatment," *Neurology India* 51:2 (2003): 254-256.

18. N. Masubuchi, H. Hakusui, and O. Okazaki, "Effects of proton pump inhibitors on thyroid hormone metabolism in rats: a comparison of UDP-glucuronyltransferase induction [abstract]," *Biochemistry and Pharmacology* 54:11 (1997): 1225-1231.

19. I. Sachmechi, D.M. Reich, M. Aninyei, F. Wibowo, G. Gupta, and P.J. Kim, "Effect of protein pump inhibitors on serum thyroid-stimulating hormone level in euthyroid patients treated with levothyroxine for hypothyroidism [abstract]," *Endocrine Practice* 13:4 (2007):345-349.

20. J. Liappas, T. Paparrigopoulos, I. Mourikis, and C. Soldatos, "Hypothyroidism induced by quetiapine: a case report," *Journal of Clinical Psychopharmacology* 26:2 (2006): 208-209.

21. D.L. Kelly and R.R. Conley, "Thyroid function in treatment-resistant schizophrenia patients treated with quetiapine, risperidone, or fluphenazine," *Journal of Clinical Psychiatry* 66:1 (2005): 80-84.

22. M.S. Benedetti, R. Whomsley, E. Baltes, and F. Tonner, "Alteration of thyroid hormone homeostasis by antiepileptic drugs in humans: involvement of glucuronosyltransferase induction [abstract]," *European Journal of Clinical Pharmacology* 61:12 (2005): 863-872.

23. Y. Dalezios and N. Matsokis, "Nuclear benzodiazepine binding: possible interaction with thyroid hormone receptors [abstract]," *Neurochemical Research* 18:3 (1993): 305-311.

24. L. Grandison, "Actions of benzodiazepines on the neuroendocrine system [abstract]," *Neuropharmacology* 22:12B (1983): 1505-1510.

Endnotes

25. S.C. Wiens and V.L. Trudeau, "Thyroid hormone and gamma-aminobutyric acid (GABA) interactions in neuroendocrine systems [abstract]," *Comparative biochemistry and physiology. Part A, Molecular & integrative physiology* 144:3 (2006): 332-344.

26. I.C. Wilson, J.C. Garbutt, C.F. Lanier, J. Moylan, W. Nelson, and A.J. Prange, Jr., "Is There a Tardive Dysmentia?," *Schizophrenia Bulletin* 9:2 (1983): 187-192.

27. M.S. Myslobodsky, "Anosognosia in Tardive Dyskinesia: 'Tardive Dysmentia' or 'Tardive Dementia'?," *Schizophrenia Bulletin* 12:1 (1986): 1-6.

28. B.D. Jones, "Tardive Dysmentia: Further Comments," *Schizophrenia Bulletin* 11:2 (1985): 187-189.

29. American Psychiatric Association, *Diagnostic and Statistical Manual of Mental Disorders, Fourth Edition – Text Revision* (Arlington, VA: American Psychiatric Association, 2000), 169.

30. M.A. Gallo, "History and Scope of Toxicology." Contributed chapter in C.D. Klassen. *Casarett & Doull's Toxicology: the Basis Science of Poisons*, 7th Ed. (New York: McGraw Hill Medical, 2008), 3.

31. L.D. Lehman-McKeeman, "Absorption, Distribution, and Excretion of Toxicants." Contributed chapter in C.D. Klassen. *Casarett & Doull's Toxicology: the Basis Science of Poisons*, 7th Ed. (New York: McGraw Hill Medical, 2008), 131.

32. A. Richter, P.A. Loschmann, and W. Loscher, "The novel antiepileptic drug, lamotrigine, exerts prodystonic effects in a mutant hamster model of generalized dystonia [abstract]," *European Journal of Pharmacology* 264:3 (1994): 345-351.

33. S.G. Papageorgiou, T. Kontaxis, A. Antelli, and N. Kalfakis, "Exacerbation of myoclonus by memantine in a patient with Alzheimer disease," *Journal of Clinical Psychopharmacology* 27:4 (2007): 407-408.

34. M.C. Sutter, "Assigning Causation in Disease: Beyond Koch's Postulates," *Perspectives in Biology and Medicine* 39:4 (1996): 581-592.

35. A.B. Hill, "The Environment and Disease: Association or Causation?," *Proceedings of the Royal Society of Medicine* 58 (1965): 295-300.

36. No author (January 13, 2009). "Daubert v. Merrell Dow Pharmaceuticals." *Wikipedia.* Accessed on February 19, 2009 at: http://en.wikipedia.org/wiki/Daubert_v._Merrell_Dow_Pharmaceuticals

Endnotes

Chapter 1 - Dementia

1. No author (January 3, 2009). "Jacobellis v. Ohio." Wikipedia. Accessed on January 29, 2009 at: http://en.wikipedia.org/wiki/Jacobellis_v._Ohio

2. P. Sachdev, "Is It Time to Retire the Term 'Dementia'?" *Journal of Neuropsychiatry and Clinical Neuroscience* 12:2 (2000): 276-279.

3. D.P. Simpson, *Cassell's Compact Latin Dictionary* (New York: Dell Publishing Company, 1982).

4. J. Brown and J. Hillam, *Your Questions Answered: Dementia* (New York: Churchill Livingstone, 2004), 3.

5. American Psychiatric Association, *Diagnostic and Statistical Manual of Mental Disorders, 3rd Edition* (American Psychiatric Association: Washington, DC, 1980). 124-128.

6. American Psychiatric Association, *Diagnostic and Statistical Manual of Mental Disorders, Third Edition, Revised* (American Psychiatric Association: Washington, DC, 1987), 103-198.

7. American Psychiatric Association, *Diagnostic and Statistical Manual of Mental Disorders, 4th Edition*, (American Psychiatric Association: Washington, DC, 1994), 133-155.

8. C.G. Ballard, R.N.C. Mohan, A. Patel, and C. Bannister. "Idiopathic clouding of consciousness – do the patients have cortical Lewy body disease [abstract]," *International Journal of Geriatric Psychiatry* 8:7 (2004): 571-576.

9. S.A. Factor, E.S. Molho, G.D. Podskainy, and D. Brown, "Parkinson's disease: drug-induced psychiatric states [abstract]," *Advances in Neurology* 65 (1995): 115-138.

10. C. Serrano and D. Garcia-Borreguero, "Fluctuations in cognition and alertness in Parkinson's disease and dementia [abstract]," *Neurology* 63:8 Suppl 3 (2004): S31-S34.

11. American Psychiatric Association, *Diagnostic and Statistical Manual of Mental Disorders, 4th Edition, Text Revised* (American Psychiatric Association: Arlington, VA, 2000), 135-172.

12. A. Husband and A. Worsley, "Different types of dementia," *Pharmaceutical Journal* 277 (2006): 579-582.

13. A. Mark Clarfield, "The Decreasing Prevalence of Reversible Dementias," *Archives of Internal Medicine* 163 (2003): 2219-2229.

14. E.T. Jansson, "Alzheimer disease is substantially preventable in the United States – review of risk factors, therapy, and the prospects for an expert software system [abstract]," *Medical Hypotheses* 64:5 (2005): 960-967.

15. P. Sachdev, "Is It Time to Retire the Term 'Dementia'?," *Journal of Neuropsychiatry and Clinical Neuroscience* 12:2 (2000): 276-279.

16. T. Erkinjuntti, T. Ostbye, R. Steenhuis, and V. Hachinski, "The Effect of Different Diagnostic Criteria on the Prevalence of Dementia," *NEJM* 337:23 (1997): 1667-1674.

17. A.W. Lemstra, E. Richard, W.A. Van Gool, "Cholinesterase inhibitors in dementia: yes, no, or maybe?," *Age and Ageing* 36:6 (2007): 625-627.

18. K. Landmark and A. Reikvam, "Cholinesterase inhibitors in the treatment of dementia – are they useful in clinical practice? [abstract]," *Tidsskrift for den Norske Laegeforening* 128:3 (2008): 294-297.

19. A.J. Pelosi, S.V. McNulty, and G.A. Jackson, "Role of cholinesterase inhibitors in dementia care needs rethinking," *BMJ* 333 (2006): 491-493.

20. M. Maggini, N. Vanacore, and R. Raschetti, "Cholinesterase Inhibitors: Drugs Looking for a Disease?," *PLoS Medicine* 3:4 (2006): e140.

21. R. Voelker, "Guideline: Dementia Drugs' Benefits Uncertain," *JAMA* 299:15 (2008): 1763.

22. P. Raina, P. Santaguida, A. Ismalia, C. Patterson, D. Cowan, M. Levine, et al, "Effectiveness of Cholinesterase Inhibitors and Memantine for Treating Dementia: Evidence Review for a Clinical Practice Guideline," *Annals of Internal Medicine* 148 (2008): 379-397.

23. A.J. Mitchell and M. Shiri-Feshki, "Temporal trends in the long term risk of progression of mild cognitive impairment: a pooled analysis [abstract]," *Journal of Neurology, Neurosurgery, and Psychiatry* 79:12 (2008): 1386-1391.

24. R. Raschetti, E. Albanese, N. Vanacore, and M. Maggini, "Cholinesterase Inhibitors in Mild Cognitive Impairment: A Systematic Review of Randomised Trials," *PLoS Medicine* 4:11 (2007): e338.

25. M. Maggini, N. Vanacore, and R. Raschetti, "Cholinesterase Inhibitors: Drugs Looking for a Disease?," *PLoS Medicine* 3:4 (2006): e140.

Endnotes

26. E.G. Duysen, B. Li, S. Darvesh, and O. Lockridge, "Sensitivity of butyrylcholinesterase knockout mice to (-)-huperzine A and donepezil suggest humans with butyrylcholinesterase deficiency may not tolerate these Alzheimer's drugs and indicates butyrylcholinesterase function in neurotransmission [abstract]," *Toxicology* 233:1-3 (2007): 60-69.

27. C.E. Creeley, D.F. Wozniak, A. Nardi, N.B. Farber, and J.W. Olney, "Donepezil markedly potentiates memantine neurotoxicity in the adult rat brain [abstract]," *Neurobiology of Aging* 29:2 (2008): 153-167.

28. No author. "Rivastigmine: new indication. Dementia and Parkinson's disease: no thank you!" [abstract]," *Prescrire International* 16:88 (2007): 66.

29. No author. "Cholinesterase inhibitors: tremor and exacerbation of Parkinson's disease [abstract]," *Prescrire International* 16:91 (2007): 197-198.

30. Y.T. Kwak, I.W. Han, J. Baik, and M.S. Koo, "Relation between cholinesterase inhibitors and Pisa syndrome [abstract]," *Lancet* 355:9222 (2000): 2222.

31. A. Villarejo, A. Camacho, R. Garcia-Ramos, T. Moreno, M. Penas, R. Juntas, et al, "Cholinergic-dopaminergic imbalance in Pisa syndrome [abstract]," *Clinical Neuropharmacology* 26:3 (2003): 119-121.

32. P. Glynn, "Neuropathy target esterase," *The Biochemical Journal* 344 (1999): 625-631.

33. L.G. Costa, "Current issues in organophosphate toxicology," *Clinica Chimica Acta* 366 (2006): 1-13.

34. J. Liu, T. Chakraborti, and C. Pope, "In vitro effects of organophosphate anticholinesterases on muscarinic receptor-mediated inhibition of acetylcholine release in rat striatum [abstract]," *Toxicology and Applied Pharmacology* 178:2 (2002): 102-108.

35. W. Balduini, M. Cimino, F. Reno, P. Marini, A. Princivalle, and F. Cattabeni, "Effects of postnatal or adult chronic acetylcholinesterase inhibition on muscarinic receptors, phosphoinositide turnover and m1 mRNA expression [abstract]," *European Journal of Pharmacology* 248:4 (1993): 281-288.

36. K. Kobayashi, T. Suzuki, M. Sakamoto, W. Hashimoto, K. Kashiwada, I. Sato, et al, "Brain regional acetylcholinesterase activity and muscarinic acetylcholine receptors in rats after repeated administration of cholinesterase inhibitors and its withdrawal [abstract]," *Toxicology and Applied Pharmacology* 219:2-3 (2007): 151-161.

Endnotes

37. M.G. Giovanni, C. Scali, L. Bartolini, B. Schmidt, and G. Pepeu, "Effect of subchronic treatment with metrifonate and tacrine on brain cholinergic function in aged F344 rats [abstract]," *European Journal of Pharmacology* 354:1 (1998): 17-24.

38. K.H. Haug, I.L. Bogen, H. Osmundsen, I. Walaas, F. Fonnum, "Effects of cholinergic markers in rat brain and blood after short and prolonged administration of donepezil [abstract]," *Neurochemical Research* 30:12 (2005): 1511-1520.

39. R.M. Friedenberg, "Dementia: One of the Greatest Fears of Aging," *Radiology* 229 (2003): 632-635.

40. R.J. Castellani, X. Zhu, H.G. Lee, P.I. Moreira, G. Perry, and M.A. Smith, "Neuropathology and the treatment of Alzheimer's disease: did we lose the forest for the trees [abstract]?," *Expert Review of Neurotherapeutics* 7:5 (2007): 473-485.

41. P.J. Whitehouse and D. George. *The Myth of Alzheimer's: What You Aren't Being Told About Today's Most Dreaded Diagnosis* (New York: St. Martin's Press, 2008).

42. X.L. Ji, L. Yang, Z.M. Shen, J.P. Cheng, G.W. Jin, L.Y. Qu, and W.H. Wang, "Neurotransmitter level changes in domestic ducks (Shaoxing duck) growing up in typical mercury contaminated area in China [abstract]," *Journal of Environmental Sciences* (China) 17:2 (2005): 256-258.

43. Z.X. Shen, "Brain cholinesterases I. The clinico-histopathological and biochemical basis of Alzheimer's disease [abstract]," *Medical Hypotheses* 63:2 (2004): 285-297.

44. Z.X. Shen, "Brain cholinesterases II. The molecular and cellular basis of Alzheimer's disease [abstract]," *Medical Hypotheses* 63:2 (2004): 308-321.

45. Z. Shen, "Brain cholinesterases III. Future perspectives of AD research and clinical practice [abstract]," *Medical Hypotheses* 63:2 (2004): 298-307.

46. Z.X. Shen, "CSF cholinesterase activity in demented and non-demented subjects [abstract]," *NeuroReport* 9:3 (1998): 483-488.

47. Z.X. Shen, "Acetylcholinesterase provides deeper insights into Alzheimer's disease [abstract]," *Medical Hypotheses* 43:1 (1994): 21-30.

48. R.S. Jope, H.J. Baker, and D.J. Connor, "Increased acetylcholine synthesis and release in brains of cats with GM1 gangliosidosis [abstract]," *Journal of Neurochemistry* 46:5 (1986): 1567-1572.

49. E.K. Perry, "The cholinergic system in old age and Alzheimer's disease [abstract]," *Age and Ageing* 9:1 (1980): 1-8.

Endnotes

50. J.F. Lyotard., *The Postmodern Condition: A Report on Knowledge.* Translation from French by Geoff Bennington and Brian Massumi. *Theory and History of Literature, Volume 10* (Minneapolis: University of Minnesota Press, 1984), 46.

51. A. Wimo, B. Winblad, H. Aguero-Torres, and E. von Strauss, "The Magnitude of Dementia Occurrence in the World," *Alzheimer Disease and Associated Disorders* 17:2 (2003): 63-67.

52. M. Knapp, A. Comas-Herrera, A. Somani, and S. Banerjee (May 2007). "Dementia: international comparisons." Summary Report for the National Audit Office. Accessed on September 1, 2008 at: http://www.pssru.ac/uk/pdf/dp2418.pdf

53. D.P. Chapman, S.M. Williams, T.W. Strine, R.F. Anda, and M.J. Moore, "Dementia and Its Implications for Public Health," *Preventing Chronic Disease: Public Health Research, Practice, and Policy* [serial online] April 2006. Available from: http://www.cdc.gov/pcd/issues/2006apr/05_0167.htm

54. L.E. Hebert, P.A. Scherr, J.L. Bienias, D.A. Bennett, and D.A. Evans, "Alzheimer Disease in the U.S. Population," *Archives of Neurology* 60 (2003): 1119-1122.

55. B.L. Plassman, K.M. Langa, G.G. Fisher, S.G. Heeringa, D.R. Weir, M.B. Ofstedal, et al, "Prevalence of Dementia in the United States: The Aging, Demographics, and Memory Study," *Neuroepidemiology* 29 (2007): 125-132.

56. Ibid.

57. Alzheimer's News (March 18, 2008). "10 million U.S. baby boomers will develop Alzheimer's disease." Accessed on February 2, 2009 at: http://www.alz.org/news_and_events_13106.asp

58. Alzheimer's Association. 2008 Alzheimer's Disease Facts and Figures. Accessed on September 2, 2009 at: http://www.alz.org/national/documents/report_alzfactsfigures2008.pdf

59. A. Wimo, B. Winblad, H. Aguero-Torres, and E. von Strauss, "The Magnitude of Dementia Occurrence in the World," *Alzheimer Disease and Associated Disorders* 17:2 (2003): 64.

60. M. Knapp, A. Comas-Herrera, A. Somani, and S. Banerjee (May 2007). "Dementia: international comparisons." Summary Report for the National Audit Office. Accessed on September 1, 2008 at: http://www.pssru.ac/uk/pdf/dp2418.pdf

61. U.S. Department of Health and Human Services. *Mental Health: A Report of the Surgeon General*. (Rockville, MD: U.S. Department of Health and Human Services, Substance Abuse and Mental Health Services Administration, Center for Mental Health Services, National Institutes of Health – National Institute of Mental Health, 1999), 336.

62. S. Boyles (December 23, 2008). "Survey Shows More Than 2 Million People Use Medications That Shouldn't Be Mixed." WebMD. Accessed on February 2, 2009 at: http://www.webmd.com/healthy-aging/news/20081223/older-americans-take-risky-drug-combos

63. Medicare Statistics. Accessed on February 2, 2009 at: http://www.cms.hhs.gov/MedicareMedicaidStatSupp/downloads/Table1.3.pdf

64. Kaiser Foundation (Summer 2005). *Medicare Chart Book, 3rd Edition*. http://www.kaiserfamilyfoundation.org/medicare/upload/Medicare-Chart-Book-3rd-Edition-Summer-2005-Section-7.pdf

Chapter 2 - Mechanisms of Dementia

1. J.H. Growdon and M.N. Rossor, Ed., *The Dementias 2* (Philadelphia: Butterworth Heinemann, 2007).

2. *Stedman's Pocket Medical Dictionary* (Philadelphia: Williams and Wilkins, 1987), 24.

3. A. Guberman, *Essentials of Clinical Practice: An Introduction to Clinical Neurology* (New York: Little, Brown, and Company, 1994), 63-81.

4. K.W. Lindsay and I. Bone, *Neurology and Neurosurgery Illustrated, 3rd Ed.* (New York: Churchill Livingstone, 1997), 121-128.

5. D.M. Kaufman, *Clinical Neurology for Psychiatrists, 4th Ed.* (Philadelphia: W.B. Saunders Company, 1995).

6. J. Brown and J. Hillam, *Your Questions Answered: Dementia* (New York: Churchill Livingstone, 2004).

7. J.H. Growdon and M.N. Rossor, Ed., *The Dementias 2* (Philadelphia: Butterworth Heinemann, 2007).

8. A. Bruck, T. Kurki, V. Kaasinen, T. Vahlberg, and J.O. Rinne, "Hippocampal and prefrontal atrophy in patients with early non-demented Parkinson's disease is related to cognitive impairment," *Journal of Neurology, Neurosurgery, and Psychiatry* 75 (2004): 1467-1469.

9. M. Bozzali, A. Falini, M. Cercignani, F. Baglio, E. Farina, M. Alberoni, et al, "Brain tissue damage in dementia with Lewy bodies : an in vivo diffusion tensor MRI study," *Brain* 128 (2005): 1595-1604.

10. *Dorland's Pocket Medical Dictionary, 24th Ed.* (Philadelphia: W.B. Saunders Company, 1989).

11. J.H. Growdon and M.N. Rossor, Ed., *The Dementias 2* (Philadelphia: Butterworth Heinemann, 2007).

12. E. Kapaki, K. Kilidireas, G.P. Paraskevas, M. Michalopoulou, and E. Patsouris, "Highly increased CSF tau protein and decreased B-amyloid (1-41) in sporadic CJD: a discrimination from Alzheimer's disease?," *Journal of Neurology, Neurosurgery, and Psychiatry* 71 (2001): 401-403.

13. J.H. Growdon and M.N. Rossor, Ed., *The Dementias 2* (Philadelphia: Butterworth Heinemann, 2007).

14. A.H. Ropper and R.H. Brown, *Adams and Victor's Principles of Neurology, 8th Ed.* (New York: McGraw-Hill, 2005), 369.

15. P. Demay-Wechsler, "Iatrogenic lupus-like syndromes [abstract]," *Revue de stomatologie et de chirurgie maxillo-faciale* 77:5 (1976): 727-734.

16. M. Gallien, J.P. Schnetzler, and J. Morin, "Antinuclear antibodies and lupus due to phenothiazines in 600 hospitalized patients [abstract]," *Annales médico-psychologiques* 1:2 (1975): 237-248.

17. A.F. Rami, D. Barkan, D. Mevorach, E. Leitersdorf, and Y. Caraco, "Clozapine-induced systemic lupus erythematosus," *Annals of Pharmacotherapy* 40:5 (2006): 983-985.

18. J. Wolf, A. Sartorius, B. Alm, and F.A. Henn, "Clozapine-induced lupus erythematosus," *Journal of Clinical Psychopharmacology* 24:2 (2004): 236-238.

19. V.R. Shukla and R.L. Borison, "Lithium and lupus-like syndrome [abstract]," *JAMA* 248:8 (1982): 921-922.

20. P. Amerio, C. Innocente, C. Feliciani, D. Angelucci, D. Gambi, and A. Tulli, "Drug-induced cutaneus lupus erythematosus after 5 years of treatment with carbamazepine," *European Journal of Dermatology* 16:3 (2006): 281-283.

21. M. Kajs-Wyllie, "Lupus Cerebritis: A Case Study," *Journal of Neuroscience Nursing* 34:4 (2002): 176-183.

22. A. Wiik, "Drug-induced vasculitis," *Current Opinion in Rheumatology* 20 (2008): 35-39.

23. G. Thomalia, T. Kucinski, C. Weiller, and J. Rother, "Cerebral vasculitis following oral methylphenidate intake in an adult: a case report [abstract]," *World Journal of Biological Psychiatry* 7:1 (2006): 56-58.

24. A. Schteinschnaider, L.L. Plaghos, S. Garbugino, D. Riveros, A. Lazarowski, S. Intruvini, et al, "Cerebral arteritis following methylphenidate use [abstract]," *Journal of Child Neurology* 15:4 (2000): 265-267.

25. J.M. Trugman, "Cerebral arteritis and oral methylphenidate [abstract]," *Lancet* 1988 1:8585 (1988): 584-585.

26. D.G. Bostwick, "Amphetamine induced cerebral vasculitis [abstract]," *Human Pathology* 12:11 (1981): 1031-1033.

27. D.J. Feeney and W.M. Klykylo, "Medication-Induced Seizures," *Journal of the American Academy of Child and Adolescent Psychiatry* 36:8 (1997): 1018-1019.

28. M. Schertz and T. Steinberg, "Seizures induced by the combination of methylphenidate and sertraline [abstract]," *Journal of Child and Adolescent Psychopharmacology* 18:3 (2008): 301-303.

29. F. Pisani, G. Oteri, C. Costa, G. Di Raimondo, and R. Di Perri, "Effects of psychotropic drugs on seizure threshold [abstract]," *Drug Safety* 25:2 (2002): 91-110.

30. D.M. White and A.C. Van Cott, "Clozapine (Clozaril), seizures, and EEG abnormalities [abstract]," *American Journal of Electroneurodiagnostic Technology* 47:3 (2007): 190-197.

31. K.O. Nakken and S.I. Johannessen, "Seizure exacerbation caused by antiepileptic drugs [abstract]," *Tidsskrift for den Norske lægeforening: tidsskrift for praktisk medicin, ny række* 128:18 (2008): 2052-2055.

32. K. Alper, K.A. Schwartz, R.L. Kolts, and A. Khan, "Seizure incidence in psychopharmacological clinical trials: an analysis of Food and Drug Administration (FDA) summary basis of approval reports [abstract]," *Biological Psychiatry* 62:4 (2007): 345-354.

33. S. Thomas, and H. Upadhyaya, "Adderall and Seizures," *Journal of the American Academy of Child and Adolescent Psychiatry* 41:4 (2002): 365.

34. A. Ickowicz, "Bupropion-methylphenidate combination and grand mal seizures [abstract]," *Canadian Journal of Psychiatry* 47:8 (2002): 790-791.

35. D. Taylor, "Antidepressant drugs and cardiovascular pathology: a clinical overview of effectiveness and safety [abstract]," *Acta Psychiatrica Scandinavica* 118:6 (2008): 434-442.

36. D.C. Henderson, T.B. Daley, L. Kunkel, M. Rodriguez-Scott, P. Koul, and D. Hayden, "Clozapine and hypertension: a chart review of 82 patients [abstract]," *Journal of Clinical Psychiatry* 65:5 (2004): 686-689.

37. Y.S. Woo, W. Kim, J.H. Chae, B.H. Yoon, and W.M. Bahk, "Blood pressure changes during clozapine or olanzapine treatment in Korean schizophrenic patients [abstract]," *World Journal of Biological Psychiatry* 6:1-6 (2008) epub.

38. T.E. Wilens, P.G. Hammerness, J. Biederman, A. Kwon, T.J. Spencer, S. Clark, et al, "Blood pressure changes associated with medication treatment of adults with attention-deficit/hyperactivity disorder [abstract]," *Journal of Clinical Psychiatry* 66:2 (2005): 253-259.

39. A. Kiev, H.L. Masco, T.L. Wenger, J.A. Johnston, S.R. Batey, and L.C. Holloman, "The cardiovascular effects of bupropion and nortriptyline in depressed outpatients [abstract]," *Annals of Clinical Psychiatry* 6:2 (1994): 107-115.

40. S.P. Roose, G.W. Dalack, A.H. Glassman, S. Woodring, B.T. Walsh, and E.G. Giardina, "Cardiovascular effects of bupropion in depressed patients with heart disease [abstract]," *American Journal of Psychiatry* 148:4 (1991): 512-516.

41. F.O. Bastian and J.W. Foster, "Spiroplasma sp. 16S rDNA in Creutzfeldt-Jakob disease and scrapie as shown by PCR and DNA sequence analysis [abstract]," *Journal of Neuropathology and Experimental Neurology* 60:6 (2001): 613-620.

42. C. Farber, *Serious Adverse Events: An Uncensored History of AIDS* (Hoboken, NJ: Melville House Publishing, 2006).

43. R. Root-Bernstein, *Rethinking AIDS: The Tragic Cost of Premature Consensus* (New York: The Free Press, 1993).

44. P.H. Duesberg, *Inventing the AIDS Virus* (Washington, D.C.: Regnery Publishing, Inc., 1996).

45. http://www.rethinkingaids.com/

46. http://www.theperthgroup.com/

47. J. Ishikawa, A. Ishikawa, S. Nakamura, "Interferon-alpha reduces the density of monoaminergic axons in the rat brain [abstract]," *NeuroReport* 18:2 (2007): 137-140.

Endnotes

48. A.R. Sas, H.A. Bimonte-Nelson, and W.R. Tyor, "Cognitive dysfunction in HIV encephalitic SCID mice correlates with levels of Interferon-alpha in the brain [abstract]," *AIDS* 21:16 (2007): 2151-2159.

49. S. Saracchini, E. Vaccher, E. Covezzi, G. Tortorici, A. Carbone, and U. Tirelli, "Lethal neurotoxicity associated with azidothymidine therapy," *Journal of Neurology, Neurosurgery, and Psychiatry* 52:4 (1989): 544-545.

50. G. Famularo, S. Moretti, S. Marcellin, V. Trinchieri, S. Tzantzoglu, G. Santini, et al, "Acetyl-carnitine deficiency in AIDS patients with neurotoxicity on treatment with antiretroviral nucleoside analogues [abstract]," *AIDS* 11:2 (1997): 185-190.

51. E. Papadopulos-Eleopulos, V.F. Turner, J.M. Papadimitriou, D. Causer, H. Alphonso, and T. Miller, "A critical analysis of the pharmacology of AZT and its use in AIDS [abstract]," *Current Medical Research and Opinion* 15 Suppl 1 (1999): S1-S45.

52. R.R. Riedel, L. Bader, L. Gurtler, J. Eberle, W. Gunther, and D. Naber, "Azidothymidine and the nervous system [abstract]," *Fortschritte der Medezin* 109:14 (1991): 302-306.

53. S. Stubner, R. Grohmann, R. Engel, B. Bandelow, W.D. Ludwig, G. Wagner, et al, "Blood dyscrasias induced by psychotropic drugs [abstract]," *Pharmacopsychiatry* 37 Suppl 1 (2004): S70-S78.

54. C.F. Yen, M.Y. Chong, M.C. Kuo, and C.S. Chang, "Severe granulocytopenia secondary to chlorpromazine despite concurrent lithium treatment: a case report [abstract]," *Kaohsiung Journal of Medical Science* 13:10 (1997): 635-638.

55. F.D. Vilinksy and A. Lubin, "Severe neutropenia associated with fluoxetine hydrochloride [abstract]," *Annals of Internal Medicine* 127:7 (1997): 573-574.

56. T. Ozcanli, B. Unsalver, S. Ozdemir, and M. Ozmen, "Sertraline- and mirtazapine-induced severe neutropenia [abstract]," *American Journal of Psychiatry* 162:7 (2005): 1386.

57. D.S. Gravenor, J.R. Leclerc, and G. Blake, "Tricyclic antidepressant agranulocytosis [abstract]," *Canadian Journal of Psychiatry* 31:7 (1986): 661.

58. A. Ahmed, "Neutropenia associated with mirtazapine use: is a drop in the neutrophil count in a symptomatic older adult a cause for concern [abstract]?," *Journal of the American Geriatric Society* 50:8 (2002): 1461-1463.

59. R. Perrichot, O. Reman, B. Hurault de Ligny, E. Albengres, A. Batho, J.P. Ryckelynck, et al, "Agranulocytosis induced by diazepam and then midazolam [abstract]," *Presse Medicale* 19:16 (1990): 764.

60. F. Lechin, B. van der Dijs, and M. Benaim, "Benzodiazepines: tolerability in elderly patients [abstract]," *Psychotherapy and Psychosomatics* 65:4 (1996): 171-182.

61. F. Lechin, B. van der Dijs, G. Vitelli-Flores, S. Baez, M.E. Lechin, A.E. Lechin, et al, "Peripheral blood immunological parameters in long-term benzodiazepine users [abstract]," *Clinical Neuropharmacology* 17:1 (1994): 63-72.

62. H.B. Solvason, "Agranulocytosis associated with lamotrigine," *American Journal of Psychiatry* 157:10 (2000): 1704.

63. M.S. Keshavan and J.S. Kennedy, Ed., *Drug-Induced Dysfunction in Psychiatry* (New York: Hemisphere Publishing Company 1992), 287-292, 311-324.

64. M.E. Sobaniec-Lotowska and W. Sobaniec, "Morphological features of encephalopathy after chronic administration of the antiepileptic drug valproate to rats. A transmission electron microscopic study of capillaries in the cerebellar cortex [abstract]," *Experimental Toxicology and Pathology* 48:1 (1996): 65-75.

65. H.V. Tran, E. Ravinet, C. Schnyder, M. Reichhart, and Y. Guex-Crosier, "Blood-brain barrier disruption associated with topiramate-induced angle-closure glaucoma of acute onset [abstract]," *Klinische Monatsblätter für Augenheilkunde* 223:5 (2006): 425-427.

66. D. Ben-Shachar, E. Livne, I. Spanier, K.L. Leenders, and M.B. Youdim, "Typical and atypical neuroleptics induce alteration in blood-brain barrier and brain 59FeCl3 uptake," *Journal of Neurochemistry* 62:3 (1994): 1112-1118.

67. S.H. Preskorn, M.E. Raichle, and B.K. Hartman, "Antidepressants alter cerebrovascular permeability and metabolic rates in primates," *Science* 217:4556 (1982): 250-252.

68. S.H. Preskorn, B.K. Hartman, M.E. Raichle, and H.B. Clark, "The effect of dibenzazepines (tricyclic antidepressants) on cerebral capillary permeability in the rat in vivo [abstract]," *Journal of Pharmacology and Experimental Therapeutics* 213:2 (1980): 313-320.

69. A. Sarmento, A. Albino-Teixeira, and I. Azevedo, "Amitriptyline-induced morphological alterations of the rat blood-brain barrier [abstract]," *European Journal of Pharmacology* 176:1 (1990): 69-74.

70. R. Sankar, E. Blossom, K. Clemons, and P. Charles, "Age-associated changes in the effects of amphetamine on the blood-brain barrier of rats [abstract]," *Neurobiology of Aging* 4:1 (1983): 65-68.

71. U. Braun, G. Braun, and T. Sargent 3rd, "Changes in blood-brain permeability resulting from D-amphetamine, 6-hydroxydopamine and pimozide measured by a new technique [abstract]," *Experientia* 36:2 (1980): 207-209.

72. B.B. Johansson and C. Carlsson, "Amphetamine-induced increase of the cerebrovascular permeability to protein [abstract]," *Acta Neurologica Scandinavica Supplementum* 64 (1977): 62-63.

73. M.S. Keshavan and J.S. Kennedy, Ed., *Drug-Induced Dysfunction in Psychiatry* (New York: Hemisphere Publishing Company 1992), 209-218.

74. C.T. Gualtieri and L.G. Johnson, "Comparative neurocognitive effects of 5 psychotropic anticonvulsants and lithium," *Medscape General Medicine* 8:3 (2006): 46.

75. G.J. Dumont, S.J. de Visser, A.F. Cohen, and J.M. van Gerven, "Biomarkers for the effects of selective serotonin reuptake inhibitors (SSRIs) in healthy subjects," *British Journal of Clinical Pharmacology* 59:5 (2005): 495-510.

76. M. Fava, L.M. Graves, F. Benazzi, M.J. Scalia, D.V. Iosifescu, J.E. Alpert, and G.I. Papakostas, "A Cross-Sectional Study of the Prevalence of Cognitive and Physical Symptoms During Long-Term Antidepressant Treatment," *Journal of Clinical Psychiatry* 67 (2006): 1754-1759.

77. A. Brunnauer, G. Laux, E. Geiger, M. Soyka, and H.J. Moller, "Antidepressants and driving ability: results from a clinical study [abstract]," *Journal of Clinical Psychiatry* 67:11 (2006): 1776-1781.

78. M. Wingen, J.G. Ramaekers, and J.A. Schmitt, "Driving impairment in depressed patients receiving long-term antidepressant treatment [abstract]," *Psychopharmacology* 188:1 (2006): 84-91.

79. F. Ridout, R. Meadows, S. Johnsen, and I. Hindmarch, "A placebo controlled investigation into the effects of paroxetine and mirtazapine on measures related to car driving performance," *Human Psychopharmacology* 18 (2003): 261-269.

80. S.A. Chong, Mythily, and R. Mahendran, "Cardiac effects of psychotropic drugs [abstract]," *Annals of the Academy of Medicine Singapore* 30:6 (2001): 625-631.

81. R. Constans, A. Chapelet, J. Marco, P. Dardenne, "Reversible sinus dysfunction during treatment with lithium carbonate [abstract]," *Acta Cardiologica* 33:5 (1978): 315-322.

82. A. Hagman, K. Arnman, and L. Ryden, "Syncope caused by lithium treatment. Report on two cases and a prospective investigation of the prevalence of lithium-induced sinus node dysfunction [abstract]," *Acta Medica Scandinavica* 205:6 (1979): 467-471.

83. K.A. Hewetson, A.E. Ritch, and R.D. Watson, "Sick sinus syndrome aggravated by carbamazepine therapy for epilepsy," *Postgraduate Medical Journal* 62:728 (1986): 497-498.

84. S. Stone and L.S. Lange, "Syncope and sudden unexpected death attributed to carbamazepine in a 20-year old epileptic," *Journal of Neurology, Neurosurgery, and Psychiatry* 49:12 (1986): 1460-1461.

85. R. Feder, "Bradycardia and syncope induced by fluoxetine [abstract]," *Journal of Clinical Psychiatry* 52:3 (1991) 139.

86. L.E. McAnally, K.R. Threlkeld, and C.A. Dreyling, "Case report of a syncopal episode associated with fluoxetine [abstract]," *Annals of Pharmacotherapy* 26:9 (1992): 1090-1091.

87. P.M. Haddad and S.M. Dursun, "Neurological complications of psychiatric drugs: clinical features and management [abstract]," *Human Psychopharmacology* 23 Suppl 1 (2008): 15-26.

88. H. Allain, D. Bentue-Ferrer, E. Polard, Y. Akwa, and A. Patat, "Postural instability and consequent falls and hip fractures associated with the use of hypnotics in the elderly: a comparative review [abstract]," *Drugs and Aging* 22:9 (2005): 749-765.

89. M. Rudzinksa, S. Bukowszan, K. Banaszkiewicz, J. Stozek, K. Zajdel, and A. Szczudlik, "Causes and risk factors of falls in patients with Parkinson's disease [abstract]," *Neurologia i neurochirurgia polska* 42:3 (2008): 216-222.

90. F. Landi, G. Onder, M. Cesari, C. Barillaro, A. Russo, R. Bernabei, et al, "Psychotropic medications and risk for falls among community-dwelling frail older people: an observational study [abstract]," *The Journals of Gerontology. Series A, Biological Sciences and Medical Sciences* 60:5 (2005): 622-626.

91. R.M. Leipzig, R.G. Cumming, and M.E. Tinetti, "Drugs and falls in older people: a systematic review and meta-analysis: I. Psychotropic drugs [abstract]," *Journal of the American Geriatric Society* 47:1 (1999): 30-39.

92. M.S. Keshavan and J.S. Kennedy, Ed. *Drug-Induced Dysfunction in Psychiatry* (New York: Hemisphere Publishing Company 1992), 293-310.

93. E. Arvat, R. Giordano, S. Grottoli, and E. Ghigo, "Benzodiazepines and anterior pituitary function [abstract]," *Journal of Endocrinological Investigation* 25:8 (2002): 735-747.

94. T. Humbert, "Neuroendocrine effects of benzodiazepines [abstract]," *Annales médico-psychologiques* 152:3 (1994): 161-171.

95. R.J. Urban, P. Harris, and B. Masel, "Anterior hypopituitarism following traumatic brain injury," *Brain Injury* 19:5 (2005): 349-358.

96. J.R. Dusick, C. Wang, P. Cohan, R. Swerloff, and D.F. Kelly, "Chapter 1: pathophysiology of hypopituitarism in the setting of brain injury," *Pituitary* 2008 (epub).

97. F. Tanriverdi, H. Ulutabanca, K. Unluhizarci, A. Selcuklu, F.F. Casaneuva, and F. Kelestimur, "Three years prospective investigation of anterior pituitary function after traumatic brain injury: a pilot study," *Clinical Endocrinology* 68 (2008): 573-579.

98. O. Dogru, R. Koken, A. Bukulmez, H. Melek, F. Ovali, and R. Albayrak, "Delay in diagnosis of hypopituitarism after traumatic head injury: A case report and review of the literature," *Neuroendocrinology Letters* 4:26 (2005): 311-313.

99. A. McDonald, M. Lindell, D.B. Dunger, and C.L. Acerini, "Traumatic Brain Injury is a Rarely Reported Cause of Growth Hormone Deficiency," *Journal of Pediatrics* 152 (2008): 590-593.

100. A.Villarejo, J. Diaz-Guzman, and F. Bermejo-Pareja, "Dementia in hypopituitarism," *European Journal of Neurology* 15:6 (2008): e53-354.

101. L.A. Behan, J. Phillips, C.J. Thompson, and A. Agha, "Neuroendocrine disorders after traumatic brain injury [abstract]," *Journal of Neurology, Neurosurgery, and Psychiatry* 79:7 (2008): 753-759.

102. E.H. Nielsen, J. Lindholm, and P. Laurberg, "Excess mortality in women with pituitary disease: a meta-analysis," *Clinical Endocrinology* 67 (2007): 693-697.

103. J.W. Tomlinson, N. Holden, R.K. Hills, K. Wheatley, R.N. Clayton, A.S. Bates, et al, "Association between premature mortality and hypopituitarism," *Lancet* 357 (2001): 425-431.

104. E.M.T. Erfurth, B. Bulow, and LE. Hagmar, "Is Vascular Mortality Increased in Hypopituitarism?," *Pituitary* 3 (2000): 77-81.

Endnotes

105. A.S. Bates, W. Van't Hoff, P.J. Jones, and R.N. Clayton, "The Effect of Hypopituitarism on Life Expectancy," *Journal of Clinical Endocrinology and Metabolism* 81:3 (1996): 1169-1172.

106. VA/DoD Clinical Practice Guidelines (May 2000). Major Depressive Disorder. Accessed on June 8, 2008 at:
http://www.oqp.med.va.gov/cpg/MDD/MDD_cpg/content/MDDParticipants_1.htm
http://www.oqp.med.va.gov/cpg/MDD/MDD_cpg/content/appendices/mdd_app3.htm
http://www.oqp.med.va.gov/cpg/MDD/MDD_cpg/content/appendices/mdd_app5.htm
http://www.oqp.med.va.gov/cpg/MDD/MDD_cpg/content/appendices/mdd_app8.htm

107. American Psychiatric Association (November 2004). Treatment of Patients with Acute Stress Disorder and Posttraumatic Stress Disorder. Accessed on June 8, 2008 at:
http://www.psychiatryonline.com/pracGuide/pracGuideTopic_11.aspx

108. VA/DoD Clinical Practice Guideline for the Management of Post-Traumatic Stress (January 2004). Accessed on June 10, 2008 at:
http://www.oqp.med.va.gov/cpg/PTSD/G/PTSD_about.htm

109. National Institute of Mental Health (revised 2008). Post-Traumatic Stress Disorder (PTSD). Accessed on February 9, 2009 at:
http://www.nimh.nih.gov/health/publications/post-traumatic-stress-disorder-ptsd/summary.shtml

110. No author. Department of Defense / Tricare. Transforming Strategy into Action – 2007 MHS Conference – Partners in Excellence. "Traumatic Brain Injuries: Pathophysiology, Treatment, and Prevention." Accessed on February 9, 2009 at:
www.tricare.mil/conferences/2007/Mon/M213-(1).ppt

111. Department of Veterans Affairs. Quick Guide – Patient/Family: Traumatic Brain Injury. Accessed on February 9, 2009 at:
http://www1.va.gov/environagents/docs/TBI-handout-patients.pdf

112. Department of Veterans Affairs. Quick Guide – Provider: Traumatic Brain Injury. Accessed on February 9, 2009 at:
http://www1.va.gov/environagents/docs/TBI-handout-physicians.pdf

113. N. Carney, H. Maynard, P. Patterson, N.C. Mann, and M. Helfand (February 1999). Health Services Technology Assessment. Agency for Health Care Policy and Research. "Rehabilitation for Traumatic Brain Injury." Accessed on February 9, 2009 at:
http://www.ncbi.nlm.nih.gov/books/bv.fcgi?rid=hstat1.chapter.1280

114. C.W. Hoge, D. McGurk, J.L. Thomas, A.L. Cox, C.C. Engel, and C.A. Castro, "Mild Traumatic Brain Injury in U.S. Soldiers Returning from Iraq," *NEJM* 358:2 (2008): 453-463.

115. D.L. Warden, B. Gordon, T.W. McAllister, J.M. Silver, J.T. Barth, J. Bruns, et al, "Guidelines for the Pharmacologic Treatment of Neurobehavioral Sequelae of Traumatic Brain Injury," *Journal of Neurotrauma* 23:10 (2006): 1468-1501.

116. T.A. Elhadd, T.A. Abdu, J. Oxtoby, G. Kennedy, M. McLaren, R. Neary, et al, "Biochemical and Biophysical Markers of Endothelial Dysfunction in Adults with Hypopituitarism and Severe GH Deficiency," *Journal of Clinical Endocrinology and Metabolism* 86:9 (2001): 4223-4232.

117. H. Oflaz, F. Sen, A. Elitok, A.O. Cimen, I. Onur, E. Kasicioglu, et al, "Coronary flow reserve is impaired in patients with adult growth hormone (GH) deficiency," *Clinical Endocrinology* 66 (2007): 524-529.

118. B. Merola, A. Cittadini, A. Colao, S. Longobardi, S. Fazio, D. Sabatini, et al, "Cardiac Structural and Functional Abnormalities in Adult Patients with Growth Hormone Deficiency," *Journal of Clinical Endocrinology and Metabolism* 77:6 (1993): 1658-1661.

119. S. Bird, "Failure to diagnose: Addison disease," *Australian Family Physician* 36:10 (2007): 859-861.

120. J. Ringstad, S. Rodge, W. Loland, and L. Rode, "Rapidly fatal Addison's disease: three case reports [abstract]," *Journal of Internal Medicine* 230:5 (1991): 465-467.

121. G.R. Heninger, P.L. Delgado, and D.S. Charney, "The revised monoamine theory of depression: A modulatory role for monoamines, based on new findings from monoamine depletion experiments in humans," *Pharmacopsychiatry* 29 (1996): 2-11.

122. H.G. Ruhe, N.S. Mason, and A.H. Schene, "Mood is indirectly related to serotonin, norepinephrine and dopamine levels in humans: a meta-analysis of monoamine depletion studies [abstract]," *Molecular Psychiatry* 12:4 (2007): 331-359.

123. M.K. Spillmann, A.J. Van der Does, M.A. Rankin, R.D. Vuolo, J.E. Alpert, A.A. Nierenberg, et al, "Tryptophan depletion in SSRI-recovered depressed outpatients [abstract]," *Psychopharmacology* 155:2 (2001): 123-127.

124. P.L. Delgado (2003). "The Alphabet Soup of Antidepressant Pharmacology: From TCAs and MAOIs to SSRIs, SNRIs, and Beyond." Presented as part of Medscape CME, Postgraduate Institute for Medicine. Optimizing Efficacy and Tolerability of Antidepressant Therapy: Does Selectivity of Action Matter? Accessed on August 23, 2007 at: http://www.medscape.com/viewarticle/458647_4

125. P.L. Haynes, J.R. McQuaid, J. Kelsoe, M. Rapaport, and J.C. Gillin, "Affective state and EEG sleep profile in response to rapid tryptophan depletion in recently recovered nonmedicated depressed individuals," *Journal of Affective Disorders* 83 (2004): 253-262.

126. A.J.W. Van der Does and L. Booij, "Cognitive Therapy Does Not Prevent a Response to Tryptophan Depletion in Patients also Treated with Antidepressants," *Biological Psychiatry* 58 (2005): 913-915.

127. J.P. O'Reardon, M.P. Chopra, A. Bergan, R. Gallop, R.J. DeRubeis, and P. Crits-Christoph, "Response to tryptophan depletion in major depression treated with either cognitive therapy or selective serotonin reuptake inhibitor antidepressants [abstract]," *Biological Psychiatry* 55:9 (2004): 957-959.

128. S. Caccia, S. Confalonieri, A. Bergami, C. Fracasso, M. Anelli, and S. Garattini, "Neuropharmacological Effects of Low and High Doses of Repeated Oral Dexfenfluramine in Rats: A Comparison with Fluoxetine," *Pharmacology Biochemistry and Behavior* 57:4 (1997): 851-856.

129. P.V. Hrdina, "Regulation of high- and low-affinity [^{3}H]imipramine recognition sites in rat brain by chronic treatment with antidepressants," *European Journal of Pharmacology* 138 (1987): 159-168.

130. R.J. Baldessarini, E.R. Marsh, and N.S. Kula, "Interactions of fluoxetine with metabolism of dopamine and serotonin in rat brain regions," *Brain Research* 579 (1992): 152-156.

131. J.B. Long, W.Y. Youngblood, and J.S. Kizer, "Regional differences in the Response of Serotonergic Neurons in Rat CNS to Drugs," *European Journal of Pharmacology* 88 (1983): 89-97.

132. D.T. Wong and F.P. Bymaster, "Subsensitivity of Serotonin Receptors After Long-Term Treatment of Rats With Fluoxetine," *Research Communications in Chemical Pathology and Pharmacology* 32:1 (1981): 41-51.

133. J.F. Rodriguez-Landa, C.M. Contreras, A..G. Gutierrez-Garcia, and B. Bernal-Morale, "Chronic, but Not Acute, Clomipramine or Fluoxetine Treatment Reduces the Spontaneous Firing Rate in the Mesoaccumbens Neurons of the Rat," *Neuropsychobiology* 48 (2003): 116-123.

134. E.C. Hwang, I. Magnussen, and M.H. Van Woert, "Effects of Chronic Fluoxetine Administration on Serotonin Metabolism," *Research Communications in Chemical Pathology and Pharmacology* 29:1 (1980): 79-98.

Endnotes

135. J.H. Trouvin, A.M. Gardier, E. Chanut, N. Pages, and C. Jacquot, "Time Course of Brain Serotonin Metabolism After Cessation of Long-Term Fluoxetine Treatment in the Rat," *Life Sciences* 52 (1993): PL187-PL192.

136. M.R. Thompson, K.M. Li, K.J. Clemens, C.G. Gurtman, G.E. Hunt, J.L. Cornish, et al, "Chronic Fluoxetine Treatment Partly Attenuates the Long-Term Anxiety and Depressive Symptoms Induced by MDMA ("Ecstasy") in Rats," *Neuropsycho-pharmacology* 29 (2004): 694-704.

137. T.D. Smith, R. Kuczenski, K. George-Friedman, J.D. Malley, and S.L. Foote, "In Vivo Microdialysis Assessment of Extracellular Serotonin and Dopamine Levels in Awake Monkeys During Sustained Fluoxetine Administration," *Synapse* 38 (2000): 460-470.

138. J.C. Alvarez, M. Sanceaume, C. Advenier, and O. Spreux-Varoquaux, "Differential changes in brain and platelet 5-HT concentrations after steady-state achievement and repeated administration of antidepressant drugs in mice," *European Neuropsycho-pharmacology* 10 (1999): 31-36.

139. A. Carlsson and M. Lindqvist, "Effects of Antidepressant Agents on the Synthesis of Brain Monoamines," *Journal of Neural Transmission* 43 (1978): 73-91.

140. K. Fuxe, S.O. Ogren, L.F. Agnati, K. Andersson, and P. Eneroth, "Effects of Subchronic Antidepressant Drug Treatment on Central Serotonergic Mechanisms in the Male Rat," *Advances in Biochemistry and Psychopharmacology* 31 (1982): 91-107.

141. K. Nakayama, H. Katsu, T. Ando, and R. Nakajo, "Possible alteration of tryptophan metabolism following repeated administration of sertraline in the rat brain," *Brain Research Bulletin* 59:4 (2003): 293-297.

142. T. Segawa and T. Mizuta, "Effect of Imipramine on Central 5-Hydroxytryptamine Turnover and Metabolism in Rats," *Japanese Journal of Pharmacology* 30 (1980): 789-793.

143. T. Pringsheim, M. Diksic, C. Dobson, K. Nguyen, and E. Hamel, "Selective decrease in serotonin synthesis rate in rat brainstem raphe nuclei following chronic administration of low doses of amitriptyline: an effect compatible with an anti-migraine effect," *Cephalalgia* 23:5 (2003): 367-375.

144. S. Caccia, M. Anelli, A.M. Codegoni, C. Fracasso, and S. Garattini, "The effects of single and repeated anorectic doses of 5-hydroxytryptamine uptake inhibitors on indole levels in rat brain," *British Journal of Pharmacology* 110 (1993): 355-359.

145. F. Yamane, H. Okazawa, P. Blier, and M. Diksic, "Reduction in serotonin synthesis following acute and chronic treatments with paroxetine, a selective serotonin reuptake inhibitor, in rat brain: An autoradiographic study with alpha-[^{14}C]methyl-L-tryptophan," *Biochemical Pharmacology* 62 (2001): 1481-1489.

146. H. Miura, T. Kitagami, and N. Ozaki, "Suppressive Effect of Paroxetine, a Selective Serotonin Reuptake Inhibitor, on Tetrahydrobiopterin Levels and Dopamine as Well as Serotonin Turnover in the Mesoprefrontal System of Mice," *Synapse* 61 (2007): 698-706.

147. T. Goto, F. Kuzuya, H. Endo, T. Tajima, and H. Ikari, "Some effects of CNS cholinergic neurons on memory [abstract]," *Journal of Neural Transmission Supplement* 30 (1990): 1-11.

148. K. Mishima, K. Iwasaki, H. Tsukikawa, Y. Matsumoto, N. Egashira, K. Abe, et al, "The scopolamine-induced impairment of spatial cognition parallels the acetylcholine release in the ventral hippocampus in rats [abstract]," *Japanese Journal of Pharmacology* 84:2 (2000): 163-173.

149. H. Watanabe and H. Shimizu, "Effect of anticholinergic drugs on striatal acetylcholine release and motor activity in freely moving rats studied by brain microdialysis [abstract]," *Japanese Journal of Pharmacology* 51:1 (1989): 75-82.

150. M.J. Stillman, B. Shukitt-Hale, R.L. Galli, A. Levy, and H.R. Lieberman, "Effects of M2 antagonists on in vivo hippocampal acetylcholine levels [abstract]," *Brain Research Bulletin* 41:4 (1996): 221-226.

151. K.L. Marek, D.M. Bowen, N.R. Sims, and A.N. Davison, "Stimulation of acetylcholine synthesis by blockade of presynaptic muscarinic inhibitory autoreceptors: observations in rat and human brain preparations and comparison with the effect of choline [abstract]," *Life Sciences* 30:18 (1982): 1517-1524.

152. M.L. Catterson and R.L. Martin, "Anticholinergic toxicity masquerading as neuroleptic malignant syndrome: a case report and review [abstract]," *Annals of Clinical Psychiatry* 6:4 (1994): 267-269.

153. S. Singh and N. Sharma, "Neurological syndromes following organophosphate poisoning," *Neurology India* 48:4 (2000): 308-313.

154. A. Moretto and L. Marcello, "Poisoning by organophosphorus insecticides and sensory neuropathy," *Journal of Neurology, Neuropsychiatry, and Psychiatry* 64 (1998):463-468.

155. G. Ochi, K. Watanabe, H. Tokuoka, S. Hatakenaka, and T. Arai, "Neuroleptic Malignant-like syndrome: a complication of acute organophosphate poisoning," *Canadian Journal of Anaesthesia* 42:11 (1995): 1027-1030.

156. C.J. Pavlis, E.C. Kutscher, R.M. Carnahan, W.K. Kennedy, S. Van Gerpen, and E. Schlenker, "Rivastigmine-induced dystonia," *American Journal of Health-System Pharmacy* 64 (2007): 2468-2470.

157. D.L. Stevens, M.R. Lee, and Y. Padua, "Olanzapine-associated neuroleptic-malignant syndrome in a patient receiving concomitant rivastigmine therapy [abstract]," *Pharmacotherapy* 28:3 (2008): 403-405.

158. K.D. Katz, D.E. Brooks, M.C. Furtado, and L. Chan, "Toxicity, Organophosphate." eMedicine. Accessed on February 14, 2009 at: http://emedicine.medscape.com/article/167726-overview

159. M.C. Mancini and G.G. Deshpande, "Neuroleptic Malignant Syndrome." eMedicine. Accessed on February 14, 2009 at: http://emedicine.medscape.com/article/907949-print

160. N. Vanacore, G. Suzzareddu, M. Maggini, A. Casula, P. Capelli, and R. Raschetti, "Pisa syndrome in a cohort of Alzheimer's disease patients [abstract]," *Acta Neurologica Scandinavica* 111:3 (2005): 199-201.

161. Y. Wang, D. Zhou, B. Wang, H. Li, H. Chai, Q. Zhou, et al, "A kindling model of pharmaceutical temporal lobe epilepsy in Sprague-Dawley rats induced by Coriaria lactone and its possible mechanism," *Epilepsia* 44:4 (2003): 475-488.

162. C. Schaub, M. Uebachs, and H. Beck, "Diminished Response of CA1 Neurons to Antiepileptic Drugs in Chronic Epilepsy," *Epilepsia* 48:7 (2007): 1339-1350.

163. W. Loscher, "Drug Transporters in the Epileptic Brain," *Epilepsia* 48 Supplement 1 (2007): 8-13.

164. H. Beck, "Plasticity of Antiepileptic Drug Targets," *Epilepsia* 48 Supplement 1 (2007): 14-18.

165. W. Loscher and D. Schmidt, "Experimental and Clinical Evidence for Loss of Effect (Tolerance) during Prolonged Treatment with Antiepileptic Drugs," *Epilepsia* 47:8 (2006): 1253-1284.

166. C. Allison and J.A. Pratt, "Neuroadaptive processes in GABAergic and glutamatergic systems in benzodiazepine dependence," *Pharmacology and Therapeutics* 98 (2003): 171-195.

Endnotes

167. D.J. Roca, I. Rozenberg, M. Farrant, and D.H. Farb, "Chronic agonist exposure induces down-regulation and allosteric uncoupling of the gamma-aminobutyric acid/benzodiazepine receptor complex [abstract]," *Molecular Pharmacology* 37:1 (1990): 37-43.

168. S.M. Aburawi, A.S. Elhuwuegi, S.S. Ahmed, S.F. Saad, and A.S. Attia, "Effects of acute and chronic triazolam treatments on brain GABA levels in albino rats," *Acta neurobiologiae experimentalis* 60 (2000): 447-455.

169. E. Barnes, "Use Dependent Regulation of GABA-A Receptors," *International Review of Neurobiology* 39 (1996): 53-76.

170. G.R. Siggins and J.E. Schultz, "Chronic treatment with lithium or desipramine alters discharge frequency and norepinephrine responsiveness of cerebellar Purkinje cells," *Proceedings of the National Academy of Sciences* USA 76:11 (1979): 5987-5991.

171. M.A. Banchaabouchi, S.P. de Ortiz, R. Menendez, K. Ren, and C.S. Maldonado-Vlaar, "Chronic lithium decreases Nurr1 expression in the rat brain and impairs spatial discrimination," *Pharmacology Biochemistry and Behavior* 79 (2004): 607-621.

172. O. Hetmar, M. Nielsen, and C. Braestrup, "Decreased Number of Benzodiazepine Receptors in Frontal Cortex of Rat Brain Following Long-Term Lithium Treatment," *Journal of Neurochemistry* 41 (1983): 217-221.

173. A.N. Samaha, P. Seeman, J. Stewart, H. Rajabi, and S. Kapur, "'Breakthrough' Dopamine Supersensitivity during Ongoing Antipsychotic Treatment Leads to Treatment Failure Over Time," *Journal of Neuroscience* 27:11 (2007): 2979-2986.

174. S.M. Boye and P.P. Rompre, "Behavioral Evidence of Depolarization Block of Dopamine Neurons after Chronic Treatment with Haloperidol and Clozapine," *Journal of Neuroscience* 20:3 (2000): 1229-1239.

175. L.A. Chiodo and B.S. Bunney, "Population response of midbrain dopaminergic neurons to neuroleptics: further studies on time course and nondopaminergic neuronal influences [abstract]," *Journal of Neuroscience* 7:3 (1987): 629-633.

176. H. Moore, C.L. Todd, and A.A. Grace, "Striatal Extracellular Dopamine Levels in Rats with Haloperidol-Induced Depolarization Block of Substantia Nigra Dopamine Neurons," *Journal of Neuroscience* 18:13 (1998): 5068-5077.

177. C.D. Blaha and R.F. Lane, "Chronic treatment with classical and atypical antipsychotic drugs differentially decreases dopamine release in striatum and nucleus accumbens in vivo," *Neuroscience Letters* 78:2 (1987): 199-204.

178. G. Grunder, I. Vernaleken, M.J. Muller, E. Davids, N. Heydari, H.G. Buccholz, et al, "Subchronic Haloperidol Downregulates Dopamine Synthesis Capacity in the Brain of Schizophrenic Patients *In Vivo*," *Neuropsychopharmacology* 28 (2003): 787-794.

179. T.P. Giordano, S.S. Satpute, J. Striessnig, B.E. Kosofksy, and A.M. Rajadhyasksha, "Up-regulation of dopamine D(2)L mRNA levels in the ventral tegmental area and dorsal striatum of amphetamine-sensitized C57BL/6 mice: role of Ca(v)1.3 L-type CA(2+) channels [abstract]," *Journal of Neurochemistry* 99:4 (2006): 1197-1206.

180. P.K. Thanos, M. Michaelides, H. Benveniste, G.J. Wang, and N.D. Volkow, "Effects of chronic oral methylphenidate on cocaine self-administration and striatal dopamine D2 receptors in rodents [abstract]," *Pharmacology Biochemistry and Behavior* 87:4 (2007): 426-433.

181. P. Seeman, T. Tallerico, F. Ko, C. Tenn, and S. Kapur, "Amphetamine-sensitized animals show a marked increase in dopamine D2 high receptors occupied by endogenous dopamine, even in the absence of acute challenges [abstract]," *Synapse* 46:4 (2002): 235-239.

182. P. Seeman, P.N. McCormick, and S. Kapur, "Increased dopamine D2(High) receptors in amphetamine-sensitized rats, measured by the agonist [^{3}H]+PHNO [abstract]," *Synapse* 61:5 (2007): 263-267.

183. A. Imperato, M.C. Obinu, G. Carta, M.S. Mascia, M.A. Casu, and G.L. Gessa, "Reduction of dopamine release and synthesis by repeated amphetamine treatment: role in behavioral sensitization [abstract]," *European Journal of Pharmacology* 317:1-2 (1996): 231-237.

184. R.Y. Shen, K.C. Choong, and A.C. Thomson, "Long-Term Reduction in Ventral Tegmental Area Dopamine Neuron Population Activity Following Repeated Stimulant or Ethanol Treatment," *Biological Psychiatry* 61 (2007): 93-100.

185. Y. Schmitz, C.J. Lee, C. Schmauss, F. Gonon, and D. Sulzer, "Amphetamine Distorts Stimulation-Dependent Dopamine Overflow: Effects on D2 Autoreceptors, Transporters, and Synaptic Vesicle Stores," *Journal of Neuroscience* 21:16 (2001): 5916-5924.

186. P.B. Yang, A.C. Swann, and N. Dafny, "Dose-response characteristics of methylphenidate on locomotor behavior and on sensory evoked potentials recorded from the VTA, NAc, and PFC in freely behaving rats," *Behavioral and Brain Functions* 2:3 (2006).

187. M. Federici, R. Geracitano, G. Bernardi, and N.B. Mercuri, "Actions of Methylphenidate on Dopaminergic Neurons of the Ventral Midbrain," *Biological Psychiatry* 57 (2005): 361-365.

188. H. Forssberg, E. Fernell, S. Waters, N. Waters, and H. Tedroff, "Altered pattern of brain dopamine synthesis in male adolescents with attention deficit hyperactivity disorder," *Behavioral and Brain Functions* 2:40 (2006).

189. P.H. Lipkin, I.J. Goldstein, and A.R. Adesman, "Tics and dyskinesias associated with stimulant treatment in attention-deficit hyperactivity disorder [abstract]," *Archives of Pediatric and Adolescent Medicine* 148:8 (1994): 859-861.

190. C.U. Correll and H.E. Carlson, "Endocrine and Metabolic Adverse Effects of Psychotropic Medications in Children and Adolescents," *Journal of the American Academy of Child and Adolescent Psychiatry* 45:7 (2006): 771-791.

191. M. Gitlin, L.L. Altshuler, M.A. Frye, R. Suri, E.L. Huynh, L. Fairbanks, et al, "Peripheral thyroid hormones and response to selective reuptake inhibitors," *Journal of Psychiatry and Neuroscience* 29:5 (2004): 383-386.

192. A. Bereket, S. Turan, M.G. Karaman, G. Haklar, F. Ozbay, and M.Y. Yazgan, "Height, Weight, IGF-1, IGFB-3 and Thyroid Functions in Prepubertal Children with Attention Deficit Hyperactivity Disorder: Effect of Methylphenidate Treatment," *Hormone Research* 63 (2005): 159-164.

193. M.S. Benedetti, R. Whomsley, E. Baltes, and F. Tonner, "Alteration of thyroid hormone homeostasis by antiepileptic drugs in humans: involvement of glucuronosyltransferase induction," *European Journal of Clinical Pharmacology* 61 (2005): 863-872.

194. C. Constatinou, S. Bolaris, T. Valcana, and M. Margarity, "Diazepam affects the nuclear thyroid hormone receptor density and their expression levels in adult rat brain," *Neuroscience Research* 52 (2005): 269-275.

195. A. Greenspan, G. Gharabawi, and J. Kwentus, "Thyroid Dysfunction During Treatment With Atypical Antipsychotics," *Journal of Clinical Psychiatry* 66:10 (2005): 1334-1335.

196. T. Baptista, D. Reyes, and L. Hernandez, "Antipsychotic Drugs and Reproductive Hormones: Relationship to Body Weight Regulation," *Pharmacology Biochemistry and Behavior* 62:3 (1999): 409-417.

197. A.H.V. Schapira, "The 'new' mitochondrial disorders," *Journal of Neurology, Neurosurgery, and Psychiatry* 72 (2002): 144-149.

198. J.C. von Kleist-Retzow, U. Schauseil-Zipf, D.V. Michalk, and W.S. Kunz, "Mitochondrial diseases – an expanding spectrum of disorders and affected genes," *Experimental Physiology* 88:1 (2003): 155-166.

199. J. Finsterer, "Cognitive decline as a manifestation of mitochondrial disorders (mitochondrial dementia)," *Journal of the Neurological Sciences* 272:1-2 (2008): 20-33.

200. J. Finsterer, "Central nervous system manifestations of mitochondrial disorders," *Acta Neurologica Scandinavica* 114 (2006): 217-238.

201. J. Neustadt and S.R. Pieczenik, "Medication-induced mitochondrial damage and disease," *Molecular Nutrition and Food Research* 52 (2008): 780-788.

202. J.A. Dykens, J.D. Jamieson, L.D. Marroquin, S. Nadanaciva, J.J. Xu, M.C. Dunn, et al, "In vitro assessment of mitochondrial dysfunction and cytotoxicity of nefazodone, trazodone, and buspirone [abstract]," *Toxicological Sciences* 103:2 (2008): 335-345.

203. C. Curti, F.E. Mingatto, A.C.M. Polizello, L.O. Galastri, S.A. Uyemura, and A.C. Santos, "Fluoxetine interacts with the lipid bilayer of the inner membrane in isolated rat brain mitochondria, inhibiting electron transport and F_1F_0-APTase activity," *Molecular and Cellular Biochemistry* 199 (1999): 103-109.

204. S. Ponchaut, F. van Hoof, and K. Veitch, "Cytochrome aa3 depletion is the cause of deficient mitochondrial respiration induced by chronic valproate administration [abstract]," *Biochemistry and Pharmacology* 43:3 (1992): 644-647.

205. S. Ponchaut, F. van Hoof, and K. Veitch, "Valproate and cytochrome c oxidase deficiency," *European Journal of Pediatrics* 154:1 (1995): 79.

206. R. Haas, D.A. Stumpf, J.K. Parks, and L. Eguren, "Inhibitory effects of sodium valproate on oxidative phosphorylation [abstract]," *Neurology* 31:11 (1981): 1473-1476.

207. M.F.B. Silva, J.P.N. Ruiter, L. Ijlst, C. Jakobs, M. Duran, I. Tavares, et al, "Valproate inhibits the mitochondrial pyruvate-driven oxidative phosphorylation in vitro," *Journal of Inherited Metabolic Disorders* 20 (1997): 397-400.

208. M. Nag and N. Nandia, "Antidepressants and Brain Respiration," *Bioscience Reports* 11:1 (1991): 11-14.

209. E.L. Streck, G.T. Rezin, L.M. Barbosa, L.C. Assis, E. Grandi, and J. Quevedo, "Effect of antipsychotics on succinate dehydrogenase and cytochrome oxidase activities in rat brain [abstract]," *Naunyn Schmiedebergs Archives of Pharmacology* 376:1-2 (2007):127-133.

210. C. Burkhardt, J.P. Kelly, Y.H. Lim, C.M. Filley, and W.D. Parker, Jr., "Neuroleptic medication inhibit complex I of the electron transport chain [abstract]," *Annals of Neurology* 33:5 (1993): 512-517.

211. J.A. Prince, M.S. Yassin, and L. Oreland, "Neuroleptic-induced mitochondrial enzyme alterations in the rat brain [abstract]," *Journal of Pharmacology and Experimental Therapeutics* 280:1 (1997): 261-267.

212. M. Colleoni, B. Costa, E. Gori, and A. Santagostino, "Biochemical characterization of the effects of the benzodiazepine, midazolam, on mitochondrial electron transfer," *Pharmacology and Toxicology* 78:2 (1996): 69-76.

213. F. di Jeso, A. Truscello, G. Mattinotti, B. di Jeso, B. Magnani, and A. Martinotti, "Effect of diazepam on mitochondrial respiration [abstract]," *Comptes rendus des séances de la Société de biologie et de ses filiales* 184:1 (1990): 37-40.

214. I.A. Vorobjev and D.B. Zorov, "Diazepam inhibits cell respiration and induces fragmentation of mitochondrial reticulum," *Federation of European Biochemical Societies Letters* 163:2 (1983): 311-314.

215. H. Gonzalez-Pardo, N.M. Conejo, and J.L. Arias, "Oxidative metabolism of limbic structures after acute administration of diazepam, alprazolam, and zolpidem," *Progress in Neuro-Psychopharmacology and Biological Psychiatry* 30 (2006): 1020-1026.

216. J.D. Hirsch, C.F. Beyer, L. Malkowitz, B. Beer, and A.J. Blume, "Mitochondrial Benzodiazepine Receptors Mediate Inhibition of Mitochondrial Respiratory Control," *Molecular Pharmacology* 34 (1988):157-163.

217. M. Nag and N. Nandia, "Antidepressants and Brain Respiration," *Bioscience Reports* 11:1 (1991): 11-14.

218. Z. Xia, B. Lundgren, A. Bergstrand, J.W. DePierre, and L. Nassberger, "Changes in the Generation of Reactive Oxygen Species and in Mitochondrial Membrane Potential during Apoptosis Induced by Antidepressants Imipramine, Clomipramine, and Citalopram and the Effects on These Changes by Bcl-2 and Bcl-X_L," Biochemical *Pharmacology* 57 (1999): 1199-1208.

219. E. El-Demerdash and A.M. Mohamadin, "Does oxidative stress contribute in tricyclic antidepressants-induced cardiotoxicity?," *Toxicology Letters* 152 (2004): 159-166.

220. M. Bilici, H. Efe, A. Koroglu, H.A. Uydu, M. Bekaroglu, and O. Deger, "Antioxidative enzyme activities and lipid peroxidation in major depression: alterations by antidepressant treatments," *Journal of Affective Disorders* 64 (2001): 43-51.

221. A.Verrotti, A. Scardapane, E. Franzoni, R. Manco, and F. Chiarelli, "Increased oxidative stress in epileptic children treated with valproic acid [abstract]," *Epilepsy Research* 78:2-3 (2008): 171-177.

222. P. Maertens, P. Dyken, W. Graf, C. Pippenger, R. Chronister, and A. Shah, "Free radicals, anticonvulsants, and the neuronal ceroid-lipofuscinoses [abstract]," *American Journal of Medical Genetics* 57:2 (1995): 225-228.

223. A. Aycicek and A. Iscan, "The effects of carbamazepine, valproic acid, and phenobarbital on the oxidative and antioxidative balance in epileptic children [abstract]," *European Neurology* 57:2 (2007): 65-69.

224. N.A. Santos, W.S. Medina, N.M. Martins, M.A. Rodrigues, C. Curti, and A.C. Santos, "Involvement of oxidative stress in the hepatotoxicity induced by aromatic antiepileptic drugs [abstract]," *Toxicology In Vitro* 22:8 (2008): 1820-1824.

225. I.M. Araujo, A.F. Ambrosio, E.C. Leal, M.J. Verdasca, J.O. Malva, P. Soares-da-Silva, et al, "Neurotoxicity induced by antiepileptic drugs in cultured hippocampal neurons: a comparative study between carbamazepine, oxcarbazepine, and two new putative antiepileptic drugs, BIA 2-024 and BIA 2-093," *Epilepsia* 45:12 (2004): 1498-1505.

226. M.R. Martins, F.C. Petronilho, K.M. Gomes, F. Dal-Pizzol, E.L. Streck, and J. Quevedo, "Antipsychotic-induced oxidative stress in rat brain [abstract]," *Neurotoxicity Research* 13:1 (2008): 63-69.

227. X.Y. Zhang, Y.L. Tan, L.Y. Cao, G.Y. Wu, Q. Xu, Y. Shen, et al, "Antioxidant enzymes and lipid peroxidation in different forms of schizophrenia treated with typical and atypical antipsychotics [abstract]," *Schizophrenia Research* 81:2-3 (2006): 291-300.

228. G.S. Dhaunsi, B. Singh, A.K. Singh, D.A. Kirschner, and I. Singh, "Thioridazine induces lipid peroxidation in myelin of rat brain [abstract]," *Neuropharmacology* 32:2 (1993): 157-167.

229. S. Kropp, V. Kern, K. Lange, D. Degner, G. Hajak, J. Kornhuber, et al, "Oxidative stress during treatment with first- and second-generation antipsychotics [abstract]," *Journal of Neuropsychiatry and Clinical Neurosciences* 17:2 (2005): 227-231.

230. M. Polydoro, N. Schroder, M.N. Lima F. Caldana, D.C. Laranja, E. Bromberg, et al, "Haloperidol- and clozapine-induced oxidative stress in the rat brain [abstract]," *Pharmacology Biochemistry and Behavior* 78:4 (2004): 751-756.

231. S. Musavi and P. Kakkar, "Effect of diazepam treatment and its withdrawal on pro/antioxidative processes in rat brain," *Molecular and Cellular Biochemistry* 245 (2003): 51-56.

232. S. Musavi and P. Kakkar, "Diazepam induced early oxidative changes at the subcellular level in rat brain," *Molecular and Cellular Biochemistry* 178 (1998): 41-46.

Endnotes

233. K.M. Gomes, F.C. Petronilho, M. Mantovani, T. Garbelotto, C.R. Boeck, F. Dal-Pizzol, et al, "Antioxidant Enzyme Activities Following Acute or Chronic Methylphenidate Treatment in Young Rats," *Neurochemistry Research* 33 (2008): 1024-1027.

234. T. Cunha-Oliveira, A.C. Rego, S.M. Cardoso, F. Borges, R.H. Swerdlow, T. Macebo, et al, "Mitochondrial dysfunction and caspase activation in rat cortical neurons treated with cocaine or amphetamine," *Brain Research* 1089 (2006): 44-54.

235. A. Rodriguez-Casado, I. Alvarez, A. Toledano, E. de Miguel, and P. Carmona, "Amphetamine Effects on Brain Protein Structure and Oxidative Stress as Revealed by FTIR Microspectroscopy," *Biopolymers* 86:5-6 (2007): 437-446.

236. B.N. Frey, M.R. Martins, F.C. Petronilho, F. Dal-Pizzol, J. Quevedo, and F. Kapczinski, "Increased oxidative stress after repeated amphetamine exposure: possible relevance as a model of mania," *Bipolar Disorder* 8:3 (2006): 275-280.

237. B.N. Frey, S.S. Valvassoi, K.M. Gomes, M.R. Martins, F. Dal-Pizzol, F. Kapczinski, et al, "Increased oxidative stress in submitochondrial particles after chronic amphetamine exposure," *Brain Research* 1097 (2006): 224-229.

238. F.J. Wan, H.C. Lin, K.L. Huang, C.J. Tseng, and C.S. Wong, "Systemic administration of d-amphetamine induces long-lasting oxidative stress in the rat striatum," *Life Sciences* 66:15 (2000): PL205-PL212.

239. M.R. Martins, A. Reinke, F.C. Petronilho, K.M. Gomes, F. Dal-Pizzol, and J. Quevedo, "Methylphenidate treatment induces oxidative stress in young rat brain," *Brain Research* 1078:1 (2006): 189-197.

240. J.L. Cadet, S. Jayanth, and X. Deng, "Speed kills: cellular and molecular bases of methamphetamine-induced nerve terminal degeneration and neuronal apoptosis," *FASEB Journal* 17 (2003): 1775-1788.

241. N.D. Volkow, G.J. Wang, J.S. Fowler, J. Logan, M. Gerasimov, L. Maynard, et al, "Therapeutic Doses of Oral Methylphenidate Significantly Increase Extracellular Dopamine in the Human Brain," *Journal of Neuroscience* 21:2 (2001): RC121.

242. M. Cyr, J.M. Beaulieu, A. Laakso, T.D. Sotnikova, W.D. Yao, L.M. Bohn, et al, "Sustained elevation of extracellular dopamine causes motor dysfunction and selective degeneration of striatal GABAergic neurons," *Proceedings of the National Academy of Sciences of the USA* 100:19 (2003): 11035-11040.

243. L. Chen, Y. Ding, B. Cagniard, A.D. Van Laar, A. Mortimer, W. Chi, et al, "Unregulated Cytosolic Dopamine Causes Neurodegeneration Associated with Oxidative Stress in Mice," *Neurobiology of Disease* 28:2 (2008): 425-433.

244. J.M. Brown and B.K. Yamamoto, "Effects of amphetamines on mitochondrial function: role of free radicals and oxidative stress," *Pharmacology and Therapeutics* 99 (2003): 45-53.

245. L. Iacovelli, F. Fulceri, A. De Blasi, F. Nicoletti, S. Ruggieri, and F. Fornai, "The neurotoxicity of amphetamines: Bridging drugs of abuse and neurodegenerative disorders," *Experimental Neurology* 201:1 (2006): 24-31.

246. T. Kita, G.C. Wagner, and T. Nakashima, "Current Research on Methamphetamine-Induced Neurotoxicity: Animal Models of Monoamine Disruption," *Journal of Pharmacological Sciences* 92 (2003): 178-195.

247. A.L. Miller, "The Methionine-Homocysteine Cycle and Its Effects on Cognitive Diseases," *Alternative Medicine Review* 8:1 (2003): 7-19.

248. M.P. Mattson, S.L. Chan, and W. Duan, "Modification of Brain Aging and Neurodegenerative Disorders by Genes, Diet, and Behavior," *Physiological Reviews* 82 (2002): 637-672.

249. R. Diaz-Arrastia, "Homocysteine and Neurologic Disease," *Archives of Neurology* 57 (2000): 1422-1428.

250. F. Vaillant, F. Turrel, M. Bost, G. Bricca, J. Descotes, B. Bui-Xuzn, et al, "Role of selenium in heart lesions produced by neuroleptics in the rabbit," *Journal of Applied Toxicology* 28:2 (2008): 212-216.

251. U. Schweizer, A.U. Brauer, J. Kohrle, R. Nitsch, and N.E. Savaskan, "Selenium and brain function: a poorly recognized liaison," *Brain Research and Brain Research Review* 45:3 (2004): 164-178.

252. K.S. Vaddai, E. Soosai, and G. Vaddadi, "Low blood selenium concentrations in schizophrenic patients on clozapine," *British Journal of Clinical Pharmacology* 55 (2003): 307-309.

253. M. Anil, M. Helvaci, E. Ozbal, O. Kalenderer, A.B. Anil, and M. Dilek, "Serum and muscle carnitine levels in epileptic children receiving sodium valproate [abstract]," *Journal of Child Neurology* 24:1 (2009): 80-86.

254. J. Nicolai, S.J. Smith, and R.W. Keunen, "Simultaneous side effects of both clozapine and valproate [abstract]," *Intensive Care Medicine* 27:5 (2001): 943.

255. S. Zierz and S. Neumann-Schmid, "Inhibition of carnitine palmitoyltransferase (CPT) by chlorpromazine in muscle of patients with CPT deficiency [abstract]," *Journal of Neurology* 236:4 (1989): 251-252.

256. M. McCall and J.A. Bourgeois, "Valproic Acid-Induced Hyperammonemia," *Journal of Clinical Psychopharmacology* 24:5 (2004): 521-526.

257. N. Segura-Bruna, A. Rodriguez-Campello, V. Puente, and J. Roquer, "Valproate-induced hyperammonemic encephalopathy," *Acta Neurological Scandinavica* 114 (2006): 1-7.

258. M.K. Sun and D.L. Alkon, "Carbonic anhydrase gating of attention: memory therapy and enhancement," *Trends in Pharmacological Sciences* 23:2 (2002): 83-89.

259. P. Latour, A. Biraben, E. Polard, D. Bentue-Ferrer, A. Beauplet, O. Tribut, et al, "Drug induced encephalopathy in six epileptic patients: topiramate? valproate? Or both [abstract]?," *Human Psychopharmacology* 19:3 (2004): 193-203.

260. H. Bailleiux, W. Versiegers, P. Paquier, P.P. De Deyn, and P. Marien, "Cerebellar cognitive affective syndrome associated with topiramate [abstract]," *Clinical Neurology and Neurosurgery* 110:5 (2008): 496-499.

261. M. Sadowski and E.H. Kolodny, "Dementia in Young Adults." Contributed chapter in J.H. Growdon and M.N. Rossor, Ed. *The Dementias 2* (Philadelphia: Butterworth Heineman 2007), 313-328.

262. W.H. Halliwell, "Cationic Amphiphilic Drug-Induced Phospholipidosis," *Toxicologic Pathology* 25:1 (1997): 53-60.

263. N. Anderson and J. Borlak, "Drug-induced phospholipidosis," *FEBS Letters* 580 (2006): 5533-5540.

264. T. Nonoyama and R. Fukuda, "Drug-induced Phospholipidosis – Pathological Aspects and Its Prediction," *Journal of Toxicology and Pathology* 21 (2008): 9-24.

265. R.J. Gonzalez-Rothi, D.S. Zander, and P.R. Ros, "Fluoxetine Hydrochloride (Prozac)-Induced Pulmonary Disease," *Chest* 107:6 (1995): 1763-1765.

266. S.L. Dexter, "Zimelidine induced neuropathies [abstract]," *Human Toxicology* 3:2 (1984): 141-143.

267. H. Bockhardt and R. Lullmann-Rauch, "Zimelidine-induced lipidosis in rats [abstract]," *Acta pharmacologica et toxicological* 47:1 (1980): 45-48.

268. M.J. Reasor and S. Kacew, "Drug-Induced Phospholipidosis: Are There Functional Consequences?," *Experimental Biology and Medicine* 226:9 (2001): 825-830.

269. M.J. Reasor, K.L. Hastings, and R.G. Ulrich, "Drug-induced phospholipidosis: issues and future directions," *Expert Opinion* 5:4 (2006): 567-583.

Endnotes

270. Z. Hruban, "Pulmonary and Generalized Lysosomal Storage Induced by Amphiphilic Drugs," *Environmental Health Perspectives* 55 (1984): 53-76.

271. U.P. Kodavanti and H.M. Mehendale, "Cationic Amphiphilic Drugs and Phospholipid Storage Disorder," *Pharmacological Reviews* 42:4 (1990): 327-354.

272. W.H. Halliwell, "Cationic Amphiphilic Drug-Induced Phospholipidosis," *Toxicologic Pathology* 25:1 (1997): 53-60.

273. H. Sawada, K. Takami, and S. Asahi, "A Toxicogenomic Approach to Drug-Induced Phospholipidosis: Analysis of Its Induction Mechanism and Establishment of a Novel *in vitro* Screening System," *Toxicological Sciences* 83 (2005): 282-292.

274. G. Schmitz and G. Muller, "Structure and function of lamellar bodies, lipid-protein complexes involved in storage and secretion of cellular lipids," *Journal of Lipid Research* 32 (1991): 1539-1570.

275. V.W. Fischer and H. Barner, "Cardiomyopathic Findings Associated with Methylphenidate," *JAMA* 238:14 (1977): 1497.

276. T.A. Henderson and V.W. Fischer, "Effects of Methylphenidate (Ritalin) on Mammalian Myocardial Ultrastructure," *American Journal of Cardiovascular Pathology* 5:1 (1994): 68-78.

277. J.R. Piccotti, M.S. LaGattuta, S.A. Knight, A.J. Gonzales, and M.R. Bleavins, "Induction of Apoptosis by Cationic Amphiphilic Drugs Amiodarone and Imipramine," *Drugs and Chemical Toxicology* 1 (2005): 117-133.

278. A. Vejux, E. Kahn, F. Menetrier, T. Montange, J. Lherminier, J.M. Riedinger, et al, "Cytotoxic oxysterols induce caspase-independent myelin figure formation and caspase-dependent polar lipid accumulation," *Histochemistry and Cell Biology* 127:6 (2007): 609-624.

279. M. Koike, M. Shibata, S. Waguri, K. Yoshimura, I. Tanida, E. Kominami, et al, "Participation of Autophagy in Storage of Lysosomes in Neurons from Mouse Models of Neuronal Ceroid-Lipofuscinoses (Batten Disease)," *American Journal of Pathology* 167:6 (2005): 1713-1728.

280. K. Kiselyov, J.J. Jennigs, Jr., Y. Rbaibi, and C.T. Chu, "Autophagy, mitochondria, and cell death in lysosomal storage diseases," *Autophagy* 3:3 (2007): 259-262.

281. C.T. Chu, "Autophagic Stress in Neuronal Injury and Disease," *Journal of Neuropathology and Experimental Neurology* 65:5 (2006): 423-432.

282. P. Codogno and A.J. Meijer, "Autophagy and signaling: their role in cell survival and cell death," *Cell Death and Differentiation* 12 (2005): 1509-1518.

283. S.J. Cherra, III, and C.T. Chu, "Autophagy in neuroprotection and neurodegeneration: A question of balance," *Future Neurology* 3:3 (2008): 309-323.

284. M.J. Reasor, K.L. Hastings, and R.G. Ulrich, "Drug-induced phospholipidosis: issues and future directions," *Expert Opinion* 5:4 (2006): 567-583.

285. M.E. Charness, F. Morady, and M.M. Scheinman, "Frequent neurologic toxicity associated with amiodarone therapy [abstract]," *Neurology* 34:5 (1984): 669-671.

286. J.A. Maciel Junior and L. Queiroz, "Neuropathy caused by amiodarone: clinico-pathologic study of 2 cases [abstract]," *Arquivos de neuro-psiquiatria* 47:4 (1989): 474-478.

287. N.E. Anderson, N.M. Lynch, and K.P. O'Brien, "Disabling neurological complications of amiodarone [abstract]," *Australia and New Zealand Journal of Medicine* 15:3 (1985): 300-304.

288. C. Masson, P. Boulu, and D. Henin, "Iatrogenic neuropathies [abstract]," *La Revue de médecine interne* 13:3 (1992): 225-232.

289. J.M. Jacobs and F.R. Costa-Jussa, "The pathology of amiodarone neurotoxicity II. Peripheral neuropathy in man [abstract]," *Brain* 108:Pt3 (1985): 753-769.

290. A. Arnaud, J.P. Neau, T. Rivasseau-Jonveaux, R. Marechaud, and R. Gil, "Neurological toxicity of amiodarone – 5 case reports [abstract]," *La Revue de médecine interne* 13:6 (1992): 419-422.

291. P.R. Palakurthy, V. Iyer, and R.J. Meckler, "Unusual neurotoxicity associated with amiodarone therapy [abstract]," *Archives of Internal Medicine* 147:5 (1987): 881-884.

292. L.N. Johnson, G.B. Krohel, and E.R. Thomas, "The clinical spectrum of amiodarone-associated with optic neuropathy [abstract]," *Journal of the National Medical Association* 96:11 (2004): 1477-1491.

293. M.J. Reasor, K.L. Hastings, and R.G. Ulrich, "Drug-induced phospholipidosis: issues and future directions," *Expert Opinion* 5:4 (2006): 567-583.

294. H.A. Jensen and M. Dalsgaard, "Neuromuscular blocking effect of gentamicin as a possible cause of respiratory insufficiency [abstract]," *Ugeskrift for Laeger* 134:35 (1972): 1855-1856.

295. A. Bischoff, C. Meier, and F. Roth, "Gentamicin neurotoxicity – polyneuropathy – encephalopathy [abstract]," *Schweizerische medizinische Wochenschrift* 107:1 (1977): 3-8.

296. W.B. Wadlington, H. Hatcher, and D.J. Turner, "Osteomyelitis of the patella. Gentamicin therapy associated with encephalopathy [abstract]," *Clinical Pediatrics* 10:10 (1971): 577-580.

297. G.J. Byrd, "Acute organic brain syndrome associated with gentamicin therapy [abstract]," *JAMA* 238:1 (1977): 53-54.

298. F. Robert and T.K. Hevor, "Abnormal organelles in cultured astrocytes are largely enhanced by streptomycin and intensively by gentamicin [abstract]," *Neuroscience* 144:1 (2007): 191-197.

299. G.E. Jackson, "Chemo brain – A Psychotropic Drug Phenomenon?," *Medical Hypotheses* 70 (2008): 572-577.

300. D.C. Williams, Jr., G.V. Massey, E.C. Russell, R.S. Riley, and J. Ben-Ezra, "Translocation-positive acute myeloid leukemia associated with valproic acid therapy [abstract]," *Pediatric Blood Cancer* 50:3 (2008): 641-643.

301. T.E. Coyle, A.K. Bair, C. Stein, N. Vajpayee, S. Mehdi, and J. Wright, "Acute leukemia associated with valproic acid treatment: a novel mechanism for leukemogenesis?," *American Journal of Hematology* 78:4 (2005): 256-260.

302. L.E. Orr and J.F. McKernan, "Lithium reinduction of acute myeloblastic leukemia [abstract]," *Lancet* 1:8113 (1979): 449-450.

303. J.L. Nielsen, "Development of acute myeloid leukaemia during lithium treatment [abstract]," *Acta Haematologica* 63:3 (1980): 172-173.

304. B.B. Sethi, R. Prakash, and N. Sethi, "Chronic myeloid leukemia during lithium therapy: case report [abstract]," *Journal of Clinical Psychiatry* 43:7 (1982): 296.

305. J. Moreb and C. Hershko, "Increased leucocyte alkaline phosphatase and transcobalamin III in chronic myeloid leukaemia associated with lithium therapy [abstract]," *Scandinavian Journal of Haematology* 34:3 (1985): 238-241.

306. W.A. Wallace, M. Balsitis, and B.J. Harrison, "Male breast neoplasia in association with selective serotonin reuptake inhibitor therapy: a report of three cases," *European Journal of Surgical Oncology* 27:4 (2001): 429-431.

307. P.W. Harvey, D.J. Everett, and C.J. Springall, "Adverse effects of prolactin in rodents and humans: breast and prostate cancer [abstract]," *Journal of Psychopharmacology* 22:2 Suppl (2008): 20-27.

308. J. Cox-Hippisley, Y. Vinogradova, C. Coupland, and C. Parker, "Risk of malignancy in patients with schizophrenia or bipolar disorders: nested case-control study [abstract]," *Archives of General Psychiatry* 64:12 (2007): 1368-1376.

309. P.M. Doraiswamy, G. Schott, K. Star, R. Edwards, and B. Mueller-Oerlinghausen, "Atypical antipsychotics and pituitary neoplasms in the WHO database [abstract]," *Psychopharmacology Bulletin* 40:1 (2007): 74-76.

310. J.K. Pal and W.A. Sarino, "Effect of risperidone on prolactinoma growth in a psychotic woman," *Psychosomatic Medicine* 62:5 (2000): 736-738.

311. G.E. Jackson, "Chemo brain – A Psychotropic Drug Phenomenon?," *Medical Hypotheses* 70 (2008): 572-577.

312. L. Rumbach, G. Cremel, C. Marescaux, J.M. Warter, and A. Waksman, "Succinate transport inhibition by valproate in rat renal mitochondria [abstract]," *European Journal of Pharmacology* 164:3 (1989): 577-581.

313. S. Ponchaut, F. van Hoof, and K. Veitch, "Valproate and cytochrome c oxidase deficiency," *European Journal of Pediatrics* 154:1 (1995): 79.

314. R. Haas, D.A. Stumpf, J.K.Parks, and L. Eguren, "Inhibitory effects of sodium valproate on oxidative phosphorylation [abstract]," *Neurology* 31:11 (1981): 1473-1476.

315. M.F.B. Silva, J.P.N. Ruiter, L. Ijlst, C. Jakobs, M. Duran, I. Tavares, et al, "Valproate inhibits the mitochondrial pyruvate-driven oxidative phosphorylation in vitro," *Journal of Inherited Metabolic Disorders* 20 (1997): 397-400.

316. H. Karabiber, E. Sonmezgoz, E. Ozerol, C. Yakinci, B. Otlu, and S. Yologlu, "Effects of valproate and carbamazepine on serum levels of homocysteine, vitamin B12, and folic acid [abstract]," *Brain Development* 25:2 (2003): 113-115.

317. J. Hendel, M. Dam, L. Gram, P. Winkel, and I. Jorgensen, "The effects of carbamazepine and valproate on folate metabolism in man [abstract]," *Acta Neurologica Scandinavica* 69:4 (1984): 226-231.

318. E. Alonso-Aperte, N. Ubeda, M. Achon, J. Perez-Miguelsanz, and G. Varela-Moreiras, "Impaired methionine synthesis and hypomethylation in rats exposed to valproate during gestation [abstract]," *Neurology* 52:4 (1999): 750-756.

319. M. Gottlicher, S. Minucci, P. Zhu, O.H. Kramer, A. Schimpf, S. Giavara, et al, "Valproic acid defines a novel class of HDAC inhibitors inducing differentiation of transformed cells," *EMBO Journal* 20:24 (2001): 6969-6978.

320. N. Gurvich, M.G. Berman, B.S. Wittner, R.C. Gentleman, P.S. Klein, and J.B.A. Green, "Association of valproate-induced teratogenesis with histone deacetylase inhibition in vivo," *FASEB Journal* 19:9 (2005): 1166-1168.

321. G.R. Cannell, M.J. Bailey, and R.G. Dickinson, "Inhibition of tubulin assembly and covalent binding to microtubular protein by valproic acid glucuronide in vitro [abstract]," *Life Sciences* 71:22 (2002): 2633-2643.

322. P.S. Walmod, G. Skladchikova, A. Kawa, V. Berezin, and E. Bock, "Antiepileptic teratogen valproic acid (VPA) modulates organisation and dynamics of the actin cytoskeleton [abstract]," *Cell Motility Cytoskeleton* 42:3 (1999): 241-255.

323. M. McCall and J.A. Bourgeois, "Valproic Acid-Induced Hyperammonemia," *Journal of Clinical Psychopharmacology* 24:5 (2004): 521-526.

324. N. Segura-Bruna, A. Rodriguez-Campello, V. Puente, and J. Roquer, "Valproate-induced hyperammonemic encephalopathy," *Acta Neurologica Scandinavica* 114 (2006): 1-7.

325. V. Felipo and R.F. Butterworth, "Neurobiology of ammonia," *Progress in Neurobiology* 67 (2002): 259-279.

326. M. Steiner, A. Attarbaschi, U. Kastner, M. Dworzak, O.A. Haas, H. Gadner, et al, "Distinct fluctuations of ammonia levels during asparaginase therapy for childhood acute leukemia [abstract]," *Pediatric Blood Cancer* 49:5 (2007): 640-642.

327. J.V. Leonard and J.D. Kay, "Acute encephalopathy and hyperammonemia complicating treatment of acute lymphoblastic leukaemia with asparaginase [abstract]," *Lancet* 1:847 (1986): 162-163.

328. R.A. El-Zein, S.Z. Abdel-Rahman, M.J. Hay, M.S. Lopez, M.L. Bondy, D.L. Morris, et al, "Cytogenetic effects in children treated with methylphenidate," *Cancer Letters* 230 (2005): 284-291.

329. R.A. El-Zein, M.J. Hay, M.S. Lopez, M.L. Bondy, D.L. Morris, M.S. Legator, "Response to comments on 'Cytogenetic effects in children treated with methylphenidate' by El-Zein et al," *Cancer Letters* 231 (2006): 146-148.

330. H. Norppa, S. Bonassi, I.L. Hansteen, L. Hagmar, U. Stromberg, P. Rossner, et al, "Chromosomal aberrations and SCEs as biomarkers of cancer risk," *Mutation Research* 600 (2006): 37-45.

Endnotes

331. A.C. Andreazza, B.N. Frey, S.S. Valvassori, C. Zanotto, K.M. Gomes, C.M. Comim, et al, "DNA damage in rats after treatment with methylphenidate," *Progress in Neuro-Psychopharmacology and Biological Psychiatry* 31 (2007): 1282-1288.

332. L. Smigan and C. Perris, "Cortisol changes in long-term lithium therapy [abstract]," *Neuropsychobiology* 11:4 (1984): 219-23.

333. M. Zatz and T.D. Reisine, "Lithium induces corticotrophin secretion and desensitization in cultured anterior pituitary cells [abstract]," *Proceedings of the National Academy of Sciences of the USA* 82:4 (1985): 1286-1290.

334. D.K. Raap and L.D. Van de Kar, "Selective Serotonin Reuptake Inhibitors and Neuroendocrine Function," *Life Sciences* 65:12 (1999): 1217-1235.

335. Z. Zemishlany, R. McQueeney, S.M. Gabriel, and M. Davidson, "Neuroendocrine and monoaminergic responses to acute administration of alprazolam in normal subjects [abstract]," *Neuropsychobiology* 23:3 (1990): 124-128.

336. M.A. Schuckit, R. Hauger, and J.L. Klein, "Adrenocorticotropin hormone response to diazepam in health young men [abstract]," *Biological Psychiatry* 31:7 (1992): 661-669.

337. P.P. Roy-Byrne, D.S. Cowley, D. Hommer, J. Ritchie, D. Greenblatt, and C. Nemeroff, "Neuroendocrine effects of diazepam in panic and generalized anxiety disorders [abstract]," *Biological Psychiatry* 30:1 (1991): 73-80.

338. N. Pomara, L.M. Willoughby, J.J. Sidtis, T.B. Cooper, and D.J. Greenblatt, "Cortisol response to diazepam: its relationship to age, dose, duration of treatment, and presence of generalized anxiety disorder [abstract]," *Psychopharmacology* 178:1 (2005): 1-8.

339. H. Aggernaes, C. Kirkegaard, and G. Magelund, "The effect of sodium valproate on serum cortisol levels in healthy subjects and depressed patients [abstract]," *Acta Psychiatrica Scandinavica* 77:2 (1988): 170-174.

340. C. Invitti, L. Danesi, A. Dubini, and F. Cavagnin, "Neuroendocrine effects of chronic administration of sodium valproate in epileptic patients [abstract]," *Acta Endocrinologica* 118:3 (1988): 381-388.

341. P. Putignano, G.A. Kaltsas, M.A. Satta, and A.B. Grossman, "The effects of anti-convulsant drugs on adrenal function [abstract]," *Hormone and Metabolic Research* 30:6-7 (1998): 389-397.

342. N. Weintrob, D. Cohen, Y. Klipper-Aurbach, Z. Zadik, and Z. Dickerman, "Decreased Growth During Therapy With Selective Serotonin Reuptake Inhibitors," *Archives of Pediatric and Adolescent Medicine* 156 (2002): 696-701.

343. D.K. Raap and L.D. Van de Kar, "Selective serotonin reuptake inhibitors and neuroendocrine function," *Life Sciences* 65:12 (1999): 1217-1235.

344. D.A. Schmid, A. Wichniak, M. Uhr, M. Ising, H. Brunner, K. Held, et al, "Changes of sleep, architecture, spectral composition of sleep EEG, the nocturnal secretion of cortisol, ACTH, GH, prolactin, melatonin, ghrelin, and leptin, and the DEX-CRH test in depressed patients during treatment with mirtazapine [abstract]," *Neuropsycho-pharmacology* 31:4 (2006): 832-844.

345. C. Constatinou, S. Bolaris, T. Valcana, and M. Margarity, "Diazepam affects the nuclear thyroid hormone receptor density and their expression levels in the adult rat brain [abstract]," *Neuroscience Research* 52:3 (2005): 269-275.

346. P.M. Haddad and A. Wieck, "Antipsychotic-Induced Hyperprolactinemia: Mechanisms, Clinical Features and Management," *Drugs* 64:20 (2004): 2291-2314.

347. G.I. Papakostas, K.K. Miller, T. Petersen, K.G. Sklarsky, S.E. Hilliker, A. Klibanski, et al, "Serum prolactin levels among outpatients with major depressive disorder during the acute phase of treatment with fluoxetine [abstract]," *Journal of Clinical Psychiatry* 67:6 (2006): 952-957.

348. A.B.F. Emiliano and J.L. Fudge, "From Galactorrhea to Osteopenia: Rethinking Serotonin-Prolactin Interactions," *Neuropsychopharmacology* 29 (2004): 833-846.

349. M. Gitlin, L.L. Altshuler, M.A. Frye, R. Suri, E.L. Huynh, L. Fairbanks, et al, "Peripheral thyroid hormones and response to selective serotonin reuptake inhibitors," *Journal of Psychiatry and Neuroscience* 29:5 (2004): 383-386.

350. M. Eravci, G. Pinna, H. Meinhold, and A. Baumgartner, "Effects of Pharmacological and Nonpharmacological Treatments on Thyroid Hormone Metabolism and Concentrations in Rat Brain," *Endocrinology* 141:3 (2000): 1027-1040.

351. M.S. Benedetti, R. Whomsley, E. Baltes, and F. Tonner, "Alteration of thyroid hormone homeostasis by antiepileptic drugs in humans: involvement of glucuronosyltransferase induction," *European Journal of Clinical Pharmacology* 61 (2005): 863-872.

352. D.L. Kelly and R.R. Conley, "Thyroid Function in Treatment-Resistant Schizophrenia Patients Treated With Quetiapine, Risperidone, or Fluphenazine," *Journal of Clinical Psychiatry* 66 (2005): 80-84.

353. A. Bereket, S. Turan, M.G. Karaman, G. Haklar, F. Ozbay, and M.Y. Yazgan, "Height, weight, IGF-1, IFGBP-3 and thyroid functions in prepubertal children with attention deficit hyperactivity disorder: effect of methylphenidate treatment," *Hormone Research* 63:4 (2005): 159-164.

354. C. Valerde, L.S. Pastrana, J.A. Ruiz, H. Solis, J.L Jurado, C.M. Sordo, et al, "Neuroendocrine and electroencephalographic sleep changes due to acute amphetamine ingestion in human beings," *Neuroendocrinology* 22:1 (1976): 57-71.

355. G.H. Greeley, Jr., G. Jahnke, G.F. Nicholson, and J.S. Kizer, "Decreased serum 3,5,3'-triiodothyronine and thyroxine levels accompanying acute and chronic Ritalin treatment of developing rats [abstract]," *Endocrinology* 106:3 (1980): 898-904.

356. T. Humbert, "Neuroendocrine effects of benzodiazepines [abstract]," *Annales médico-psychologiques* 152:3 (1994): 161-171.

357. V. Popovic, M. Doknic, N. Maric, S. Pekic, A. Damjanovic, D. Miljic, et al, "Changes in neuroendocrine and metabolic hormones induced by atypical antipsychotics in normal weight patients with schizophrenia [abstract]," *Neuroendocrinology* 85:4 (2007): 249-256.

358. M. Jakovljevic, N. Pivac, A. Mihljevic-Peles, M. Mustapic, M. Relja, D. Ljubicic, et al, "The effects of olanzapine and fluphenazine on plasma cortisol, prolactin and muscle rigidity in schizophrenic patients: a double blind study [abstract]," *Progress in Neuropsychopharmacology and Biological Psychiatry* 31:2 (2007): 399-402.

359. Y. Kaneda, A. Fujii, and T. Ohmori, "The hypothalamic-pituitary-adrenal axis in chronic schizophrenic patients long-term treated with neuroleptics [abstract]," *Progress in Neuropsychopharmacology and Biological Psychiatry* 26:5 (2002): 935-938.

360. S. Cohrs, C. Roher, W. Jordan, A. Meier, G. Huether, W. Wuttke, et al, "The atypical antipsychotics olanzapine and quetiapine, but not haloperidol, reduce ACTH and cortisol secretion in healthy subjects [abstract]," *Psychopharmacology* 185:1 (2006): 11-18.

361. R.H. Syed and T.L. Moore, "Methylphenidate and dextroamphetamine-induced peripheral vasculopathy [abstract]," *Journal of Clinical Rheumatology* 14:1 (2008): 30-33.

362. W. Goldman, R. Seltzer, and P. Reuman, "Association between treatment with central nervous system stimulants and Raynaud's syndrome in children: a retrospective case-control study of rheumatology patients [abstract]," *Arthritis and Rheumatism* 58:2 (2008): 563-566.

363. G. Thomalia, T. Kucinski, C. Weiller, and J. Rother, "Cerebral vasculitis following oral methylphenidate intake in an adult: a case report [abstract]," *World Journal of Biological Psychiatry* 7:1 (2006): 56-58.

364. A. Schteinschnaider, L.L. Plaghos, S. Garbugino, D. Riveros, A. Lazarowski, S. Intruvini, et al, "Cerebral arteritis following methylphenidate use [abstract]," *Journal of Child Neurology* 15:4 (2000): 265-267.

365. S. Bird, "Failure to diagnose: Addison disease," *Australian Family Physician* 36:10 (2007): 859-861.

366. S. Diederich, N.F. Franzen, V. Bahr, and W. Oelkers, "Severe hyponatremia due to hypopituitarism with adrenal insufficiency: report on 28 cases," *European Journal of Endocrinology* 148:6 (2003): 609-617.

367. C.M. Brosnan and N.F. Gowing, "Addison's disease," *British Medical Journal* 312:7038 (1996): 1085-1087.

368. T.C. Harvey, "Addison's disease and the regulation of potassium: the role of insulin and aldosterone [abstract]," *Medical Hypotheses* 69:5 (2007): 1120-1126.

369. T.H. Elhadd, T.A. Abdu, J. Oxtoby, G. Kennedy, M. McLaren, R. Neary, et al, "Biochemical and Biophysical Markers of Endothelial Dysfunction in Adults with Hypopituitarism and Severe GH Deficiency," *Journal of Clinical Endocrinology and Metabolism* 86:9 (2001): 4223-4232.

370. H. Oflaz, F. Sen, A. Elitok, A.O. Cimen, I. Onur, E. Kasikcioglu, et al, "Coronary flow reserve is impaired in patients with adult growth hormone (GH) deficiency," *Clinical Endocrinology* 66 (2007): 524-529.

371. B. Merola, A. Cittadini, A. Colao, S. Longobardi, S. Fazio, D. Sabatini, et al, "Cardiac Structural and Functional Abnormalities in Adult Patients with Growth Hormone Deficiency," *Journal of Clinical Endocrinology and Metabolism* 77:6 (1993): 1658-1661.

372. H. Wallaschofski, M. Donne, M. Eigenthaler, B. Hentschel, R. Faber, H. Stepan, et al, "PRL as a Novel Potent Cofactor for Platelet Aggregation," *Journal of Clinical Endocrinology and Metabolism* 86:12 (2001): 5912-5919.

373. D. Hilfiker-Kleiner, K. Kaminski, E. Podewski, T. Bonda, A. Schaefer, K. Sliwa, et al, "A Cathepsin D-Cleaved 16 kDA Form of Prolactin Mediates Postpartum Cardiomyopathy," *Cell* 128 (2007): 589-600.

374. G. Curtareli and C. Ferrari, "Cardiomegaly and heart failure in a patient with prolactin-secreting pituitary tumor," *Thorax* 34 (1979): 328-331.

375. C. Molinari, E. Grossini, D.A.S.G. Mary, F. Uberti, E. Ghigo, F. Ribichini, et al, "Prolactin induces regional vasoconstriction through the B2-adrenergic and nitric oxide mechanisms," *Endocrinology* 148:8 (2007): 4080-4090.

376. J.A. Moolman, "Thyroid hormone and the heart," *Cardiovascular Journal of South Africa* 13:4 (2002): 159-163.

377. P. Dahl, S. Danzi, and I. Klein, "Thyrotoxic cardiac disease [abstract]," *Current Heart Failure Report* 5:3 (2008): 170-176.

378. B. Biondi, "Cardiovascular effects of mild hypothyroidism [abstract]," *Thyroid* 17:7 (2007): 625-630.

379. J. Duggal, S. Singh, C.P. Barsano, and R. Arora, "Cardiovascular risk with subclinical hyperthyroidism and hypothyroidism: pathophysiology and management [abstract]," *Journal of the Cardiometabolic Syndrome* 2:3 (2007): 198-206.

Chapter 3 - Epidemiology

1. R.S. Greenberg, S.R. Daniels, W.D. Flanders, J.W. Eley, and J.R. Boring, III, *Medical Epidemiology, 3rd Ed.* (New York: Lange Medical Books – McGraw Hill, 2001), 192.

2. R.S. Greenberg, S.R. Daniels, W.D. Flanders, J.W. Eley, and J.R. Boring, III, *Medical Epidemiology, 3rd Ed.* (New York: Lange Medical Books – McGraw Hill, 2001).

3. B. Dawson and R.G. Trapp, *Basic & Clinical Biostatistics, 4th Ed.* (New York: Lange Medical Books – McGraw Hill, 2004).

4. J.P. Geyman, R.A. Deyo, and S.D. Ramsey, *Evidence-Based Clinical Practice: Concepts and Approaches* (Boston: Butterworth Heinemann, 2000).

5. S.B. Hulley, S.R. Cummings, W.S. Browner, D. Grady N. Hearst, and T.B. Newman, *Designing Clinical Research, 2nd Ed.* (Philadelphia: Lippincott Williams & Wilkins, 2001).

6. C.W. Colton and R.W. Manderscheid, "Congruencies in Increased Mortality Rates, Years of Potential Life Lost, and Causes of Death Among Public Mental Health Clients in Eight States," *Preventing Chronic Disease: Public Health Research, Practice, and Policy* 3:2 (2006): 1-14.

7. S. Saha, D. Chant, and J. McGrath, "A Systematic Review of Mortality in Schizophrenia: Is the Differential Mortality Gap Worsening Over Time?," *Archives of General Psychiatry* 64:10 (2007): 1123-1131.

Endnotes

8. M. Wachterman, D.K. Kiely, and S.L. Mitchell, "Reporting Dementia on the Death Certificates of Nursing Home Residents Dying with End-Stage Dementia," *JAMA* 300:22 (2008): 2608-2610.

9. B. Cooper and C. Holmes, "Previous psychiatric history as a risk factor for late-life dementia: a population-based case-control study," *Age and Ageing* 27 (1998): 181-188.

10. L.V. Kessing, E.W. Olsen, P.B. Mortensen, and P.K. Andersen. "Dementia in affective disorder: a case-register study," *Acta Psychiatrica Scandinavica* 100 (1999): 176-185.

11. M. Brandt-Christensen, K. Kvist, F.M. Nilsson, P.K. Andersen, and L.V. Kessing, "Treatment with antidepressants and lithium is associated with increased risk of treatment with antiparkinsonian drugs: a pharmacoepidemiological study," *Journal of Neurology*, Neurosurgery, and Psychiatry 77 (2006): 781-783.

12. Y. Chen, J.J. Guo, H. Li, L. Wulsin, and N.C. Patel, "Risk of Cerebrovascular Events Associated with Antidepressant Use in Patients with Depression: A Population-Based, Nested Case-Control Study," *Annals of Pharmacotherapy* 42 (2008): 177-184.

13. L.V. Kessing, L. Sondergard, J.L. Forman, and P.K. Andersen, "Antidepressants and dementia," *Journal of Affective Disorders* (2009): doi:10.1016/j.jad.2008.11.020

14. No author. "Emil Kraepelin." Accessed on January 23, 2009 at: http://en/wikipedia.org/wiki/Emil_Kraepelin

15. U.S. Food and Drug Administration (April 11, 2005). "FDA Talk Paper: FDA Issues Public Health Advisory for Antipsychotic Drugs used for Treatment of Behavioral Disorders in Elderly Patients." Accessed on January 19, 2009 at: http://www.fda.gov/bbs/topics/ANSWERS/2005/ANS01350.html

16. U.S. Food and Drug Administration (April 11, 2005). "FDA Public Health Advisory: Deaths with Antipsychotics in Elderly Patients with Behavioral Disturbances." Accessed on January 19, 2009 at: http://www.fda.gov/CDER/drug/advisory/antipsychotics.htm

17. U.S. Food and Drug Administration (June 16, 2008). "Information for Health Care Professionals: Antipsychotics." Accessed on January 19, 2009 at: http://www.fda.gov/CDER/drug/InfoSheets/HCP/antipsychotics_conventional.htm

18. H.C. Kales, M. Valenstein, H.M. Kim, J.F. McCarthy, D. Ganoczy, F. Cunningham, and F.C. Blow, "Mortality Risk in Patients With Dementia Treated With Antipsychotics Versus Other Psychiatric Medications," *American Journal of Psychiatry* 164 (2007): 1568-1576.

19. C. Ballard, M.L. Hanney, M. Theodoulou, S. Douglas, R. McShane, K. Kossakowski, et al, "The dementia antipsychotic withdrawal trial (DART-AD): long-term follow-up of a randomised placebo-controlled trial," *Lancet* 8 (2009): 151-157.

20. M.S. Myslobodsky, "Anosognosia in Tardive Dyskinesia: 'Tardive Dysmentia' or 'Tardive Dementia'?," *Schizophrenia Bulletin* 12:1 (1986): 1-6.

21. I.C. Wilson, J.C. Garbutt, C.F. Lanier, J. Moylan, W. Nelson, and A.J. Prange, Jr., "Is There a Tardive Dysmentia?," *Schizophrenia Bulletin* 9:2 (1983): 187-192.

22. B.D. Jones, "Tardive Dysmentia: Further Comments," *Schizophrenia Bulletin* 11:2 (1985): 187-189.

23. A.S. DeWolfe, J.J. Ryan, and M.E. Wolf, "Cognitive sequelae of tardive dyskinesia [abstract]," *Journal of Nervous and Mental Disorders* 176:5 (1988): 270-274.

24. S. Saddicha, N. Manjunatha, S. Ameen, and S. Akhtar, "Diabetes and schizophrenia – effect of disease or drug? Results from a randomized, double-blind, controlled prospective study in first-episode schizophrenia," *Acta Psychiatrica Scandinavica* 117 (2008): 342-347.

25. J. Suvisaari, J. Perala, S.I. Saarni, T. Harkanen, S. Pirkola, M. Joukamaa, et al, "Type 2 diabetes among persons with schizophrenia and other psychotic disorders in a general population survey," *European Archives of Psychiatry and Clinical Neuroscience* 258 (2008): 129-136.

26. D.C. Henderson, E. Cagliero, C. Gray, R.A. Nasrallah, D.L. Hayden, D.A. Schoenfeld, et al, "Clozapine, diabetes mellitus, weight gain, and lipid abnormalities: A five-year naturalistic study," *American Journal of Psychiatry* 157:6 (2000): 975-981.

27. D.C. Henderson, D.D. Nguyen, P.M. Copeland, D.L. Hayden, C.P. Borba, P.M. Louie, et al, "Clozapine, diabetes mellitus, hyperlipidemia, and cardiovascular risks and mortality: results of a 10-year naturalistic study [abstract]," *Journal of Clinical Psychiatry* 66:9 (2005): 1116-1121.

28. M.W.J. Strachan, R.M. Reynolds, B.M. Frier, R.J. Mitchell, and J.F. Price, "The relationship between type 2 diabetes and dementia," *British Medical Bulletin* 88:1 (2008): 131-146.

29. C.L. Leibson, W.A. Rocca, V.A. Hanson, R. Cha, E. Kokman, P.C. O'Brien, et al, "Risk of Dementia among Persons with Diabetes Mellitus: A Population-based Cohort Study," *American Journal of Epidemiology* 145:4 (1997): 301-308.

30. A. Akomolafe, A. Beiser, J.B. Meigs, R. Au, R.C. Green, L.A. Farrer, et al, "Diabetes Mellitus and Risk of Developing Alzheimer Disease," *Archives of Neurology* 63 (2006): 1551-1555.

31. S. Gebhardt, F. Hartling, M. Hanke, M. Mittendorf, F.M. Theisen, K. Wolf-Ostermann, et al, "Prevalence of movement disorders in adolescent patients with schizophrenia and in relationship to predominantly atypical antipsychotic treatment [abstract]," *European Child and Adolescent Psychiatry* 15:7 (2006): 371-382.

32. S. Janno, M.M. Holi, K. Tuisku, and K. Wahlbeck, "Neuroleptic-induced movement disorders in a naturalistic schizophrenia population: diagnostic value of actometric movement patterns," *BMC Neurology* 8:10 (2008): doi:10.1186/1471-2377-8-10

33. R.G. McCreadie, L.J.Robertson, and D.H. Wiles, "The Nithsdale schizophrenia surveys. IX: Akathisia, parkinsonism, tardive dyskinesia and plasma neuroleptics levels [abstract]," *British Journal of Psychiatry* 160 (1992): 793-799.

34. D. Aarsland, K. Andersen, J.P Larsen, A. Lolk, and P. K. Sorensen, "Prevalence and Characteristics of Dementia in Parkinson Disease," *Archives of Neurology* 60 (2003): 378-392.

35. T.A. Hughes, H.F. Ross, S. Musa, S. Bhattacherjee, R.N. Nathan, R.H.S. Mindham, et al, "A 10-year study of the incidence of and factors predicting dementia in Parkinson's disease," *Neurology* 54: 8 (2000): 1596-1603.

36. E. Sacchetti, G. Trifiro, A. Caputi, C. Turrina, E. Spina, C. Cricelli, et al, "Risk of stroke with typical and atypical anti-psychotics: a retrospective cohort study including unexposed subjects [abstract]," *Journal of Psychopharmacology* 22:1 (2008): 39-46.

37. H.C. Lin, F.H. Hsaio, S. Pfeiffer, Y.T. Hwang, and H.C. Lee, "An increased risk of stroke among young schizophrenic patients [abstract]," *Schizophrenia Research* 101:1-3 (2008): 234-241.

38. M. Percudani, C. Barbui, I. Fortino, M. Tansella, and L. Petrovich, "Second-generation antipsychotics and risk of cerebrovascular accidents in the elderly," *Journal of Clinical Psychopharmacology* 25:5 (2005): 468-470.

39. D. Leys, H. Henon, M.A. Mackowiak-Cordoliani, and F. Pasquier, "Poststroke dementia [abstract]," *Lancet Neurology* 4:11 (2005): 752-759.

40. B. Tamam, H. Tasdemir, and Y. Tamam, "The prevalence of dementia three months after stroke and its risk factors [abstract]," *Türk psikiyatri dergisi* 19:1 (2008): 46-56.

41. O.A. Seines and H.V. Vinters, "Vascular cognitive impairment [abstract]," *Nature Clinical Practice Neurology* 2:10 (2006): 538-547.

42. W. Byne, L. White, M. Parella, R. Adams, P.D. Harvey, and K.L. Davis, "Tardive dyskinesia in a chronically institutionalized population of elderly schizophrenic patients: prevalence and association with cognitive impairment [abstract]," *International Journal of Geriatric Psychiatry* 13:7 (1998): 473-479.

43. M.G. Woerner, J.M.J. Alvir, B.L. Saltz, J.A. Lieberman, and J.M. Kane, "Prospective Study of Tardive Dyskinesia in the Elderly: Rates and Risk Factors," *American Journal of Psychiatry* 155 (1998): 1521-1528.

44. J. de Leon, "The effect of atypical versus typical antipsychotics on tardive dyskinesia: a naturalistic study," *European Archives of Psychiatry and Clinical Neuroscience* 257:3 (2007): 169-172.

45. A.S. DeWolfe, J.J. Ryan, and M.E. Wolf, "Cognitive sequelae of tardive dyskinesia [abstract]," *Journal of Nervous and Mental Disorders* 176:5 (1988): 270-274.

46. P. Sachdev, F. Hume, P. Toohey, and C. Doutney, "Negative symptoms, cognitive dysfunction, tardive akathisia and tardive dyskinesia [abstract]," *Acta Psychiatrica Scandinavica* 93:6 (1996): 451-459.

47. J.L. Waddington and H.A. Youssef, "Cognitive dysfunction in chronic schizophrenia followed prospectively over 10 years and its longitudinal relationship to the emergence of tardive dyskinesia [abstract]," *Psychological Medicine* 26:4 (1996): 681-688.

48. G. Tunnicliff, N.L. Schindler, G.J. Crites, R. Goldenberg, A. Yochum, and E. Malatynska, "The $GABA_A$ Receptor Complex as a Target for Fluoxetine Action," *Neurochemical Research* 24:10 (1999): 1271-1276.

49. Z. Bhagwagar, M. Wylezinska, M. Taylor, P. Jezzard, P.M. Matthews, and P.J. Cowen, "Increased Brain GABA Concentrations Following Acute Administration of a Selective Serotonin Reuptake Inhibitor," *American Journal of Psychiatry* 161 (2004): 368-370.

50. R.T. Robinson, B.C. Drafts, and J.L. Fisher, "Fluoxetine Increases $GABA_A$ Receptor Activity through a Novel Modulatory Site," *Journal of Pharmacology and Experimental Therapeutics* 304:3 (2003): 978-984.

51. M.S. Goren, E. Kucukibrahimoglu, K. Berkman, and B. Terzioglu, "Fluoxetine Partly Exerts its Actions Through GABA: A Neurochemical Evidence," *Neurochemical Research* 32 (2007): 1559-1565.

52. I.T. Uzbay, "Serotonergic Anti-Depressants and Ethanol Withdrawal Syndrome: A Review," *Alcohol & Alcoholism* 43:1 (2008): 15-24.

Endnotes

53. S. Paterniti, C. Dufouil, and A. Alperovitch, "Long-Term Benzodiazepine Use and Cognitive Decline in the Elderly: The Epidemiology of Vascular Aging Study," *Journal of Clinical Psychopharmacology* 22:3 (2002): 285-293.

54. R. Lanaoui, B. Begaud, N. Moore, A. Chaslerie, A. Fourrier, L. Letenneur, et al, "Benzodiazepine use and risk of dementia: A nested case-control study," *Journal of Clinical Epidemiology* 55 (2002): 314-318.

55. M.J. Barker, K.M. Greenwood, M. Jackson, and S.F. Crowe, "Cognitive Effects of Long-Term Benzodiazepine Use," *CNS Drugs* 18:1 (2004): 37-48.

56. M.J. Barker, K.M. Greenwood, M. Jackson, S.F. Crowe, "Persistence of cognitive effects after withdrawal from long-term benzodiazepine use: a meta-analysis," *Archives of Clinical Neuropsychology* 19 (2004): 437-454.

57. B. Dawson and R.G. Trapp, *Basic & Clinical Biostatistics, 4th Ed.* (New York: Lange Medical Books, 2004).

58. L. Becker. "Effect Size." Accessed on January 25, 2009 at: http://web.uccs.edu/lbecker/Psy590/es.htm

59. N. Dunn, C. Holmes, and M. Mullee, "Does Lithium Protect Against the Onset of Dementia?," *Alzheimer Disease and Associated Disorders* 19:1 (2005): 20-22.

60. P.V. Nunes, O.V. Forlenza, and W.F. Gattaz, "Lithium and risk for Alzheimer's disease in elderly patients with bipolar disorder," *British Journal of Psychiatry* 190 (2007): 359-360.

61. L.V. Kessing L. Sondergard, J.L. Forman, and P.K. Andersen, "Lithium Treatment and Risk of Dementia," *Archives of General Psychiatry* 65:11 (2008): 1331-1335.

62. E.R. Garwood, W. Bekele, C.E. McCulloch, and C.W. Christine, "Amphetamine exposure is elevated in Parkinson's disease," *NeuroToxicology* 27 (2006): 1003-1006.

Chapter Four - Antidepressants

1. K. Davison, "Historical aspects of mood disorders," *Psychiatry* 5:4 (2006): 115-118.

2. No author (October 22, 2008). "Humorism." Accessed on October 23, 2008 at: http://en.wikipedia.org/wiki/Humorism

3. M.A. Kazlev (July 3, 2004). "The Four Humours." Accessed on October 24, 2008 at: http://www.kheper.net/topics/typology/four_humours.html

Endnotes

4. C. Colarusso (May 9, 1995). "The Presocratic Influence Upon Hippocratic Medicine." Accessed on October 24, 2008 at:
http://www.perseus.tufts.edu/GreekScience/Students/Chad/pre-soc.html

5. A. Coppen, A.J. Prange, P.C. Whybrow, and R. Noguera, "Abnormalities of Indoleamines in Affective Disorders," *Archives of General Psychiatry* 26 (1972): 474-478.

6. H.M. van Praag, "Central Monoamine Metabolism in Depressions. I. Serotonin and Related Compounds," *Comprehensive Psychiatry* 21:1 (1980): 30-43.

7. J.J. Schildkraut, "The Catecholamine Hypothesis of Affective Disorders," *International Journal of Psychiatry* 4:3 (1967): 203-217.

8. D. Luchins, "Biogenic Amines and Affective Disorders: A Critical Analysis," *International Pharmacopsychiatry* 11 (1976): 135-149.

9. M. Kalia, J.P. O'Callaghan, D.B. Miller, and M. Kramer, "Comparative study of fluoxetine, sibutramine, sertraline and dexfenfluramine on the morphology of serotonergic nerve terminals using serotonin immunohistochemistry," *Brain Research* 858 (2000): 92-105.

10. G. Bonanno, A. Fassio, P. Severi, A. Ruelle, and M. Raiteri, "Fenfluramine Releases Serotonin from Human Brain Nerve Endings by a Dual Mechanism," *Journal of Neurochemistry* 63 (1994): 1164-1166.

11. S.M. Wolniak. "Principles of Microscopy." Accessed on September 27, 2008 at:
http://www.life.umd/edu/CBMG/faculty/wolniak/wolniakmicro.html

12. C.K. Omoto and J. Folwell. "Using Darkfield Microscopy To Enhance Contrast: An Easy and Inexpensive Method." Accessed on September 27, 2008 at:
http://www.wsu.edu/~omoto/papers/darkfield.html

13. No author. (September 3, 2008). "Dark field microscopy," Wikipedia. Accessed on September 27, 2008 at:
http://en.wikipedia.org/wiki/Darkfield_microscope

14. S. Caccia, M. Cappi, C. Fracasso, and S. Garattini, "Influence of dose and route of administration on the kinetics of fluoxetine and its metabolite norfluoxetine in the rat. *Psychopharmacology* 100 (1990): 509-514.

15. S. Caccia, S. Confalonieri, A. Bergami, C. Fracasso, M. Anelli, and S. Garattini, "Neuropharmacological Effects of Low and High Doses of Repeated Oral Dexfenfluramine in Rats: A Comparison with Fluoxetine." *Pharmacology Biochemistry and Behavior* 57:4 (1997): 851-856.

Endnotes

16. J.F. Czachura and K. Rasmussen, "Effects of acute and chronic administration of fluoxetine on the activity of serotonergic neurons in the dorsal raphe nucleus of the rat," *Naunyn-Schmiedeberg's Archives of Pharmacology* 362 (2000): 266-275.

17. S.H. Preskorn, B. Silkey, J. Beber, and C. Dorey, "Antidepressant Response and Plasma Concentrations of Fluoxetine [abstract]" *Annals of Clinical Psychiatry* 3:2 (1991): 147-151.

18. No author. "Selected laboratory tests with reference ranges and conversion factors." Accessed on September 26, 2008 at: http://www.oup.com/us/pdf/9780195176339/table_2.pdf

19. L.M. Tremaine, W.M. Welch, and R.A. Ronfeld, "Metabolism and Disposition of the 5-Hydroxytryptamine Uptake Blocker Sertraline in the Rat and Dog," *Drug Metabolism and Disposition* 17:5 (1989): 542-550.

20. R.N. Gupta and S.A. Dziurdzy, "Therapeutic Monitoring of Sertraline," *Clinical Chemistry* 40:3 (1994): 498-499.

21. Pfizer Canada, Inc. (November 10, 2004). Product Monograph – Zoloft. Accessed on September 26, 2008 at: http://74.125.47.132/search?q=cache:pG2fYgagJFMJ:www.pfizer.ca/english/our%2520products/prescription%2520pharmaceuticals/default.asp%3Fs%3D1%26id%3D20%26doc%3Denmonograph+Pfizer+Canada+zoloft+product+monograph&hl=en&ct=clnk&cd=1&gl=us

22. C.L. DeVane, H.L. Liston, and J.S. Markowitz, "Clinical Pharmacokinetics of Sertraline," *Drug Disposition* 41:15 (2002): 1247-1266.

23. E.C. Azmitia and R. Nixon, "Dystrophic serotonergic axons in neurodegenerative diseases," *Brain Research* 1217 (2008): 185-194.

24. B. Czeh, M. Simon, B. Schmelting, C. Hiemke, and E. Fuchs, "Astroglial Plasticity in the Hippocampus is Affected by Chronic Psychosocial Stress and Concomitant Fluoxetine Treatment," *Neuropsychopharmacology* 31 (2006): 1616-1626.

25. B. Lundrigan and L. Cisneros (2005). "Tupaia glis," Animal Diversity Web. Accessed on September 30, 2008 at: http://animaldiversity.ummz.umich.edu/site/accounts/information/Tupaia_glis.html

26. D. O'Neil (March 14, 2008). "Classification of Living Things: An Introduction to the Principles of Taxonomy with a Focus on Human Classification Categories." Accessed on August 6, 2008 at: http://anthro.palomar.edu/animal/

Endnotes

27. S.H. Preskorn, B. Silkey, J. Beber, and C. Dorey, "Antidepressant Response and Plasma Concentrations of Fluoxetine [abstract]," *Annals of Clinical Psychiatry* 3:2 (1991): 147-151.

28. M. Broe, J. Kril, and G.M. Halliday, "Astrocytic degeneration relates to the severity of disease in frontotemporal dementia," *Brain* 127 (2004): 2214-2220.

29. R.M. Julien, C.D. Advokat, and J.E. Comaty, *A Primer of Drug Action: A comprehensive guide to the actions, uses, and side effects of psychoactive drugs, 11th Ed.* (New York, NY: Worth Publishers, 2008), 203.

30. S.M. Stahl, *Stahl's Essential Psychopharmacology: Neuroscientific Basis and Practical Applications, 3rd Ed.* (New York: Cambridge University Press, 2008), p. 24.

31. P. Taupin, "BrdU immunohistochemistry for studying adult neurogenesis: paradigms, pitfalls, limitations, and validation," *Brain Research Reviews* 53 (2007): 198-214.

32. J.J. Breunig, Jon I. Arellano, J.D. Macklis, and P. Rakic, "Everything that Glitters Isn't Gold: A Critical Review of Postnatal Neural Precursor Analyses," *Cell Stem Cell* 1 (2007): 612-627.

33. P. Rakic, "Adult Neurogenesis in Mammals: An Identity Crisis," *Journal of Neuroscience* 22:3 (2002): 614-618.

34. R.S. Nowakowski, "Stable neuron numbers from cradle to grave," *Proceedings of the New York Academy of Sciences* 103:33 (2006): 12219-12220.

35. M. Sairanen, G. Lucas, P. Ernfors, M. Castren, and E. Castren, "Brain-Derived Neurotrophic Factor and Antidepressant Drugs Have Different But Coordinated Effects on Neuronal Turnover, Proliferation, and Survival in the Adult Dentate Gyrus," *Journal of Neuroscience* 25:5 (2005): 1089-1094.

36. Dr. Karl Herrup, (personal communication), August 7, 2008.

37. E. Castren, "Neurotrophic effects of antidepressant drugs," *Current Opinion in Pharmacology* 4 (2004): 58-64.

38. D.S. Cowen, L.F. Takase, C.A. Fornal, B.L. Jacobs, "Age-dependent decline in hippocampal neurogenesis is not altered by chronic treatment fluoxetine," *Brain Research* 1228 (2008): 14-19.

39. R.M. Sapolsky, "Depression, antidepressants, and the shrinking hippocampus," *Proceedings of the New York Academy of Sciences* 98:22 (2001): 12320.

Endnotes

40. Ibid., p. 12320.

41. Y.I. Sheline, P.W. Wang, M.H. Gado, J.G. Csernansky, and M.W. Vannier, "Hippocampal atrophy in recurrent major depression," *Proceedings of the New York Academy of Science* 93 (1996): 3908-3913.

42. Y.I. Sheline, M. Sanghavi, M.A. Mintun, M.H. Gado, "Depression Duration But Not Age Predicts Hippocampal Volume Loss in Medically Healthy Women with Recurrent Major Depression," *Journal of Neuroscience* 19:12 (1999): 5034-5043.

43. J.D. Bremner, M. Narayan, E.R. Anderson, L.H. Staib, and D.S. Charney, "Hippocampal Volume Reduction in Major Depression," *American Journal of Psychiatry* 157:1 (2000): 115-117.

44. S. Campbell, M. Marriott, C. Nahmias, G.M. MacQueen, "Lower Hippocampal Volume in Patients Suffering From Depression: A Meta-Analysis," *American Journal of Psychiatry* 161 (2004): 598-607.

45. G.M. MacQueen, S. Campbell, B.S. MacEwen, K. Macdonald, S. Amano, R.T. Joffe, et al, "Course of illness, hippocampal function, and hippocampal volume in major depression," *Proceedings of the New York Academy of Sciences* 100:3 (2003): 1387-1392.

46. T. Frodl, E.M. Meisenzahl, T. Zetzsche, T. Hohne, S. Banac, C. Schorr, "Hippocampal and Amygdala Changes in Patients With Major Depressive Disorder and Healthy Controls During a 1-Year Follow-Up," *Journal of Clinical Psychiatry* 65:4 (2004): 492-499.

47. D.C. Steffens, H. Chung, R.R. Krishnan, W.T. Longstreth, M. Carlson, and G.L. Burke, "Antidepressant Treatment and Worsening White Matter on Serial Cranial Magnetic Resonance Imaging in the Elderly," *Stroke* 39 (2008): 857-862.

48. P.J. Lucassen, M.B. Muller, F. Holsboer, J. Bauer, A. Holtrop, J. Wouda, et al, "Hippocampal Apoptosis in Major Depression Is a Minor Event and Absent from Subareas at Risk for Glucocorticoid Overexposure," *American Journal of Pathology* 158:2 (2001): 453-468.

49. J.F. Kerr, A.H. Wyllie, and A.R. Currie, "Apoptosis: a basic biological phenomenon with wide-ranging implications in tissue kinetics," *British Journal of Cancer* 26:4 (1972): 239-257.

50. R.A. Lockshin and A. Zakeri, "Introduction," in *When Cells Die II* (Hoboken, NJ: John Wiley & Sons, Inc., 2004), 3-23.

51. W. Bursch, A. Ellinger, C. Gerner, and R. Schulte-Hermann, "Caspase-Independent and Autophagic Programmed Cell Death," in *When Cells Dies II* (Hoboken, NJ: John Wiley & Sons, Inc., 2004), 275-309.

52. S.Y. Proskuryakov, A.G. Konoplyannikov, and V.L. Gabai "Necrosis: a specific form of programmed cell death?," *Experimental Cell Research* 283 (2003): 1-16.

53. A.L. Edinger and C.B. Thompson, "Death by design: apoptosis, necrosis, and autophagy," *Current Opinion in Cell Biology* 16 (2004): 663-669.

Chapter 5 – Antipsychotics

1. P.R. Breggin, *Brain Disabling Treatments in Psychiatry* (New York: Springer Publishing Company, 1997), 67-69.

2. D. Cohen, "A Critique of the Use of Neuroleptics Drugs in Psychiatry," in S. Fisher and R.P. Greenberg, Ed., *From Placebo to Panacea: Putting Psychiatric Drugs to the Test* (New York: John Wiley & Sons, 1997), 173-185.

3. O. Sacks, *Awakenings* (New York: Vintage Books, 1990).

4. D. Healy, *The Creation of Psychopharmacology* (Cambridge, Massachusetts: Harvard University Press, 2002), 114-117.

5. D. Cohen, "A Critique of the Use of Neuroleptics Drugs in Psychiatry," in S. Fisher and R.P. Greenberg, Ed., *From Placebo to Panacea: Putting Psychiatric Drugs to the Test* (New York: John Wiley & Sons, 1997), 182.

6. K. Jellinger, "Neuropathologic Findings after Neuroleptic Long-Term Therapy," in L. Roizin, H. Shiraki, and N. Grcevic, Ed. *Neurotoxicology* (New York: Raven Press, 1977), 25-42.

7. Ibid., 33.

8. L. Roizin, C. True, and M. Knight, "Structural effects of tranquilizers," *Research Publications – Association for Research in Nervous and Mental Disease* 37 (1959): 285-324.

9. No author. Center for Neurodegenerative Research. "Amyloid plaques." Accessed on October 15, 2008 at:
http://wiki.iop.kcl.ac.uk/default.aspx/Neurodegeneration/Amyloid%20plaques.html

10. William I. Rosenblum, "Neuropathology Mini-Course for Residents," VCU Department of Pathology. Accessed on December 31, 2006 at:
http://www.pathology.vcu.edu/WirSelfInst/neur&proc.html

Endnotes

11. K. Okamoto, S. Hirai, T. Iizuka, T. Yanagisawa, and M. Watanabe, "Reexamination of granulovacuolar degeneration [abstract]," *Acta Neuropathologica* 82:5 (1991): 340-345.

12. N. Ghoshal, J.F. Smiley, A.J. DeMaggio, M.F. Hoekstra, E.J. Cochran, L.I. Binder, et al, "A new molecular link between the fibrillar and granulovacuolar lesions of Alzheimer's disease," *American Journal of Pathology* 155:4 (1999): 1163-1172.

13. P.J. Whitehouse and D. George, *The Myth of Alzheimer's: What You Aren't Being Told About Today's Most Dreaded Diagnosis* (New York: St. Martin's Press, 2008).

14. A. Korczy, "The underdiagnosis of the vascular contribution to dementia [abstract]," *Journal of the Neurological Sciences* 229 (2005): 3-6.

15. I. Prohovnik, A.J. Dwork, M.A. Kaufman, and N. Wilson, "Alzheimer-type pathology in elderly schizophrenia patients," *Schizophrenia Bulletin* 19:4 (1993): 805-816.

16. D.P. Purohit, D.P. Perl, V. Haroutinian, P. Powchik, M. Davidson, and K.L. Davis, "Alzheimer Disease and Related Neurodegenerative Diseases in Elderly Patients With Schizophrenia," *Archives of General Psychiatry* 55 (1998): 205-211.

17. I. McKeith, A. Fairbairn, R. Perry, P. Thompson, and E. Perry, "Neuroleptic Sensitivity in patients with senile dementia of Lewy body type," *BMJ* 305 (1992): 673-678.

18. A. Baskys, "Lewy Body Dementia: The Litmus Test for Neuroleptic Sensitivity and Extrapyramidal Symptoms," *Journal of Clinical Psychiatry* 65:suppl 11 (2004): 16-22.

19. A.W. Lemstra, N. Schoenmaker, A.J. Rozemuller-Kwakkel, and W.A. van Gool, "The association of neuroleptic sensitivity in Lewy body disease with a false positive clinical diagnosis of Creutzfeldt-Jakob disease [abstract]," *International Journal of Geriatric Psychiatry* 21:11 (2006): 1031-1035.

20. C.G. Ballard, R.H. Perry, I.G. McKeith, and E.K. Perry, "Neuroleptics are associated with more severe tangle pathology in dementia with Lewy bodies," *International Journal of Geriatric Psychiatry* 20 (2005): 872-875.

21. K.A. Jellinger and E. Gabriel, "No increased incidence of Alzheimer's disease in elderly schizophrenics [abstract]," *Acta Neuropathologica* 97:2 (1999): 165-169.

22. K. Niizato and K. Ikeda, "Long-term antipsychotic medication of schizophrenics does not promote the development of Alzheimer's disease brain pathology [abstract]," *Journal of Neurological Science* 138:1-2 (1996): 165-167.

23. S.E. Arnold, B.R. Franz, and J.Q. Trojanowski, "Elderly patients with schizophrenia exhibit infrequent neurodegenerative lesions [abstract]," *Neurobiology of Aging* 15:3 (1994): 299-303.

24. R.J. Baldessarini, J.D. Hegarty, E.D. Bird, and F.M. Benes, "Meta-Analysis of Postmortem Studies of Alzheimer's Disease-Like Neuropathology in Schizophrenia," *American Journal of Psychiatry* 154 (1997): 861-863.

25. G.K. Wilcock, S.M. Matthews, and T. Moss, "Comparison of three silver stains for demonstrating neurofibrillary tangles and neuritic plaques in brain tissue stored for long periods [abstract]," *Acta Neuropathologica* 79:5 (1990): 566-568.

26. T. Yamamoto and A. Hirano, "A Comparative Study of Modified Bielschowsky, Bodian and Thioflavin S Stains on Alzheimer's Neurofibrillary Tangles [abstract]," *Neuropathology and Applied Neurobiology* 12:1 (1986): 3-9.

27. R.J. Oken and P.L. McGeer, "Schizophrenia, Alzheimer's Disease, and Anti-inflammatory Agents," *Schizophrenia Bulletin* 22:1 (1996): 1-3.

28. I. Prohovnik, A.J. Dwork, M.A. Kaufman, and N. Wilson, "Alzheimer-type pathology in elderly schizophrenia patients," *Schizophrenia Bulletin* 19:4 (1993): 805-816.

29. H.M. Wisniewski, J. Constantinidis, J. Wegiel, M. Bobinski, and M. Tarnawski, "Neurofibrillary pathology in brains of elderly schizophrenics treated with neuroleptics [abstract]," *Alzheimer Disease and Associated Disorders* 8:4 (1994): 211-227.

30. V.P. Bozikas, E. Kovari, C. Bouras, and A. Karavatos, "Neurofibrillary tangles in elderly patients with late onset schizophrenia," *Neuroscience Letters* 324 (2002): 109-112.

31. M. Lesort, J. Tucholski, M.L. Miller, and G.V.W. Johnson, "Tissue transglutaminase: a possible role in neurodegenerative diseases," *Progress in Neurobiology* 61 (2000): 439-463.

32. R.M. Bonelli, A. Aschoff, G. Niederwieser, C. Heuberger, and G. Jirikowski, "Cerebrospinal Fluid Tissue Transglutaminase as a Biochemical Marker for Alzheimer's Disease," *Neurobiology of Disease* 11 (2002): 106-110.

33. G.J. Ho, E.J. Gregory, I.V. Smirnova, M.N. Zoubine, and B.W. Festoff, "Cross-linking of beta-amyloid protein precursor catalyzed by tissue transglutaminase," *FEBS Letters* 349 (1994): 151-154.

34. S.Y. Kim, T.M. Jeitner, and P.M. Steinert, "Transglutaminase in disease," *Neurochemistry International* 40 (2005): 85-103.

35. R.M. Bonelli, P. Hofmann, A. Aschoff, G. Niederwieser, C. Heuberger, G. Jirikowski, et al, "The influence of psychotropic drugs on cerebral cell death: female neurovulnerability to antipsychotics," *International Clinical Psychopharmacology* 20:3 (2005): 145-149.

36. G. Zanusso, M. Fiorini, A. Farinazzo, M. Gelati, M.D. Benedetti, S. Ferrari, et al, "Phosphorylated 14-3-3ζ protein in the CSF of neuroleptic-treated patients," *Neurology* 64 (2005): 1618-1620.

37. E. Rassart, A. Bedirian, S. Do Carmo, O. Guinard, J. Sirois, L. Terrisse, et al, "Apolipoprotein D," *Biochimica et Biophysica Acta* 1482 (2000): 185-198.

38. E.A Thomas, D.L. Copolov, and J.G. Sutcliffe, "From Pharmacotherapy to Pathophysiology: Emerging Mechanisms of Apolipoprotein D in Psychiatric Disorders," *Current Molecular Medicine* 3 (2003): 408-418.

39. W.Y. Ong, Y. He, S. Suresh, and S.C. Patel, "Differential expression of apolipoprotein D and apolipoprotein E in the kainic-acid-lesioned rat hippocampus [abstract]," *Neuroscience* 79:2 (1997): 359-367.

40. G. Franz, M. Reindl, S.C. Patel, R. Beer, I. Unterrichter, T. Berger, et al, "Increased Expression of Apolipoprotein D Following Experimental Traumatic Brain Injury," *Journal of Neurochemistry* 73 (1999): 1615-1625.

41. L. Terrisse, J. Poirier, P. Bertrand, A. Merched, S. Visvikis, G. Siest, et al, "Increased levels of apolipoprotein D in cerebrospinal fluid and hippocampus of Alzheimer's patients," *Journal of Neurochemistry* 71:4 (1998): 1643-1650.

42. B. Belloir, E. Kovari, M.S. Demiri, and A. Savioz, "Altered Apolipoprotein D Expression in the Brain of Patients with Alzheimer Disease," *Journal of Neuroscience Research* 64 (2001): 61-69.

43. F. Glockner and T.G. Ohm, "Hippocampal Apolipoprotein D Level Depends on Braak Stage and ApoE Genotype," *Neuroscience* 122 (2003): 103-110.

44. P.P. Desai, M.D. Ikonomovic, E.E. Abrahamson, R.L. Hamilton, B.A. Isanski, C.E. Hope, et al, "Apolipoprotein D is a component of compact but not diffuse amyloid-beta plaques in Alzheimer's disease temporal cortex," *Neurobiology of Disease* 20 (2005): 574-582.

45. M.M. Khan, V.V. Parikh, and S.P. Mahadik, "Antipsychotic drugs differentially modulate apolipoprotein D in rat brain," *Journal of Neurochemistry* 86:5 (2003): 1089-1100.

46. E.A. Thomas, R.C. George, P.E. Danielson, P.A. Nelson, A.J. Warren, D. Lo, et al, "Antipsychotic drug treatment alters expression of mRNAs encoding lipid metabolism-related proteins," *Molecular Psychiatry* 8 (2003): 983-993.

47. E.A. Thomas, P.E. Danielson, P.A. Nelson, T.M. Pribyl, B.S. Hilbush, K.W. Hasel, et al, "Clozapine increases apolipoprotein D expression in rodent brain: towards a mechanism for neuroleptic pharmacotherapy," *Journal of Neurochemistry* 76 (2001): 789-796.

48. E.A. Thomas, B. Dean, G. Pavey, and J.G. Sutcliffe, "Increased CNS levels of apolipoprotein D in schizophrenic and bipolar subjects: Implications for the pathophysiology of psychiatric disorders," *Proceedings of the New York Academy of Sciences* 98:7 (2001): 4066-4071.

49. S.P. Mahadik, M.M. Khan, D.R. Evans, and V.V. Parikh, "Elevated plasma level of apolipoprotein D in schizophrenia and its treatment and outcome," *Schizophrenia Research* 58 (2002: 55-62.

50. W.Y. Ong, C.Y. Hu, and S.C. Patel, "Apolipoprotein D in the Niemann-Pick type C disease mouse brain: An ultrastructural immunocytochemical analysis," *Journal of Neurocytology* 31 (2002): 121-129.

51. H. Li, J.J. Repa, M.A. Valasek, E.P. Beltroy, S.D. Turley, D.C. German, "Molecular, Anatomical, and Biochemical Events Associated with Neurodegeneration in Mice with Niemann-Pick Type C Disease," *Journal of Neuropathology and Experimental Neurology* 64:4 (2005): 323-333.

52. E. Rassart, A. Bedirian, S. Do Carmo, O. Guinard, J. Sirois, L. Terrisse, et al, "Apolipoprotein D," *Biochimica et Biophysica Acta* 1482 (2000): 185-198.

53. N.D. Volkow, J.D. Brodie, A.P. Wolf, B. Angrist, J. Russell, and R. Cancro, "Brain metabolism in patients with schizophrenia before and after acute neuroleptic administration," *Journal of Neurology, Neurosurgery, and Psychiatry* 49 (1986): 1199-1202.

54. Ibid., 1201.

55. A.L. Madsen, N. Keiding, A. Karle, S. Esbjerg, and R. Hemmingsen, "Neuroleptics in progressive structural brain abnormalities in psychiatric illness," *Lancet* 352 (1998): 784-785.

56. A.L. Madsen, A. Karle, P. Rubin, M. Cortsen, H.S. Andersen, and R. Hemmingsen, "Progressive atrophy of the frontal lobes in first-episode schizophrenia: interaction with clinical course and neuroleptic treatment," *Acta Psychiatrica Scandinavica* 100 (1999): 367-374.

Endnotes

57. R.E. Gur, P. Cowell, B.I. Turetsky, F. Gallacher, T. Cannon, W. Bilker, et al, "A Follow-up Magnetic Resonance Imaging Study of Schizophrenia: Relationship of Neuroanatomical Changes to Clinical and Neurobehavioral Measures," *Archives of General Psychiatry* 55:2 (1998): 145-152.

58. W. Cahn, H.E. Hulshoff, E.B.T.E. Lems, E.M. Neeltje, M.S. van Haren, H.G. Schnack, et al, "Brain Volume Changes in First-Episode Schizophrenia," *Archives of General Psychiatry* 59 (2002): 1002-1010.

59. B.C. Ho, N.C. Andreasen, P. Nopoulos, S. Arndt, V. Magnotta, and M. Flaum, "Progressive Structural Brain Abnormalities and Their Relationship to Clinical Outcome," *Archives of General Psychiatry* 60 (2003): 585-594.

60. Ibid., 593.

61. J.A. Lieberman, G.D. Tollefson, C. Charles, R. Zipursky, T. Sharma, R.S. Kahn, et al, "Antipsychotic Drug Effects on Brain Morphology in First-Episode Psychosis," *Archives of General Psychiatry* 62 (2005): 361-370.

62. L.E. DeLisi, W. Tew, S. Xie, A.L. Hoff, M. Sakuma, M. Kushner, et al, "A prospective follow-up study of brain morphology and cognition in first-episode schizophrenic patients: preliminary findings [abstract]," *Biological Psychiatry* 38:6 (1997): 349-360.

63. P.M. Thompson, C. Vidal, J.N. Giedd, P. Gochman, J. Blumenthal, R. Nicolson, et al, "Mapping adolescent brain change reveals dynamic wave of accelerated gray matter loss in very early-onset schizophrenia," *Proceedings of the New York Academy of Sciences* 90:20 (2001): 11650-11655.

64. D.H. Mathalon, E.V. Sullivan, K.O. Lim, and A. Pfefferbaum, "Progressive Brain Volume Changes and the Clinical Course of Schizophrenia in Men," *Archives of General Psychiatry* 58 (2001): 148-157.

65. J.R. Bustillo, J. Lauriello, L.M. Rowland, L.M. Thomson, H. Petropoulos, R. Hammond, et al, "Longitudinal follow-up of neurochemical changes during the first year of antipsychotic treatment in schizophrenia patients with minimal previous medication exposure [abstract]," *Schizophrenia Research* 58:2-3 (2002): 313-321.

66. K. Kasai, M.E. Shenton, D.F. Salisbury, Y. Hirayasu, C.U. Lee, A.A. Ciszweski, et al, "Progressive decrease of left superior temporal gyrus gray matter volume in patients with first-episode schizophrenia [abstract]," *American Journal of Psychiatry* 160:1 (2003): 156-164.

67. A.B. Whitworth, G. Kemmler, M. Honeder, C. Kremser, S. Felber, A. Hausmann, et al, "Longitudinal volumetric MRI study in first- and multiple-episode male schizophrenia patients [abstract]," *Psychiatry Research* 140:3 (2005): 225-237.

68. T.J. Whitford, S.M. Grieve, T.F.D. Farrow, L. Gomes, J. Brennan, A.W.F. Harris, et al, "Progressive grey matter atrophy over the first 1-2 years of illness in first-episode schizophrenia: A tensor-based morphometry study," *NeuroImage* 32 (2006): 511-519.

69. N.E. van Haren, H.E.H. Pol, H.G. Schnack, W. Cahn, R.C. Mandl, D.L. Collins, et al, "Focal gray matter changes in schizophrenia across the course of the illness: a 5-year follow-up study," *Neuropsychopharmacology* 32:10 (2007): 2057-2066.

70. J. Theberge, K.E. Williamson, N. Aoyama, D.J. Drost, R. Manchandra, A.K. Malla, et al, "Longitudinal grey-matter and glutamatergic losses in first-episode schizophrenia [abstract]," *British Journal of Psychiatry* 191 (2007): 325-334.

71. M. Nakamura, D.F. Salisbury, Y. Hirayasu, S. Bouix, K.M. Pohl, T. Yoshida, et al, "Neocortical gray matter volume in first-episode schizophrenia and first-episode affective psychosis: a cross-sectional and longitudinal MRI study [abstract]," *Biological Psychiatry* 62:7 (2007): 773-783.

72. S. Reig, C. Moreno, D. Moreno, M. Burdalo, J. Janssen, M. Parellada, et al, "Progression of Brain Volume Changes in Adolescent-Onset Psychosis [abstract]," *Schizophrenia Bulletin* 35:1 (2009): 233-243.

73. M.S. Koo, J.J. Levitt, D.F. Salisbury, M. Nakamura, M.E. Shenton, and R.W. McCarley, "A cross-sectional and longitudinal magnetic resonance imaging study of cingulate gyrus gray matter volume abnormalities in first-episode schizophrenia and first-episode affective psychosis [abstract]," *Archives of General Psychiatry* 65:7 (2008): 746-760.

74. J. Mackiewicz and S. Gershon, "An Experimental Study of the Neuropathological and Toxicological Effects of Chlorpromazine and Reserpine," *Journal of Neuropsychiatry* 5 (1964): 159.

75. Ibid.

76. Ibid., 168.p.45

77. K. Jellinger, "Neuropathologic Findings after Neuroleptic Long-Term Therapy," in L. Roizin, H. Shiraki, and N. Grcevic, Ed. *Neurotoxicology* (New York: Raven Press, 1977), 28.

78. I.J. Mitchell, A.C. Cooper, M.R. Griffiths, and A.J. Cooper, "Acute Administration of Haloperidol Induces Apoptosis of Neurones in the Striatum and Substantia Nigra in the Rat," *Neuroscience* 109:1 (2002): 89-99.

79. T. Sunderland and B.M. Cohen, "Blood to brain distribution of neuroleptics," *Psychiatry Research* 20:4 (1987): 299-205.

80. J. Kornhuber, A. Schultz, J. Wiltfang, I. Meineke, C.H. Gleiter, R. Zochling, et al, "Persistence of Haloperidol in Human Brain Tissue," *American Journal of Psychiatry* 156 (1999): 885-890.

81. M.E. Burton, L.M. Shaw, J.J. Schentag, and W.E. Evans. *Applied Pharmacokinetics & Pharmacodynamics, 4th Ed.* (Baltimore, MD: Lippincott Williams & Wilkins, 2006), 828.

82. KA. Dorph-Petersen, J.N. Pierri, J.M. Perel, Z. Sun, A.R. Sampson, and D.A. Lewis, "The Influence of Chronic Exposure to Antipsychotic Medications on Brain Size before and after Tissue Fixation: A Comparison of Haloperidol and Olanzapine in Macaque Monkeys," *Neuropsychopharmacology* 30 (2005): 1649-1661.

83. G.T. Konopaske, K.A. Dorph-Petersen, J.N. Pierri, Q. Wu, A.R. Sampson, and D.A. Lewis, "Effect of Chronic Exposure to Antipsychotic Medication on Cell Numbers in the Parietal Cortex of Macaque Monkeys," *Neuropsychopharmacology* 32 (2007): 1216-1223.

84. G.T. Konopaske, K.A. Dorph-Petersen, R.A. Sweet, J.N. Pierri, W. Zhang, A.R. Sampson et al, "Effect of Chronic Antipsychotic Exposure on Astrocyte and Oligodendrocyte Numbers in Macaque Monkeys," *Biological Psychiatry* 63 (2008): 759-765.

Chapter 6 - Anxiolytics

1. C.B. Nemeroff, "Anxiolytics: Past, Present, and Future Agents," *Journal of Clinical Psychiatry* 64 suppl 3 (2003): 3-6.

2. S.L. Speaker, "From 'happiness pills' to 'national nightmare': changing cultural assessment of minor tranquilizers in America, 1955-1980," *Journal of the History of Medicine and Allied Sciences* 52:3 (1997): 338-376.

3. Ibid., 345.

4. F. Lopez-Munoz, R. Ucha-Udabe, and C. Alamo, "The history of barbiturates a century after their clinical introduction," *Neuropsychiatric Disease and Treatment* 1:4 (2005): 329-343.

Endnotes

5. L.H. Sternbach, "The Benzodiazepine Story," *Journal of Medicinal Chemistry* 22:1 (1979): 1-7.

6. U.S. FDA. "Buspar," Label and Approval History. Accessed on October 22, 2008 at: http://www.accessdata.fda.gov/Scripts/cder/DrugsatFDA/index.cfm?fuseaction=Search.Overview&DrugName=BUSPAR&CFID=11808942&CFTOKEN=9b68c2d2a391444c-26D4FD01-1372-5AE1-6461D9B0271E83D6

7. U.S. FDA. "Effexor," Label and Approval History. Accessed on October 22, 2008 at: http://www.accessdata.fda.gov/Scripts/cder/DrugsatFDA/index.cfm?fuseaction=Search.Label_ApprovalHistory#apphist

8. S.H. Snyder, *Drugs and the Brain* (New York: Scientific American Library, 1999).

9. D. Healy, *The Antidepressant Era* (Cambridge, Massachusetts: Harvard University Press, 1999).

10. M.H. Lader, M. Ron, and H. Petursson, "Computed axial brain tomography in long-term benzodiazepine users," *Psychological Medicine* 14 (1984): 203-206.

11. K.M.H. Perera, T. Powell, and F.A. Jenner, "Computerized axial tomographic studies following long-term use of benzodiazepines," *Psychological Medicine* 17 (1987): 775-777.

12. C. Schmauss and J.C. Krieg, "Enlargement of cerebrospinal fluid spaces in long-term benzodiazepine abusers," *Psychological Medicine* 17 (1987): 869-873.

13. P. Moodley, S. Golombok, P. Shine, and M. Lader, "Computed Axial Brain Tomograms in Long-term Benzodiazepine Users," *Psychiatry Research* 48 (1993): 135-144.

14. M. Carassiti, A. Ysasi, and L.M. Gonzalo, "Efectos de la administracion cronica de Midazolam sobre el hipocampo de ratas," *Revista de Medicina de la Universidad de Navarra* 1-3 (1998): 18-28.

15. J.T. Backman, K.T. Olkkola, M. Ojala, H. Laaksovirta, and P.J. Neuvonen, "Concentrations and effects of oral midazolam are greatly reduced in patients treated with carbamazepine or phenytoin [abstract]," *Epilepsia* 37:3 (1996): 253-257.

16. T. Nishimura, N. Amano, Y. Kubo, M. Ono, Y. Kato, H. Fujita, et al, "Asymmetric Intestinal First-Pass Metabolism Causes Minimal Oral Bioavailability of Midazolam in Cynomolgus Monkeys," *Drug Metabolism and Disposition* 35:8 (2007): 1275-1284.

17. New Zealand Medicines and Medical Devices Safety Authority (04 April 2008). "Hypnovel (midazolam) 7.5 mg tablets," Information Data Sheet for Health Professionals. Accessed on August 8, 2008 at: http://www.medsafe.govt.nz/Profs/Datasheet/h/Hynoveltab.htm

18. C.E. Lau, F. Ma, Y. Wang, and C. Smith, "Pharmacokinetics and bioavailability of midazolam after intravenous, subcutaneous, intraperitoneal and oral administration under a chronic food-limited regimen: relating DRL performance to pharmaco-kinetics," *Psychopharmacology* 126 (1996): 241-248.

19. B.S. Jortner, "The return of the dark neuron. A histological artifact complicating contemporary neurotoxicologic evaluation," *NeuroToxicology* 27:4 (2006): 628-634.

20. A. Ohtsuka and T. Murakami, "Dark Neurons in the Mouse Brain: An Investigation into the Possible Significance of Their Variable Appearance within a Day and Their Relation to Negatively Charged Cell Coats," *Archives of Histology and Cytology* 59:1 (1996): 79-85.

21. K. Ishida, H. Shimizu, H. Hida, S. Urakawa, K. Ida, and H. Nishino, "Argyrophilic dark neurons represent various states of neuronal damage in brain insults: some come to die and others survive [abstract]," *Neuroscience* 125:3 (2004): 644-644.

22. E. Kovesdi, J. Pal, and F. Gallyas, "The fate of 'dark' neurons produced by transient focal cerebral ischemia in a non-necrotic and non-excitotoxic environment: neurobiological aspects [abstract]," *Brain Research* 1147 (2007): 272-283.

23. J.L. Poirier, R. Capek, and Y. De Koninck, "Differential progression of Dark Neuron and Fluoro-Jade labeling in the rat hippocampus following pilocarpine-induced status epilepticus [abstract]," *Neuroscience* 97:1 (2000): 59-68.

24. W.W. Christie, "Gangliosides: Structure, Occurrence, Biology and Analysis," 11 August 2008. Accessed on October 12, 2008 at: http://www.lipidlibrary.co.uk/Lipids/gang/

25. K.A. Sheikh, J. Sun, Y. Liu, H. Kawai, T.O. Crawford, R.L. Proia, et al, "Mice lacking complex gangliosides develop Wallerian degeneration and myelination defects," *Proceedings of the National Academy of Sciences* 96 (1999): 7532-7537.

26. M. Thomas, J.P. Ballantyne, S. Hansen, A.I. Weir, and D. Doyle, "Anterior horn cell dysfunction in Alzheimer's disease," *Journal of Neurology, Neurosurgery, and Psychiatry* 45 (1982): 378-381.

Endnotes

27. Q. Ma, M. Kobayashi, M. Sugiura, N. Ozaki, K. Nishio, Y. Shiraishi, et al, "Morphological study of disordered myelination and the degeneration of nerve fibers in the spinal cord of mice lacking complex gangliosides [abstract]," *Archives of Histology and Cytology* 66:1 (2003): 37-44.

28. S.R. De Luka, S. Protic and S. Vraski, "Ganglioside Content and Composition in Rat Cerebellum After Prolonged Diazepam Treatment," *Physiological Research* 48 (1999): 143-148.

29. S.R. De Luka, S. Protic, and S.R. Vrbaski, "Ganglioside a/b ratio in different rat brain regions following chronic diazepam treatment," *Neurological Sciences* 23 (2002): 69-74.

30. D.J. Greenblatt, T.P. Laughren, M.D. Allen, J.S. Harmatz, and R.I. Shader, "Plasma diazepam and desmethyldiazepam concentrations during long-term diazepam therapy," *British Journal of Clinical Pharmacology* 1 (1981): 35-40.

31. D.M. Rutherford, A. Okoko, and P.J. Tyrer, "Plasma Concentrations of Diazepam and Desmethyldiazepam During Chronic Diazepam Therapy," *British Journal of Clinical Pharmacology* 6 (1978): 69-73.

32. U. Klotz, K.H. Antonin, and P.R. Bieck, "Pharmacokinetics and Plasma Binding of Diazepam in Man, Dog, Rabbit, Guinea Pig, and Rat," *Journal of Pharmacology and Experimental Therapeutics* 199:1 (1976): 67-73.

33. H. Friedman, D.R. Abernethy, D.J. Greenblatt, and R.I. Shader, "The pharmacokinetics of diazepam and desmethyldiazepam in rat brain and plasma," *Psychopharmacology* 88 (1986): 267-270.

34. J.M. Diaz-Garcia, J. Oliver-Botana, and D. Fos-Galve, "Pharmacokinetics of Diazepam in the Rat: Influence of a Carbon Tetrachloride-Induced Hepatic Injury," *Journal of Pharmaceutical Sciences* 81:8 (1992): 768-772.

35. I. Kracun, H. Rosner, V. Drnovsek, M. Heffer-Lauc, C. Cosovic, and G. Lauc, "Human brain gangliosides in development, aging, and disease," *International Journal of Developmental Biology* 35 (1991): 289-295.

36. Y. Ohtani, Y. Tamai, Y. Ohnuki, and S. Miura, "Ganglioside alterations in the central and peripheral nervous systems of patients with Creutzfeldt-Jakob disease [abstract]," *Neurodegeneration* 5:4 (1996): 331-338.

37. T. Yamashita, Y.P. Wu, R. Sandhoff, N. Werth, H. Mizukami, J.M. Ellis, et al, "Interruption of ganglioside synthesis produces central nervous system degeneration and altered axon-glial interactions," *Proceedings of the National Academy of Sciences* 102:8 (2005): 2725-2730.

38. H. Sohn, Y.S. Kim, H.T. Kim, C.H. Kim, E.W. Cho, H.Y. Kang, et al, "Ganglioside GM3 is involved in neuronal cell death," *FASEB Journal* 20 (2006): E525-R535.

39. N. Yamamoto, W.E. Van Nostrand, and E. Yanagisawa, "Further evidence of local ganglioside-dependent amyloid beta protein assembly in brain [abstract]," *Neuroreport* 17:16 (2006): 1735-1737.

Chapter 7 - Mood Stabilizers

1. M. Harris, S. Chandran, N. Chakraborty, and D. Healy, "Mood-stabilizers: the archaeology of the concept," *Bipolar Disorders* 5 (2003): 446-452.

2. R.S. El-Mallakh and J.W. Jefferson, "Prethymoleptic Use of Lithium," *American Journal of Psychiatry* 156 (1999): 129.

3. S. Snyder, *Drugs and the Brain* (New York: Scientific American Library, 1996), 113-119.

4. M. Harris, S. Chandran, N. Chakraborty, and D. Healy, "Mood-stabilizers: the archaeology of the concept," *Bipolar Disorders* 5 (2003): 446-452.

5. R.M. Post, K.D. Denicoff, M.A. Frye, R.T. Dunn, G.S. Leverich, E. Osuch, et al, "A History of the Use of Anticonvulsants as Mood Stabilizers in the Last Two Decades of the 20th Century," *Neuropsychobiology* 38 (1998): 152-166.

6. J. Moncrieff, "Lithium Revisited: A Re-Examination of the Placebo Controlled Trials of Lithium Prophylaxis in Manic-Depressive Disorder," *British Journal of Psychiatry* 167 (1995): 569-574.

7. T. Silverstone, H. McPherson, N. Hunt, S. Romans, "How effective is lithium in the prevention of relapse in bipolar disorder? A prospective naturalistic follow-up study," *Australian and New Zealand Journal of Psychiatry* 32 (1998): 61-66.

8. R.L. Symonds and P. Williams, "Lithium and the changing incidence of mania," *Psychological Medicine* 11 (1981): 193-196.

9. S. Akhondzadeh, E. Emamian, A. Ahmadi-Abhari, O. Shabestari, and M. Dadgarnejad, "Is It Time to Have Another Look at Lithium Maintenance Therapy in Bipolar Disorder?," *Progress in Neuropsychopharmacology and Biological Psychiatry* 23 (1999): 1011-1017.

Endnotes

10. F. Goodman, P. Glassman, Q. Ma, M. Maglione, S. Rhodes, C. Rolon, et al, (December 2004). "Drug Class Review on Antiepileptic Drugs in Bipolar Mood Disorder and Neuropathic Pain." OHSU. [Produced by Southern California Evidence-based Practice Center – RAND.] Accessed on February 20, 2006 at: http://www.ohsu.edu/drugeffectiveness/reports/final.cfm

11. T.R. Henry, "The History of Valproate in Clinical Neuroscience," *Psychopharmacology Bulletin* 37 suppl 2:5 (2003): 5-16.

12. N. F. Feiner (1998). "A Short History of Antiepileptic Drugs in Psychiatry." Accessed on July 3, 2004 at: http://www.psycom.net/depression.central.aed.html

13. D.F. Scott, "The discovery of anti-epileptic drugs," *Journal of the History of Neuroscience* 1 (1992): 111-118.

14. R.M. Post, K.D. Denicoff, M.A. Frye, R.T. Dunn, G.S. Leverich, E. Osuch, et al, "A History of the Use of Anticonvulsants as Mood Stabilizers in the Last Two Decades of the 20th Century," *Neuropsychobiology* 38 (1998): 152-166.

15. S.R.B. Weiss and R.M. Post, "Kindling: Separate vs. Shared Mechanisms in Affective Disorders and Epilepsy," *Neuropsychobiology* 38 (1998): 167-180.

16. M.C. Walker, H.S. White, and J.W.A.S. Sander, "Disease modification in partial epilepsy," *Brain* 125 (2002): 1937-1950.

17. H.S. White, "Animal models of epileptogenesis," *Neurology* 59:Suppl 5 (2002): S7-S14.

18. P. Kwan and J.W. Sander, "The natural history of epilepsy: an epidemiological view," *Journal of Neurology, Neurosurgery, and Psychiatry* 75 (2004): 1376-1381.

19. A.E. Watts, "The Natural History of Untreated Epilepsy in a Rural Community in Africa," *Epilepsia* 33:3 (1992): 464-468.

20. W.F.M. Arts, O.F. Brouwer, A.C.B. Peters, H. Stroink, E.A.J. Peeters, P.I.M. Schmitz, et al, "Course and prognosis of childhood epilepsy: 5-year follow-up of the Dutch study of epilepsy in childhood," *Brain* 127 (2004): 1774-1784.

21. T. Suppes, K.A. Chisholm, D. Dhavale, M.A. Frye, L.L. Altshuler, S.L. McElroy, et al, "Tiagabine in treatment refractory bipolar disorder: a clinical case series [abstract]," *Bipolar Disorders* 4:5 (2002): 283-289.

22. S.N. Ghaemi, A.A. Shirzadi, J. Klugman, D.A. Berv, T.B. Pardo, and M.M. Filkowski, "Is adjunctive open-label zonisamide effective for bipolar disorder? [abstract]," *Journal of Affective Disorders* 105:1-3 (2008): 311-314.

23. No author (06 August 2003). "Synopsis: A Randomized, Double-Blind, Multicenter, Placebo-Controlled 12-Week Study of the Safety and Efficacy of Two Doses of Topiramate for the Treatment of Acute Manic or Mixed Episodes in Subjects With Bipolar I Disorder With an Optional Open-Label Extension." Protocol CR003199. Accessed on December 6, 2008 at: http://download.veritasmedicine.com/PDF/CR003199_CSR.pdf

24. C.Y. Yung, "A review of clinical trials of lithium in neurology [abstract]," *Pharmacology, Biochemistry, and Behavior* 21 Suppl 1 (1984): 57-64.

25. R.E. Frost, F.S. Messiha, "Clinical uses of lithium salts [abstract]," *Brain Research Bulletin* 11:2 (1983): 219-231.

26. A. Wada, H. Yokoo, T. Yanagita, and H. Kobayashi, "Lithium: Potential Therapeutics Against Acute Brain Injuries and Chronic Neurodegenerative Diseases," *Journal of Pharmacological Sciences* 99 (2005): 307-321.

27. K.N. Fountoulakis, E. Vieta, C. Bouras, G. Notaridis, P. Giannakopoulos, G. Kaprinis, et al, "A systematic review of existing data on long-term lithium therapy: neuroprotective or neurotoxic?," *International Journal of Neuropsychopharmacology* 11 (2008): 269-287.

28. D.M. Chuang, "Neuroprotective and neurotrophic actions of the mood stabilizer lithium: can it be used to treat neurodegenerative diseases? [abstract]," *Critical Reviews in Neurobiology* 16:1-2 (2004): 83-90.

29. C.G. Janson, M. Assadi, J. Francis, L. Bilaniuk, D. Shera, and P. Leone, "Lithium citrate for Canavan disease [abstract]," *Pediatric Neurology* 33:4 (2005): 235-243.

30. F. Fornai, P. Longone, L. Cafaro, O. Kastsiuchenka, M. Ferrucci, M.L. Manca, et al, "Lithium delays progression of amyotrophic lateral sclerosis [abstract]," *Proceedings of the National Academy of Sciences* USA 105:6 (2008): 2052-2057.

31. J. Zhong and W.H. Lee, "Lithium: a novel treatment for Alzheimer's disease? [abstract]," *Expert Opinion on Drug Safety* 6:4 (2007): 375-383.

32. C. Livingstone and H. Rampes, "Lithium: a review of its metabolic adverse effects," *Journal of Psychopharmacology* 20:3 (2006): 347-355.

33. C. Henry, "Lithium side-effects and predictors of hypothyroidism in patients with bipolar disorders: sex differences," *Journal of Psychiatry and Neuroscience* 27:2 (2002): 104-107.

34. M. Kusalic and F. Engelsmann, "Effect of lithium maintenance therapy on thyroid and parathyroid function," *Journal of Psychiatry and Neuroscience* 24:3 (1999): 227-233.

35. M. Gitlin, "Lithium and the Kidney: An Updated Review," *Drug Safety* 20:3 (1999): 231-243.

36. G.S. Markowitz, J. Radhakrishnan, N. Kambham, A.M. Valeri, W.H. Hines, and V.D. D'Agati, "Lithium Nephrotoxicity: A Progressive Combined Glomerular and Tubulointerstitial Nephropathy," *Journal of the American Society of Nephrology* 11 (2000): 1439-1448.

37. N. Bassilios, P. Martel, V. Godard, M. Froissart, J.P. Grunfeld, B. Stengel, et al, "Monitoring of glomerular filtration rate in lithium-treated outpatients – an ambulatory laboratory database surveillance [abstract]," *Nephrology, Dialysis, Transplantation* 23:2 (2008): 562-565.

38. C. Presne, F. Fakhouri, L.H. Noel, B. Stengel, C. Even, H. Kreis, et al, "Lithium-induced nephropathy: Rate of progression and prognostic factors [abstract]," *Kidney International* 64:2 (2003): 585-592.

39. H. Bendz, M. Aurell, J. Balldin, A.A. Mathe, and I. Sjodin, "Kidney damage in long-term lithium patients: a cross-sectional study of patients with 15 years or more on lithium [abstract]," *Nephrology, Dialysis, Transplantation* 9:9 (1994): 1250-1254.

40. E. Lederer, C.C. Dasco, and M.D.T. Tran (January 30, 2007). "Lithium Nephropathy." Accessed on December 12, 2008 at: http://emedicine.medscape.com/article/242772-print

41. G.L. Sheean, "Lithium Neurotoxicity," *Clinical and Experimental Neurology* 28 (1991): 112-127.

42. M.E.G. Sansone and D.K. Ziegler, "Lithium Toxicity: A Review of Neurologic Complications," *Clinical Neuropharmacology* 8:3 (1985): 242-248.

43. G.L. Sheean, "Lithium Neurotoxicity," *Clinical and Experimental Neurology* 28 (1991): 112-127.

44. P.W. Oakley, I.M. Whyte, and G.L. Carter, "Lithium toxicity: an iatrogenic problem in susceptible individuals," *The Australian and New Zealand Journal of Psychiatry* 35 (2001): 833-840.

Endnotes

45. G.L. Sheean, "Lithium Neurotoxicity," *Clinical and Experimental Neurology* 28 (1991): 112-127.

46. P.W. Oakley, I.M. Whyte, and G.L. Carter, "Lithium toxicity: an iatrogenic problem in susceptible individuals," *The Australian and New Zealand Journal of Psychiatry* 35 (2001): 833-840.

47. W.S. Waring, W.J. Laing, A.M. Good, and D.N. Bateman, "Pattern of lithium exposure predicts poisoning severity: evaluation of referrals to a regional poisons unit," *QJM* 100 (2007): 271-276.

48. P.E. Keck, Jr., and S.L. McElroy, "Clinical Pharmacodynamics and Pharmacokinetics of Antimanic and Mood-Stabilizing Medications," *Journal of Clinical Psychiatry* 63 Suppl 4 (2002): 3-11.

49. B. von Hartitzsch, N.A. Hoenich, R.J. Leigh, R. Wilkinson, T.H. Frost, A. Weddel, and G.A. Posen, "Permanent Neurological Sequelae Despite Hemodialysis for Lithium Intoxication," *British Medical Journal* 30 (1972): 757-759.

50. L. Goldwater and M. Pollack, "Neurological sequelae after lithium intoxication [abstract]," *New Zealand Medical Journal* 84:575 (1976):356-358.

51. H.E. Hansen and A. Amdisen, "Lithium intoxication: Report of 23 cases and review of 100 cases from the literature [abstract]," *The Quarterly Journal of Medicine* 47:186 (1978): 123-144.

52. Adityanjee, "The syndrome of irreversible lithium effectuated neurotoxicity," *Journal of Neurology, Neurosurgery, and Psychiatry* 50:9 (1987): 1246-1247.

53. M. Schou, "Long-lasting neurological sequelae after lithium intoxication [abstract]," *Acta Psychiatrica Scandinavica* 70:6 (1984): 594-602.

54. M. Roy, E. Stip, D.N. Black, V. Lew, and R. Langlois, "Sequelles neurologiques secondaires a une intoxication aigue au lithium," *Canadian Journal of Psychiatry* 44 (1999): 671-679.

55. Adityanjee, K.R. Munshi, and A. Thampy, "SILENT: The Syndrome of Irreversible Lithium-Effectuated Neurotoxicity," *Clinical Neuropharmacology* 28 (2005): 38-49.

56. H. Sun, "From Leeching to Anaphrodisiacs: Treatments of Epilepsy in the Nineteenth Century," *University of Toronto Medical Journal* 84:2 (2007): 107.

57. D.F. Scott, "The discovery of anti-epileptic drugs," *Journal of the History of Neuroscience* 1 (1992): 111.

58. R.E. Hales, S.C. Yudofsky, and J.A. Talbott, Ed., *The American Psychiatric Press Textbook of Psychiatry, 3rd Ed.* (Washington, DC: American Psychiatric Press, 1999.

59. H. Blumenfeld, *Neuroanatomy through Clinical Cases* (Sunderland, Massachusetts: Sinauer Associates, Inc., 2002).

60. A.H. Ropper and R.H. Brown, *Adams and Victor's Principles of Neurology, 8th Ed.* (New York: McGraw-Hill, 2005).

61. S.M. Stahl, *Stahl's Essential Psychopharmacology, 3rd Ed.* (New York: Cambridge University Press, 2008).

62. *Dorland's Pocket Medical Dictionary, 24th Ed.* (Philadelphia: W.B. Saunders Company, 1989.

63. NIH (December 9, 2008). NINDS Dementia Information Page. Accessed on December 30, 2008 at: http://www.ninds.nih.gov/disorders/dementias/dementia.htm

64. NIH (February 12, 2007). NINDS Encephalopathy Information Page. Accessed on December 30, 2008 at:http://www.ninds.nih.gov/disorders/encephalopathy/encephalopathy.htm

65. Pseudo. Merriam-Webster on-line dictionary. Accessed on January 5, 2009 at: http://www.merriam-webster.com/dictionary/pseudo

66. J. Gardiner, "Saving Kendall," *Ladies' Home Journal*, August 2005, pp, 82-91.

67. Ibid, 91.

68. Divalproex sodium product label (updated March 24, 2008). Accessed on January 3, 2009 at: http://www.fda.gov/cder/foi/label/2008/019680s024lbl.pdf

69. C. Armon, C. Shin, P. Miller, S. Carwile, E. Brown, J.D. Edinger, and R.G. Paul, "Reversible parkinsonism and cognitive impairment with chronic valproate use," *Neurology* 47 (1996): 626-635.

70. K. Easterford, P. Clough, M. Kellett, K. Fallon, and S. Duncan, "Reversible parkinsonism with normal B-CIT-SPECT in patients exposed to sodium valproate," *Neurology* 62 (2004): 1435-1437.

71. A.J. Ristic, N. Vojvodic, S. Jankovic, A. Sindelic, and D. Sokic, "The Frequency of Reversible Parkinsonism and Cognitive Decline Associated with Valproate Treatment: A Study of 364 Patients with Different Types of Epilepsy," *Epilepsia* 47:12 (2006): 2183-2185.

Endnotes

72. C.T. Gualtieri and L.G. Johnson, "Comparative Neurocognitive Effects of 5 Psychotropic Anticonvulsants and Lithium," *Medscape General Medicine* 8:3 (2006): 46, published on-line. Accessed on February 22, 2008 at: http://www.pubmedcentral.nih.gov/articlerender.fcgi?tool=pubmed&pubmedid=17406176

73. E.C. Bell, M.C. Willson, A.H. Wilman, S. Dave, and P.H. Silverstone, "Differential effects of chronic lithium and valproate on brain activation in healthy volunteers," *Human Psychopharmacology* 20 (2005): 415-424.

74. K.J. Meador, D.W. Loring, E.E. Moore, W.O. Thompson, M.E. Nichols, R.E. Oberzan, et al, "Comparative cognitive effects of phenobarbital, phenytoin, and valproate in healthy adults," *Neurology* 45 (1995): 1494-1499.

75. P.E. Keck, Jr., and S.L. McElroy, "Clinical Pharmacodynamics and Pharmacokinetics of Antimanic and Mood-Stabilizing Medications," *Journal of Clinical Psychiatry* 63 Suppl 4 (2002): 3-11.

76. E. Hermansen, "Lithium," *UtoxUpdate* [Newsletter of the Utah Poison Control Center] 3:4 (2001).

77. R.A. Komoroski, J.E. Newton, J.R. Sprigg, D. Cardwell, P. Mohanakrishnan, and C.N. Karson, "In vivo 7Li nuclear magnetic resonance study of lithium pharmacokinetics and chemical shift imaging in psychiatric patients [abstract]," *Psychiatry Research* 50:2 (1993): 67-76.

78. P. Plenge, A. Stensgaard, H.V. Jensen, C. Thomsen, E.T. Mellerup, and O. Henriksen, "24-hour lithium concentration in human brain studied by Li-7 magnetic resonance spectroscopy [abstract]," *Biological Psychiatry* 36:8 (1994): 511-516.

79. J.M. Pearce, M. Lyon, and R.A. Komoroski, "Localized 7Li MR spectroscopy: in vivo brain and serum concentrations in the rat [abstract]," *Magnetic Resonance in Medicine* 52:5 (2004): 1087-1092.

80. O. Wraee, "The Pharmacokinetics of Lithium in the Brain, Cerebrospinal Fluid and Serum of the Rat," *British Journal of Pharmacology* 64 (1978): 273-279.

81. D. Ghoshdastidar, R.N. Dutta, and M.K. Poddar, "In vivo distribution of lithium in plasma and brain [abstract]," *Indian Journal of Experimental Biology* 27:11 (1989): 950-954.

82. S. Dethy, M. Manto, E. Bastianelli, V. Gangji, M.A. Laute, S. Goldman, et al, "Cerebellar spongiform degeneration induced by acute lithium intoxication in the rat," *Neuroscience Letters* 224 (1997): 25-28.

83. S.J.M. Smith and R.S. Kocen, "A Creutzfeldt-Jakob like syndrome due to lithium toxicity," *Journal of Neurology, Neurosurgery, and Psychiatry* 51 (1988): 120-123.

84. C.J. Kemperman and S.L. Notermans, "Creutzfeldt-Jacob like syndrome due to lithium toxicity," *Journal of Neurology, Neurosurgery, and Psychiatry* 52:2 (1989): 291.

85. A. Primavera, G. Brusa, and M.G. Poeta, "A Creutzfeldt-Jakob like syndrome due to lithium toxicity," *Journal of Neurology, Neurosurgery, and Psychiatry* 52:3 (1989): 423.

86. E. Broussolle, A. Setiey, Y. Moene, M. Trielett, and G. Chazot, "Reversible Creutzfeldt-Jakob like syndrome induced by lithium plus levodopa," *Journal of Neurology, Neurosurgery, and Psychiatry* 52:5 (1989): 686-687.

87. P.F. Finelli, "Drug-Induced Creutzfeldt-Jakob Like Syndrome," *Journal of Psychiatry and Neuroscience* 17:3 (1992): 103-105.

88. H. Kikyo and T. Furukawa, "Creutzfeldt-Jakob-like syndrome induced by lithium, levomepromazine, and phenobarbitone," *Journal of Neurology, Neurosurgery, and Psychiatry* 66 (1999): 802-803.

89. S. Mouldi, E. Le Rhun, S. Gautier, M. Devemy, A. Destee, and L. Defebvre, "Lithium-induced encephalopathy mimicking Creutzfeldt-Jakob disease [abstract]," *Revue Neurologique* 162:11 (2006): 1118-1121.

90. J.C. Antoine, J.L. Laplanche, J.F. Mosnier, P. Beaudry, J. Chatelain, and D. Michel, "Demyelinating peripheral neuropathy with Creutzfeldt-Jakob disease and mutation at codon 200 of the prion protein gene," *Neurology* 46 (1996): 1123-1127.

91. M. Niewiadomska, J. Kulczycki, D. Wochnik-Dyjas, G.M. Szpak, M. Rakowicz, W. Lojkowska, et al, "Impairment of the Peripheral Nervous System in Creutzfeldt-Jakob Disease," *Archives of Neurology* 59 (2002): 1430-1436.

92. C.J. Gibbs and D.M. Asher, "Subacute Spongiform Unconventional Virus Encephalopathies" Accessed on December 8, 2008 at: http://gsbs.utmb.edu/microbook/ch071.htm

93. R. Chitharanjan and F.P. Thomas (April 27, 2007). "Variant Creutzfeldt-Jakob Disease and Bovine Spongiform Encephalopathy." Accessed on December 8, 2008 at: http://www.emedicine.com/neuro/topic725.htm

94. C.M. De Castro Costa, J.M. Brucher, and C. Laterre, "Sporadic Creutzfeldt-Jakob Disease," *Arquivos de Neuropsiquiatria* 56:3A (1998): 356-365.

Endnotes

95. J.L. Ridet, S.K. Malhotra, A. Privat, and F.H. Gage, "Reactive astrocytes: cellular and molecular cues to biological function," *Trends in Neurosciences* 20:12 (1997): 570-577.

96. M. Pekny and M. Nilsson, "Astrocyte Activation and Reactive Gliosis," *Glia* 50 (2005): 427-434.

97. J.P. O'Callaghan and K. Sriram, "Glial fibrillary acidic protein and related glial proteins as biomarkers of neurotoxicity," *Expert Opinion on Drug Safety* 4:3 (2005): 433-442.

98. G. Yiu and Z. He, "Glial inhibition of CNS axon regeneration," *Nature Reviews: Neuroscience* 7 (2006): 617-627.

99. M.T. Tacconi, "Neuronal Death: Is There a Role for Astrocytes?," *Neurochemical Research* 23:5 (1998): 759-765.

100. M.L. Kashon, G.W. Ross, J.P. O'Callaghan, D.B. Miller, H. Petrovich, C.M. Burchfiel, et al, "Associations of cortical astrogliosis with cognitive performance and dementia status," *Journal of Alzheimer's Disease* 6 (2004): 595-604.

101. E. Rocha and R. Rodnight, "Chronic Administration of Lithium Chloride Increases Immunodetectable Glial Fibrillary Acidic Protein in the Rat Hippocampus," *Journal of Neurochemistry* 63 (1994): 1582-1584.

102. E. Rocha, M. Achaval, P. Santos, and R. Rodnight, "Lithium treatment causes gliosis and modifies the morphology of hippocampal astrocytes in rats," *NeuroReport* 9 (1998): 3971-3974.

103. No author. Wikipedia. "Camera lucida." Accessed on December 15, 2008 at: http://en.wikipedia.org/wiki/Camera_lucida

104. E. Sykova, "Extrasynaptic Volume Transmission and Diffusion Parameters of the Extracellular Space," *Neuroscience* 129 (2004): 861-876.

105. E. Sykova, "Diffusion properties of the brain in health and disease," *Neurochemistry International* 45 (2004): 453-466.

106. P. Phatak, A. Shaldivin, S. King, P. Shapiro, and W.T. Regenold, "Lithium and inositol: effects on brain water homeostasis in the rat," *Psychopharmacology* 186 (2006): 41-47.

107. W.T. Regenold, "Lithium and Increased Cortical Gray Matter – More Tissue or More Water?," *Biological Psychiatry* 63 (2008): e17.

Endnotes

108. C.E. Bearden, P.M. Thompson, M. Dalwant, K.M. Hayashi, A.D. Lee, D.C. Glahn, et al, "Reply: Lithium and Increased Cortical Gray Matter – More Tissue or More Water ?," *Biological Psychiatry* 63 (2008): e19.

109. G.J. Moore, J.M. Bebchuk, I.B. Wilds, G. Chen, and H.K. Menji, "Lithium-induced increase in human brain grey matter," *Lancet* 356:7 (2000): 1241-1242.

110. R.B. Sassi, M. Nicoletti, P. Brambilla, A.G. Mallinger, E. Frank, D.J. Kupfer, et al, "Increased gray matter volume in lithium-treated bipolar disorder patients," *Neuroscience Letters* 329:2 (2002): 243-245.

111. No author (December 1, 2008). "Aquaporin." Accessed on December 7, 2008 at: http://en.wikipedia.org/wiki/Aquaporin

112. P. Agre, M. Bonhivers, and M.J. Borgnia, "The Aquaporins, Blueprints for Cellular Plumbing Systems," *Journal of Biological Chemistry* 273:24 (1998): 14659-14662.

113. A.J. Yool, "Aquaporins: Multiple Roles in the Central Nervous System," *The Neuroscientist* 13:5 (2007): 470-485.

114. E.E. Benarroch, "Neuron-Astrocyte Interactions: Partnership for Normal Function and Disease in the Central Nervous System," *Mayo Clinic Proceedings* 80:10 (2005): 1326-1338.

115. Y. Iwasaki, M. Mimuro, M. Yoshida, Y. Hashizume, M. Ito, T. Kitamoto, Y. Wakayama, et al, "Enhanced Aquaporin-4 Immunoreactivity in sporadic Creutzfeldt-Jakob disease," *Neuropathology* 27 (2007): 314-323.

116. B. Hexom and R.P. Barthel, "Lithium and Pseudotumor Cerebri," *Journal of the American Academy of Child and Adolescent Psychiatry* 43:3 (2004): 247.

117. J. Jonnalagadda, E. Saito, and V. Kafantaris, "Lithium, Minocycline, and Pseudotumor Cerebri," *Journal of the American Academy of Child and Adolescent Psychiatry* 44:3 (2005): 209.

118. S.H. Levine and C. Puchalski, "Pseudotumor cerebri associated with lithium therapy in two patients [abstract]," *Journal of Clinical Psychiatry* 51:6 (1990): 251-253.

119. R.F. Saul, H.A. Hamburger, and J.B. Selhorst, "Pseudotumor cerebri secondary to lithium carbonate [abstract]," *JAMA* 253:19 (1985): 2869-2870.

120. F. Wang, X.C. Feng, Y.M. Li, H. Yang, and T.H. Ma, "Aquaporins as potential drug targets," *Acta Pharmacologica Sinica* 27:4 (2006): 395-401.

Endnotes

121. D. Boassa, W.D. Stamer, and A.J. Yool, "Ion Channel Function of Aquaporin-1 Natively Expressed in Choroid Plexus," *Journal of Neuroscience* 26:30 (2006): 7811-7819.

122. W. Sobaniec, E. Jankowicz, and M. Sobaniec-Lotowska. "Effect of valproic acid on the morphology of the rat cerebellum and brain stem [abstract]," *Neuropatologia polska* 27:4 (1989): 593-606.

123. M. Sobaniec-Lotowska and W. Sobaniec, "Effect of chronic administration of sodium valproate on the morphology of the rat brain hemisphere [abstract]," *Neuropatologia polska* 31:3-4 (1993): 149-158.

124. M.E. Sobaniec-Lotowska and W. Sobaniec, "Morphological features of encephalopathy after chronic administration of the antiepileptic drug valproate to rats. A transmission electron microscopic study of capillaries in the cerebellar cortex [abstract]," *Experimental and Toxicologic Pathology* 48:1 (1996): 65-75.

125. M. Sobaniec-Lotowska, W. Sobaniec, and A. Augstynowicz, "Morphometric analysis of the cerebellar cortex capillaries in the course of experimental valproate encephalopathy and after chronic exposure to sodium valproate using transmission electron microscopy [abstract]," *Folia Neuropathologica* 39:4 (2001): 277-280.

126. M.E. Sobaniec-Lotowska, "Ultrastructure of synaptic junctions in the cerebellar cortex in experimental valproate encephalopathy and after terminating chronic application of the antiepileptic [abstract]," *Folia Neuropathologica* 40:2 (2002): 87-96.

127. M.E. Sobaniec-Lotowska and J.M. Lotowska, "Ultrastructural study of cerebellar dentate nucleus astrocytes in chronic experimental model with valproate," *Folia Neuropathologica* 43:3 (2005): 166-171.

128. M.E. Sobaniec-Lotowska, "Ultrastructure of Purkinje cell perikarya and their dendritic processes in the rat cerebellar cortex in experimental encephalopathy induced by chronic application of valproate," *International Journal of Experimental Pathology* 82 (2001): 337-348.

129. M.E. Sobaniec-Lotowska, "Ultrastructure of astrocytes in the cortex of the hippocampal gyrus and in the neocortex of the temporal lobe in experimental valproate encephalopathy and after valproate withdrawal," *International Journal of Experimental Pathology* 84 (2003): 115-125.

130. M.E. Sobaniec-Lotowska, "A transmission electron microscopic study of microglia/macrophages in the hippocampal cortex and neocortex following chronic exposure to valproate," *International Journal of Experimental Pathology* 86 (2005): 91-96.

Endnotes

131. Electron Microscope. Wikipedia. Accessed on January 14, 2009 at: http://en.wikipedia.org/wiki/File:Electron_Microscope.png

132. Food and Drug Administration (April 2006). Depakene product label. http://www.fda.gov/medwatch/SAFETY/2006/Jan_PI/Depakene_PI.pdf

133. M.H. Allen, R.M. Hirschfeld, P.J. Wozniak, J.D. Baker, and C.L. Bowden, "Linear Relationship of Valproate Serum Concentration to Response and Optimal Serum Levels for Acute Mania," *American Journal of Psychiatry* 163:2 (2006): 272-275.

134. K.D.K. Adkison, G.A. Ojemann, R.L. Rapport, R.L. Dills, and D.D. Shen, "Distribution of Unsaturated Metabolites of Valproate in Human and Rat Brain – Pharmacologic Relevance?," *Epilepsia* 36:8 (1995): 772-782.

135. V. Murugesan and P. Subramanian, "Enhancement of Circulatory Antioxidants by alpha-Ketoglutarate During Sodium Valproate Treatment in Wistar Rats," *Polish Journal of Pharmacology* 55 (2003): 31-36.

136. L. Sveberg Roste, E. Tauboll, A. Berner, K.A. Berg, M. Aleksandersen, L. Gjerstad, et al, "Morphological changes in the testis after long-term valproate treatment in male Wistar rats [abstract]," *Seizure* 10:8 (2001): 559-565.

137. L. Sveberg Rose, E. Tauboll, J.I. Isojarvi, A.J. Pakarinen, I.T. Huhtaniemi, M. Knip, et al, "Effects of chronic valproate treatment on reproductive endocrine hormones in female and male Wistar rats [abstract]," *Reproductive Toxicology* 16:6 (2002): 767-773.

138. S. Bakerman (Revised by P. Bakerman and P. Strausbach), *Bakerman's ABC's of Interpretive Laboratory Data, 3rd Ed.* (Myrtle Beach, SC: Interpretive Laboratory Data, Inc., 1994), p. 40.

139. A.V. Chicharro, A.J. de Marinis, and A.M. Kanner, "The measurement of ammonia blood levels in patients taking valproic acid: looking for problems where they do not exist?," *Epilepsy and Behavior* 11:3 (2007): 361-366.

140. S. Panda and K. Radhakrishnan, "Two Cases of Valproate-induced Hyperammonemic Encephalopathy Without Hepatic Failure," *Journal of the Association of Physicians of India* 52 (2004): 746-748.

141. S. Alqahtani, P. Federico, and R.P. Myers, "A case of valproate-induced hyperammonemic encephalopathy: look beyond the liver," *Canadian Medical Association Journal* 177:6 (2007): 568-569.

142. J. Wadzinski, R. Franks, D. Roane, and M. Bayard, "Valproate-associated Hyperammonemic Encephalopathy," *Journal of the American Board of Family Medicine* 20 (2007): 499-502.

143. M.J. Kempton, J.R. Geddes, U. Ettinger, S.C.R. Williams, and P.M. Grasby, "Meta-analysis, Database, and Meta-regression of 98 Structural Imaging Studies in Bipolar Studies," *Archives of General Psychiatry* 65:9 (2008): 1017-1032.

144. E.S. Monkul, G.S. Malhi, and J.C. Soares, "Anatomical MRI abnormalities in bipolar disorder: do they exist and do they progress?," *Australian and New Zealand Journal of Psychiatry* 39 (2005): 222-226.

145. W.T. Regenold, "Lithium and Increased Cortical Gray Matter – More Tissue or More Water?," *Biological Psychiatry* 63 (2008): e17.

146. M. Niethammer and B. Ford, "Permanent Lithium-Induced Cerebellar Toxicity: Three Cases and Review of the Literature," *Movement Disorders* 22:4 (2007): 570-573.

147. A.C. Rodrigues de Cerqueira, M. Costa dos Reis, F.D. Novis, J.M.F. Bezerra, G. Canedo de Magahlaes, et al, "Cerebellar Degeneration Secondary to Acute Lithium Carbonate Intoxication," *Arquivos de Neuropsiquiatria* 66:3-A (2008): 578-580.

148. M.P. DelBello, S.M. Strakowski, M.E. Zimmerman, J.M. Hawkins, and K.W. Sax, "MRI Analysis of the Cerebellum in Bipolar Disorder: A Pilot Study," *Neuropsychopharmacology* 21:1 (1999): 63-68.

149. N.P. Mills, M.P. DelBello, C.M. Adler, and S.M. Strakowski, "MRI Analysis of Cerebellar Vermal Abnormalities in Bipolar Disorder," *American Journal of Psychiatry* 162 (2005): 1530-1532.

150. R. Guerrini, A. Belmonte, R. Canapicchi, C. Casalini, and E. Perucca, "Reversible Pseudoatrophy of the Brain and Mental Deterioration Associated with Valproate Treatment," *Epilepsia* 39:1 (1998): 27-32.

151. O. Papazian, E. Canizales, I. Alfonso, R. Archila, M. Duchowny, and J. Aicardi, "Reversible Dementia and Apparent Brain Atrophy During Valproate Therapy," *Annals of Neurology* 38 (1995): 687-691.

152. C.A. Galimberti, M. Diegoli, I. Sartori, C. Uggetti, A. Brega, A. Tartara, et al, "Brain pseudoatrophy and mental regression on valproate and a mitochondrial DNA mutation," *Neurology* 67 (2006): 1715-1717.

153. R.S. McLachlan, "Pseudoatrophy of the Brain with Valproic Acid Monotherapy," *Canadian Journal of Neurological Sciences* 14 (1987): 294-296.

Endnotes

154. U. Thirugnanasampanthan, A. Foster, and R.A. Rauch, "Reversible cerebral atrophy: a case report and literature review," *General Hospital Psychiatry* 28:5 (2006): 458-462.

155. J.P. Soares-Fernandes, A. Machado, M. Ribeiro, C. Ferreira, J. Figueiredo, and J.F. Rocha, "Hippocampal Involvement in Valproate-Induced Acute Hyperammonemic Encephalopathy," *Archives of Neurology* 63:8 (2006): 1202-1203.

156. *Dorland's Pocket Medical Dictionary, 24th Ed.* (Philadelphia: W.B. Saunders Company, 1989), p. 285.

157. P. Kaufmann, D.C. Shungu, M.C. Sano, S. Jhung, K. Engelstad, E. Mitsis, et al, "Cerebral lactic acidosis correlates with neurological impairment in MELAS," *Neurology* 62 (2004): 1297-1302.

158. C. Madhavi and M. Jacob, "Light and electron microscopic structure of choroid plexus in hydrocephalic guinea pig [abstract]," *Indian Journal of Medical Research* 101 (1995): 217-224.

159. K. Watanabe, T. Kumagai, Y. Wakayama, K. Hara, and H. Yamada, "Congenital myopathy and communicating hydrocephalus – a possible pathogenetic combination [abstract]," *Brain Development* 4:6 (1982): 455-462.

160. M. Castro-Gago, A. Alonso, E. Pintos-Martinez, A. Beiras-Iglesias, Y. Campos, J. Arenas, et al, "Congenital hydraencephalic-hydrocephalic syndrome associated with mitochondrial dysfunction [abstract]," *Journal of Child Neurology* 14:2 (1999): 131-135.

161. O.J. Castejon and H.V. de Castejon, "Structural patterns of injured mitochondria in human oedematous cerebral cortex [abstract]," *Brain Injury* 18:11 (2004): 1107-1126.

162. F. Scaglia, "MELAS Syndrome," eMedicine. Accessed on January 3, 2009 at: http://emedicine.medscape.com/article/946864-print

163. S. Shanske, J. Pancrudo, P. Kaufmann, K. Engelstad, S. Jhung, J. Lu, et al, "Varying Loads of the Mitochondrial DNA A3243G Mutation in Different Tissues: Implications for Diagnosis," *American Journal of Medical Genetics* 130A (2004): 134-137.

164. T.P. Hutchin and G.A. Cortopassi, "Mitochondrial defects and hearing loss," *Cellular and Molecular Life Sciences* 57 (2000): 1927-1937.

165. K.G. Kimata, L. Gordan, E.T. Ajax, P.H. Davis, and T. Grabowski, "A Case of Late-Onset MELAS," *Archives of Neurology* 55 (1998): 722-725.

166. J.M. Gilchrist, M. Sikirica, E. Stopa, and S. Shanske, "Adult-Onset MELAS – Evidence for Involvement of Neurons as Well as Cerebral Vasculature in Strokelike Episodes," *Stroke* 27 (1996): 1420-1423.

167. C.A. Galimberti, M. Diegoli, I. Sartori, C. Uggetti, A. Brega, A. Tartara, et al, "Brain pseudoatrophy and mental regression on valproate and a mitochondrial DNA mutation," *Neurology* 67 (2006): 1715-1717.

168. C.W. Lam, C.H. Lau, J.C. Williams, Y.W. Chan, and L.J. Wong, "Mitochondrial myopathy, encephalopathy, lactic acidosis and stroke-like episodes (MELAS) triggered by valproate therapy [abstract]," *European Journal of Pediatrics* 156:7 (1997): 562-564.

169. C.M. Lin, and P. Thajeb, "Valproic acid aggravates epilepsy due to MELAS in a patient with an A3243G mutation of mitochondrial DNA [abstract]," *Metabolic Brain Disease* 22:1(2007): 105-109

170. G.C. Kujoth, C. Leeuwenburgh, and T.A. Prolla, "Mitochondrial DNA Mutations and Apoptosis in Mammalian Aging," *Cancer Research* 66:15 (2006): 7386-7389.

171. H.C. Lee and Y.H. Wei, "Oxidative Stress, Mitochondrial DNA Mutation, and Apoptosis in Aging," *Experimental Biology and Medicine* 232 (2007): 592-606.

172. E. Ohama and F. Ikuta, "Involvement of choroid plexus in mitochondrial encephalomyopathy (MELAS) [abstract]," *Acta Neuropathologica* 75:1 (1987): 1-7.

173. S. Ponchaut, F. van Hoof, and K. Veitch, "Cytochrome aa3 depletion is the cause of the deficient mitochondrial respiration induced by chronic valproate administration [abstract]," *Biochemistry and Pharmacology* 43:3 (1992): 644-647.

174. S.A. Hamed, M.M. Abdellah, and N. El-Melegy, "Blood levels of Trace Elements, Electrolytes, and Oxidative Stress/Antioxidant Systems in Epileptic Patients," *Journal of Pharmacological Sciences* 96 (2004): 465-473.

175. V. Tong, X.W. Teng, T.K.H. Chang, and F.S. Abbott, "Valproic Acid I: Time Course of Lipid Peroxidation Biomarkers, Liver Toxicity, and Valproic Acid Metabolite Levels in Rats," *Toxicological Sciences* 86 (2005): 427-435.

176. V. Tong, X.W. Teng, T.K.H. Chang, and F.S. Abbott, "Valproic Acid II: Effects on Oxidative Stress, Mitochondrial Membrane Potential, and Cytotoxicity in Glutathione-Depleted Rat Hepatocytes," *Toxicological Sciences* 86:2 (2005): 436-443.

177. J.A. Schneider and S.S. Mirra, "Neuropathologic Correlates of Persistent Neurologic Deficit in Lithium Intoxication," *Annals of Neurology* 36 (1994): 928-931.

Endnotes

178. E.D. Louis, J.P.G. Vonsattel, L.S. Honig, A. Lawton, C. Moskowitz, B. Ford, and S. Frucht, "Essential Tremor Associated With Pathologic Changes in the Cerebellum," *Archives of Neurology* 63 (2006): 1189-1193.

179. A. Deep-Soboslay, B. Iglesias, T.M. Hyde, L.B. Bigelow, V. Imamovic, M.M. Herman, et al, "Evaluation of tissue collection for postmortem studies of bipolar disorder," *Bipolar Disorders* 10 (2008): 822-828.

180. E.A. Thomas, B. Dean, E. Scarr, D. Copolov, and J.G. Sutcliffe, "Differences in neuroanatomical sites of apoD elevation discriminate between schizophrenia and bipolar disorder," *Molecular Psychiatry* 8 (2003): 167-175.

181. A. Digney, D. Keriakous, E. Scarr, E. Thomas, and B. Dean, "Differential Changes in Apolipoprotein E in Schizophrenia and Bipolar I Disorder," *Biological Psychiatry* 57 (2005): 711-715.

182. D. Ongur, W.C. Drevets, and J.L. Price, "Glial reduction in the subgenual prefrontal cortex in mood disorders," *Proceedings of the National Academy of Sciences* 95 (1998): 13290-13295.

183. F.M. Benes, S.L. Vincent, and M. Todtenkopf, "The Density of Pyramidal and Nonpyramidal Neurons in Anterior Cingulate Cortex of Schizophrenic and Bipolar Subjects," *Biological Psychiatry* 50 (2001): 395-406.

184. G. Rajkowska, A. Halaris, and L.D. Selemon, "Reductions in Neuronal and Glial Density Characterize the Dorsolateral Prefrontal Cortex in Bipolar Disorder," *Biological Psychiatry* 49 (2001): 741-752.

185. D. Cotter, L. Hudson, and S. Landau, "Evidence for orbitofrontal pathology in bipolar disorder and major depression, but not in schizophrenia," *Bipolar Disorder* 7 (2005): 358-369.

186. W.T. Regenold, P. Phatak, C.M. Marano, L. Gearhart, C.H. Viens, and K.C. Hisley, "Myelin staining of deep white matter in the dorsolateral prefrontal cortex in schizophrenia, bipolar disorder, and unipolar major depression," *Psychiatry Research* 151 (2007): 179-188.

187. B. Baumann, P. Danos, D. Krell, S. Diekmann, A. Leschinger, R. Stauch, et al, "Reduced Volume of Limbic System-Affiliated Basal Ganglia in Mood Disorders: Preliminary Data From a Postmortem Study," *Journal of Neuropsychiatry and Clinical Neuroscience* 11:1 (1999): 71-78.

188. Y.B. Bezchlibnyk, X. Sun, J.F. Wang, G.M. MacQueen, B.S. McEwen, and T. Young, "Neuron somal size is decreased in the lateral amygdalar nucleus of subjects with bipolar disorder," *Journal of Psychiatry and Neuroscience* 32:3 (2007): 203-210.

189. H. Bielau, K. Trubner, D. Krell, M.W. Agelink, H.G. Bernstein, R. Stauch, et al, "Volume deficits of subcortical nuclei in mood disorders: A postmortem study," *European Archives of Psychiatry and Clinical Neuroscience* 255 (2005): 401-412.

190. L. Liu, S.C. Schulz, S. Lee, T.J. Reutiman, and S.H. Fatemi, "Hippocampal CA1 Pyramidal Cell Size is Reduced in Bipolar Disorder," *Cellular and Molecular Neurobiology* 27:3 (2007): 351-358.

Chapter 8 – Stimulants

1. G.E. Jackson, *Rethinking Psychiatric Drugs: A Guide for Informed Consent* (Bloomington, IN: Author House, 2005), 260-262.

2. N. Rasmussen, "Making the First Anti-Depressant Amphetamine in American Medicine, 1929-1950," *Journal of the History of Medicine and Allied Sciences* 61:3 (2006): 288-323.

3. S. Scheindlin, "Ephedra: Once a Boon, Now a Bane," *Molecular Interventions* 3 (2003): 358-360.

4. CHADD. "About Us." Accessed October 22, 2004 at: http://www.chadd.org/webpage.cfm?cat_id=2

5. No author (February 28, 1996; March 4, 1997). "United Nations' Warnings on Ritalin." Accessed on October 14, 2004: http://216.239.41.104/search?q=cache:6RroSjUcTPsJ:www.pbs.org/wgbh/pages/frontline/shows/medicating/backlash/un.html+ritalin+abuse+in+canada&hl=en

6. NAMI. "Mission and History." Accessed on October 24, 2004 at: http://madman-bbs.dyndns.org/mirrors/nami/web.nami.org/history.htm

7. T.A. Sokka and I.T. Huhtaniemi, "Functional Maturation of the Pituitary-Gonadal Axis in the Neonatal Female Rat," *Biology of Reproduction* 52 (1995): 1404-1409.

8. R. Mittendorf, M.A. Williams, C.S. Berkey, and P.F. Cotter, "The Length of Uncomplicated Human Gestation [abstract]," *Obstetrics & Gynecology* 75 (1990): 929-932.

9. B. Clancy, B.L. Finlay, R.B. Darlington, and K.J.S. Anand, "Extrapolating brain development from experimental species to humans," *NeuroToxicology* (2007) doi: 10.1016/j.neuro.2007.01.014

10. D. Rice and S. Barone, Jr., "Critical Periods of Vulnerability for the Developing Nervous System: Evidence from Humans and Animal Models," *Environmental Health Perspectives* 108 suppl 3 (2000): 511-533.

Endnotes

11. U.S. Department of Health and Human Services (August 2005). NTP-CERHR Monograph on the Potential Human Reproductive and Developmental Effects of Methylphenidate. Accesed on July 16, 2008: http://cerhr.niehs.nih.gov/chemicals/stimulants/methylphenidate/MethylphenidateMonograph.pdf

12. M.R. Gerasimov, M. Franceschi, N.D. Volkow, A. Gifford, S.J. Gatley, D. Marsteller, et al, "Comparison between Intraperitoneal and Oral Methylphenidate Administration: A Microdialysis and Locomotor Activity Study," *Journal of Pharmacology and Experimental Therapeutics* 295:1 (2000): 51-57.

13. R. Kuczenski and D.S. Segal, "Exposure of Adolescent Rats to Oral Methylphenidate: Preferential Effects on Extracellular Norepinephrine and Absence of Sensitization and Cross-Sensitization to Methamphetamine," *Journal of Neuroscience* 22:16 (2002): 7264-7271.

14. J.D. Gray, M. Punsoni, N.E. Tabori, J.T. Melton, V. Fanslow, M.J. Ward, et al, "Methylphenidate Administration to Juvenile Rats Alters Brain Areas Involved in Cognition, Motivated Behaviors, Appetite, and Stress," *Journal of Neuroscience* 27:27 (2007): 7196-7207.

15, I. Husson, B. Mesles, F. Medja, P. Leroux, B. Kosofsky, and P. Gressens, "Methylphenidate and MK-801, an N-Methyl-D-Aspartate Receptor Antagonist: Shared Biological Properties," *Neuroscience* 125 (2004): 163-170.

16. D.C. Lagace, J.K. Yee, C.A. Bolanos, and A.J. Eisch, "Juvenile Administration of Methylphenidate Attenuates Adult Hippocampal Neurogenesis," *Biological Psychiatry* 60 (2006): 1121-1130.

17. G.A. Ricaurte, A.O. Mechan, J. Yuan, G. Hatzidimitriou, T. Xie, A.H. Mayne, et al, "Amphetamine Treatment Similar to That Used in the Treatment of Adult Attention-Deficit/Hyperactivity Disorder Damages Dopaminergic Nerve Endings in the Striatum of Adult Nonhuman Primates," *Journal of Pharmacology and Experimental Therapeutics* 315:1 (2005): 91-98.

18. D.M. Grilly an A. Loveland, "What is a 'low dose' of d-amphetamine for inducing behavioral effects in laboratory rats?," *Psychopharmacology* 153 (2001): 155-169.

19. J.J. McGough, J. Biederman, L.L. Greenhill, J.T. McCracken, T.J. Spencer, K. Posner, et al, "Pharmacokinetics of SLI381 (Adderall XR), an Extended-Release Formulation of Adderall," *Journal of the American Academy of Child and Adolescent* Psychiatry 42:6 (2003): 684-691.

20. B.G. Borcherding, C.S. Keysor, T.B. Cooper, and J.L. Rapoport, "Differential Effects of Methylphenidate and Dextroamphetamine on the Motor Activity Level of Hyperactive Children," *Neuropsychopharmacology* 2:4 (1989): 255-263.

21. U.S. Department of Health and Human Services. National Toxicology Program – Center for the Evaluation of Risks to Human Reproduction (July 2005). NTP-CERHR Expert Panel Report on the Reproductive and Developmental Toxicity of Amphetamines, II-9, II-10. Accessed on July 16, 2008 at: http://cerhr.niehs.nih.gov/chemicals/stimulants/amphetamines/AmphetamineMonograph.pdf

22. V. Armstrong, C.M. Reichel, J.F. Doti, C.A. Crawford, and S.A. McDougall, "Repeated amphetamine treatment causes a persistent elevation of glial fibrillary acidic protein in the caudate-putamen," *European Journal of Pharmacology* 488 (2004): 111-115.

23. B.N. Frey, A.C. Andreazza, K.M. Cereser, M.R. Martins, F.C. Petronilho, D.F. de Souza, et al, "Evidence of astrogliosis in rat hippocampus after d-amphetamine exposure," *Progress in Neuro-Psychopharmacology & Biological Psychiatry* 30:7 (2006): 1231-1234.

24. G. Bartzokis, M. Beckson, P.H. Lu, N. Edwards, P. Bridge, and J. Mintz, "Brain Maturation May Be Arrested in Chronic Cocaine Addicts," *Biological Psychiatry* 51 (2002): 605-611.

25. G. Bartzokis, M. Beckson, P.H. Lu, N. Edwards, R. Rapoport, E. Wiseman, et al, "Age-related brain volume reductions in amphetamine and cocaine addicts and normal controls: implications for addiction research," *Psychiatry Research Neuroimaging* 98 (2000): 93-102.

26. T.R. Franklin, P.D. Acton, J.A. Maldjian, J.D. Gray, J.R. Croft, C.A. Dackis, et al, "Decreased Gray Matter Concentration in the Insular, Orbitofrontal, Cingulate, and Temporal Cortices of Cocaine Patients," *Biological Psychiatry* 51 (2002): 134-142.

27. R. Rojas, R. Riascos, D. Vargas, H. Cuellar, and J. Borne, "Neuroimaging in Drug and Substance Abuse Part I: Cocaine, Cannabis, Ecstasy," *Topics in Magnetic Resonance Imaging* 16:3 (2005): 231-2238.

28. MTA Cooperative Group, "A 14-month randomized clinical trial of treatment strategies for attention-deficit/hyperactivity disorder. The MTA Cooperative Group. Multimodal Treatment Study of Children with ADHD," *Archives of General Psychiatry* 56:12 (1999): 1073-1086.

29. MTA Cooperative Group, "Moderators and mediators of treatment response for children with attention-deficit/hyperactivity disorder: the Multimodal Treatment Study of children with Attention-deficit/hyperactivity disorder," *Archives of General Psychiatry* 56:12 (1999): 1088-1096.

30. MTA Cooperative Group, "National Institute of Mental Health Multimodal Treatment Study of ADHD Follow-up: Changes in Effectiveness and Growth After the End of Treatment," *Pediatrics* 113:4 (2004): 762-769.

31. J. Swanson, G.R. Elliott, L.L. Greenhill, T. Wigal, L.E. Arnold, B. Bitiello, et al, "Effects of Stimulant Medication on Growth Rates Across 3 Years in the MTA Follow-up," *Journal of the American Academy of Child and Adolescent Psychiatry* 46:8 (2007): 1015-1027.

32. J. Swanson, L. Greenhill, T. Wigal, S. Kollins, A. Stehli, M. Davies, et al, "Stimulant-Related Reductions of Growth Rates in the PATS," *Journal of the American Academy of Child and Adolescent Psychiatry* 45:11 (2006): 1304-1313.

33. S. Moses (November 3, 2008). "Linear Growth Velocity." Accessed on November 11, 2008 at:
http://www.fpnotebook.com/Endo/Exam/LnrGrwthVlcty.htm

34. S. Najjar, "Growth and Development." Accessed on November 12, 2008 at:
http://www.lsfm.net/5th%20Annual%20Conference/Sunday/Najjar.pps

35. S. Kemp (November 16, 2007). "Growth Failure." Accessed on April 20, 2008 at:
http://www.emedicine.com/ped/TOPIC902.HTM

36. No author (December 22, 2007). "Growth Failure." Wikipedia. Accessed on April 20, 2008 at:
http://en.wikipedia.org/wiki/Growth_failure

37. M.S. Patel and F. Elefteriou, "The New Field of Neuroskeletal Biology," *Calcified Tissue International* 80 (2007): 337-347.

38. F. Eleftriou, "Regulation of bone remodeling by the central and peripheral nervous system," *Archives of Biochemistry and Biophysics* 473:2 (2008): 231-236.

39. D.S. Robinson, "Increased Fracture Risk and Psychotropic Medications," *Primary Psychiatry* 15:10 (2008): 32-34.

40. K.B. Jones, A.V. Mollano, J.A. Morcuende, R.R. Cooper, and C.L. Saltzman, "Bone and Brain: A Review of Neural, Hormonal, and Musculoskeletal Connections," *The Iowa Orthopaedic Journal* 24 (2004): 123-132.

41. M. Bliziotes, S. McLoughlin, M. Gunness, F. Fumagalli, S.R. Jones, and M.G. Caron, "Bone histomorphometric and biochemical abnormalities in mice homozygous for deletion of the dopamine transporter gene [abstract]," *Bone* 26:1 (2000): 15-19.

42. G. Diaz-Torga, C. Feierstein, C. Libertun, D. Gelman, M.A. Kelly, M.J. Low, et al, "Disruption of the D2 Dopamine Receptor Alters GH and IGF-1 Secretion and Causes Dwarfism in Male Mice [abstract]," *Endocrinology* 143:4 (2002): 1270-1279.

43. M. Oz, L. Zhang, A. Rotondo, H. Sun, and M. Morales, "Direct Activation by Dopamine of Recombinant Human 5HT1A Receptors: Comparison With Human 5-HT2C and 5-HT3 Receptors," *Synapse* 50 (2003): 303-313.

44. S. Bhattacharyya, I. Raote, A. Bhattacharya, R. Miledi, and M.M. Panicker, "Activation, internalization, and recycling of the serotonin 2A receptor by dopamine," *PNAS* 103:41 (2006): 15248-15253.

45. H.L. Chen, P.J. Lein, J.Y. Wang, D. Gash, B.J. Hoffer, and Y.H. Chiang, "Expression of bone morphogenetic proteins in the brain during normal aging and in 6-hydroxydopamine-lesioned animals [abstract]," *Brain Research* 994:1 (2003): 81-90.

46. J. Jordan, M. Bottner, H.J. Schluesener, K. Unsicker, and K. Krieglstein, "Bone morphogenetic proteins: neurotrophic roles for midbrain dopaminergic neurons and implications of astroglial cells [abstract]," *European Journal of Neuroscience* 9:8 (1997): 1699-1709.

47. H. Shen, Y. Luo, C.C. Kuo, and Y. Wang, "BMP7 reduces synergistic injury by methamphetamine and ischemia in mouse brain," *Neuroscience Letters* 442:1 (2008): 15-18.

48. J. Chou, B.K. Harvey, C.F. Chang, H. Shen, M. Morales, and Y. Wang, "Neuroregenerative effects of BMP7 after stroke in rats [abstract]," *Journal of the Neurological Sciences* 240:1-2 (2006): 21-29.

49. A.W. Toga, P.M. Thompson, and E.R. Sowell, "Mapping brain maturation," *Trends in Neuroscience* 29:3 (2006): 148-159.

50. P. Shaw, N.J. Kabani, J.P. Lerch, K. Eckstrand, R. Lenroot, N. Gogtay, et al, "Neurodevelopmental Trajectories of the Human Cerebral Cortex," *Journal of Neuroscience* 28:14 (2008): 3586-3594.

51. J.N. Giedd, J. Blumenthal, N.O. Jeffries, F.X. Castellanos, H. Liu, A. Zijdenbos, et al, "Brain development during childhood and adolescence: a longitudinal MRI study," *Nature Neuroscience* 2:10 (1999): 861-863.

Endnotes

52. E.R. Sowell, B.S. Peterson, P.M. Thompson, S.E. Welcome, A.L. Henkenius, and A.W. Toga, "Mapping cortical change across the human life span," *Nature Neuroscience* 6:3 (2003): 309-315.

53. E.R. Sowell, P.M. Thompson, K.D. Tessner, and A.W. Toga, "Mapping Continued Brain Growth and Gray Matter Density Reduction in Dorsal Frontal Cortex: Inverse Relationships during Postadolescent Brain Maturation," *Journal of Neuroscience* 21:22 (2001): 8819-8829.

54. E.R. Sowell, P.M. Thompson, C.M. Leonard, S.E. Welcome, E. Kan, and A.W. Toga, "Longitudinal Mapping of Cortical Thickness and Brain Growth in Normal Children," *Journal of Neuroscience* 24:38 (2004): 8223-8231.

55. S. O'Donnell, M.D. Noseworthy, B. Levine, and M. Dennis, "Cortical thickness of the frontopolar area in typically developing children and adolescents," *NeuroImage* 24 (2005): 948-954.

56. P. Shaw, D. Greenstein, J. Lerch, R. Lenroot, N. Gogtay, A. Evans, et al, "Intellectual ability and cortical development in children and adolescents," *Nature* 440 (2006): 676-679.

57. M. Ashtari, S. Kumra, S.L. Bhaskar, T. Clarke, E. Thaden, K.L. Cervellione, et al, "Attention-Deficit/Hyperactivity Disorder: A Preliminary Diffusion Tensor Imaging Study," *Biological Psychiatry* 57 (2005): 448-455.

58. S. Carmona, O. Vilarroya, A. Bielsa, V. Tremols, J.C. Soliva, M. Rovira, J. Tomas, et al, "Global and regional gray matter reductions in ADHD: A voxel-based morphometric study," *Neuroscience Letters* 389 (2005): 88-93.

59. S. Durston, H.E. Hulshoff, Hg. Schnack, J.K. Buitelaar, M.P. Steenhuis, R.B. Mindera, et al, "Magnetic Resonance Imaging of Boys With Attention-Deficit/Hyperactivity Disorder and Their Unaffected Siblings," *Journal of the American Academy of Child and Adolescent Psychiatry* 43:3 (2004): 332-340.

60. S.H. Mostofsky, K.L. Cooper, W.R. Kates, M.B. Denckla, and W.E. Kaufmann, "Smaller Prefrontal and Premotor Volumes in Boys with Attention-Deficit/Hyperactivity Disorder," *Biological Psychiatry* 52 (2002): 785-794.

61. S.R. Pliszka, J. Lancaster, M. Liotti, and M. Semrud-Clikeman, "Volumetric MRI differences in treatment-naïve vs. chronically treated children with ADHD," *Neurology* 67 (2006): 1023-1027.

62. E.R. Sowell, P.M. Thompson, S.E. Welcome, A.L. Henkenius, A.W. Toga, and B.S. Peterson, "Cortical abnormalities in children and adolescents with attention-deficit hyperactivity disorder," *Lancet* 362 (2003): 1699-1707.

63. E.R. Sowell, P.M. Thompson, S.E. Welcome, A.L. Henkenius, A.W. Toga, and B.S. Peterson, "Cortical abnormalities in children and adolescents with attention-deficit hyperactivity disorder," *Lancet* 362 (2003): 1705.

64. S. Carmona, O. Vilarroya, A. Bielsa, V. Tremols, J.C. Soliva, M. Rovira, J. Tomas, et al, "Global and regional gray matter reductions in ADHD: A voxel-based morphometric study," *Neuroscience Letters* 389 (2005): 93.

65. National Institute of Mental Health, (October 8, 2002). NIMH Press Release, "Brain Shrinkage in ADHD Not Caused by Medications." Accessed on November 7, 2008 at:
http://www.nimh.nih.gov/science-news/2002/brain-shrinkage-in-adhd-not-caused-by-medications.shtml

66. Ibid.

67. F.X. Castellanos, P.P. Lee, W. Sharp, N.O. Jeffries, D.K. Greenstein, LS. Clasen, et al, "Developmental Trajectories of Brain Volume Abnormalities in Children and Adolescents With Attention-Deficit/Hyperactivity Disorder," *JAMA* 288:14 (2002): 1740-1748.

68. F.X. Castellanos, P.P. Lee, W. Sharp, N.O. Jeffries, D.K. Greenstein, LS. Clasen, et al, "Developmental Trajectories of Brain Volume Abnormalities in Children and Adolescents With Attention-Deficit/Hyperactivity Disorder," *JAMA* 288:14 (2002): 1746.

69. S. Mackie, P. Shaw, R. Lenroot, R. Pierson, D.K. Greenstein, T.F. Nugent III, et al, "Cerebellar Development and Clinical Outcome in Attention Deficit Hyperactivity Disorder," *American Journal of Psychiatry* 164 (2007): 647-655.

70. No author (September 2, 2008). Wikipedia, "Children's Global Assessment Scale." Accessed on November 22, 2008 at:
http://en.wikipedia.org/wiki/Children%E2%80%99s_Global_Assessment_Scale

71. P. Shaw, K. Eckstrand, W. Sharp, J. Blumenthal, J.P. Lerch, D. Greenstein, et al, "Attention-deficit/hyperactivity disorder is characterized by a delay in cortical maturation," *Proceedings of the National Academy of Sciences* 104:49 (2007): 19649-19654.

72. P. Shaw, J. Lerch, D. Greenstein, W. Sharp, L. Clasen, A. Evans, et al, "Longitudinal Mapping of Cortical Thickness and Clinical Outcome in Children and Adolescents With Attention-Deficit/Hyperactivity Disorder," *Archives of General Psychiatry* 63 (2006): 540-549.

73. P. Shaw, W.S. Sharp, M. Morrison, K. Eckstrand, D.K. Greenstein, L.S. Clasen, et al, "Psychostimulant Treatment and the Developing Cortex in Attention Deficit Hyperactivity Disorder," *American Journal of Psychiatry* (e-publication ahead of print edition: 2008), AiA 1-6.

Epilogue

1. C.S. Lewis, *God in the Dock,* Edited by W. Hooper (Grand Rapids, Michigan: William B. Erdmans Publishing Company, 2002), 292.

2. N. Fox (March 13-15, 2001), "Could MRI be used to measure progression in AD?," Abstract. Proceedings of the Symposium: The Second Kuopio Alzheimer Symposium, Kuopio, Finland. Accessed on February 28, 2009 at: http://www.uku.fi/neuro/ad01s4.htm

3. N.C. Fox and P.A. Freeborough, "Brain atrophy progression measured from registered serial MRI: validation and application to Alzheimer's disease [abstract]," *Journal of Magnetic Resonance Imaging* 7:6 (1997): 1069-1075.

4. N.C. Fox, E.K. Warrington, P.A. Freeborough, P. Hartikainen, A.M. Kennedy, J.M. Stevens, et al, "Presymptomatic hippocampal atrophy in Alzheimer's disease. A longitudinal MRI study," *Brain* 119 (1996): 2001-2007.

Appendix A

1. M. Zimmerman, J.I. Mattia, and M.A. Posternak, "Are Subjects in Pharmacological Treatment Trials of Depression Representative of Patients in Routine Clinical Practice?" *American Journal of Psychiatry* 159 (2002): 469-473.

2. M. Zimmerman, I. Chelminski, and M.A. Posternak, "Generalizability of Antidepressant Efficacy Trials: Differences Between Depressed Psychiatric Outpatients Who Would or Would Not Qualify for an Efficacy Trial," *American Journal of Psychiatry* 162 (2005): 1370-1372.

3. M. Zetin and C.T. Hoepner, "Relevance of exclusion criteria in antidepressant clinical trials: a replication study [abstract]," *Journal of Clinical Psychopharmacology* 27:3 (2007): 295-301.

4. Susan Stefan. "Fact Sheet: Tort Litigation Against Pharmaceutical Companies Involving Psychiatric Drugs: Lessons for Attorneys and Advocates." Accessed on February 22, 2009 at: http://psychrights.org/Research/Legal/SStefanPsychMedstortlitigationFactSheet.htm

5. D.W. Miller, Jr., and C.G. Miller, "On Evidence, Medical and Legal," *Journal of American Physicians and Surgeons* 10:3 (2005): 70-75.

Endnotes

6. A. See, "Use of human epidemiology studies in proving causation," *Defense Counsel Journal* (October 1, 2000).

7. A.B. Hill, "The Environment and Disease: Association or Causation ?," *Proceedings of the Royal Society of Medicine* 58 (1965): 295-300.

8. J.J.G. Steffensmeier, M.E. Ernst, M. Kelly, and A.J. Hartz, "Do Randomized Controlled Trials Always Trump Case Reports? A Second Look at Propranolol and Depression," *Pharmacotherapy* 26:2 (2006): 162-167.

9. W. Lenz (February 1995), "The History of Thalidomide." Accessed on February 22, 2009 at: http://www.thalidomide.ca/en/information/history_of_thalidomide.html

10. No author (February 8, 2009). William McBride. Wikipedia. Accessed on February 22, 2009 at: http://en.wikipedia.org

11. Reproductive Science Center of the Bay Area. "Diethylstilbesterol (DES) Daughters." Accessed on February 22, 2009 at: http://www.rscbayarea.com/for_patients/female_overview/des_daughters.html

12. No author (January 16, 2009). "Diethylstilbesterol (DES)." Wikipedia. Accessed on February 22, 2009 at: http://en.wikipedia.org/wiki/Diethylstilbesterol

13. A.L. Herbst, H. Ulfelder, and D.C. Poskanzer, "Adenocarcinoma of the vagina. Association of maternal stilbesterol therapy with tumor appearance in young women," *NEJM* 284:15 (1971): 878-881.

14. D. Layton, C. Key, and S.A. Shakir, "Prolongation of the QT interval and cardiac arrhythmias associated with cisapride: limitations of the pharmacoepidemiological studies conducted and proposals for the future," *Pharmacoepidemiology and Drug Safety* 12 (2003): 31-40.

15. M. Perrio, S. Voss, and S.A.W. Shakir, "Application of the Bradford Hill Criteria to Assess the Causality of Cisapride-Induced Arrhythmia: A Model for Assessing Causal Association in Pharmacovigilance," *Drug Safety* 30:4 (2007): 333-346.

16. R.W. Williams and K. Herrup, "The Control of Neuron Number," *Annual Review of Neuroscience* 11 (1988): 423-453.

17. D.A. Colon-Ramos and K. Shen, "Cellular Conductors: Glial Cells as Guideposts during Neural Circuit Development," *PLoS Biology* 6:4 (2008): e112.

18. S.E. Hyman., "Neurotransmitters," *Current Biology* 15:5 (2005): R154-R158.

Endnotes

19. J. Kornhuber, A. Schultz, J. Wiltfang, I. Meineke, C.H. Gleiter, R. Zochling, et al, "Persistence of haloperidol in human brain tissue," *American Journal of Psychiatry* 156:6 (1999): 885-890.

20. N.R. Bolo, Y. Hode, J.F. Nedelec, E. Laine, G. Wagner, and J.P. Macher, "Brain pharmacokinetics and tissue distribution in vivo of fluvoxamine and fluoxetine by fluorine magnetic resonance spectroscopy," *Neuropsychopharmacology* 23:4 (2000): 428-438.

21. N.R. Bolo, Y. Hode, and J.P. Macher, "Long-term sequestration of fluorinated compounds in tissues after fluvoxamine or fluoxetine treatment: a fluorine magnetic resonance spectroscopy study in vivo," *MAGMA* 16:6 (2004): 268-276.

22. G.E. Jackson, *Rethinking Psychiatry Drugs: A Guide for Informed Consent* (Bloomington, IN: Author House, 2005), 52-68.

23. W.Y. Tang and S.M. Ho, "Epigenetic reprogramming and imprinting in origins of disease [abstract]," *Reviews in Endocrine and Metabolic Disorders* 8:2 (2007): 173-182.

24. A.R. Isles and L.S. Wilkinson, "Epigenetics: what is it and why is it important to mental disease?," *British Medical Bulletin* 85 (2008): 35-45.

25. S.M. Reamon-Buettner and J. Borlak, "A new paradigm in toxicology and teratology: altering gene activity in the absence of DNA sequence variation [abstract]," *Reproductive Toxicology* 24:1 (2007): 20-30.

26. P. Myers and W. Hessler (April 30, 2007). "Does 'the dose make the poison?'," Environmental Health News. Accessed on February 7, 2008 at: http://www.environmentalhealthnews.org/sciencebackground/2007/2007-0415nmdrc.html

Appendix B

1. K.L. Cozza, S.C. Armstrong, and J.R. Oesterheld., *Concise Guide to Drug Interaction Principles for Medical Practice, 2nd Ed.* (Washington, D.C.: American Psychiatric Publishing, Inc., 2003).

2. K.L. Cozza, S.C. Armstrong, and J.R. Oesterheld, *Pocket Guide: Drug Interaction Principles* (Washington, D.C.: American Psychiatric Publishing, Inc., 2003).

3. A. Szarfman, J.M. Tonning, J.G. Levine, and P.M. Doraiswamy, "Atypical antipsychotics and pituitary tumors: a Pharmacovigilance study [abstract]," *Pharmacotherapy* 26:6 (2006): 748-758.

4. D.N. Mendhekar, R.C. Jiloha, and P.K. Srivastava, "Effect of risperidone on prolactinoma – a case report [abstract]," *Pharmacopsychiatry* 37:1 (2004): 41-42.

Endnotes

5. T.E. Coyle, A.K. Bair, C. Stein, N. Vajpayee, S. Mehdi, and J. Wright. Acute leukemia associated with valproic acid treatment: a novel mechanism for leukemogenesis?," *American Journal of Hematology* 78:4 (2005): 256-260.

Appendix C

1. No author. Bayh-Dole Act (January 7, 2009). Wikipedia. Accessed on March 4, 2009 at: http://en.wikipedia.org/wiki/Bayh-Dole_Act

2. D.U. Vogt (March 25, 2005). CRS Report for Congress: Direct-to-Consumer Advertising of Prescription Drugs. Congressional Research Service. The Library of Congress, Washington, DC.

3. C. Lewis (March-April 2003). The Impact of Direct-to-Consumer Advertising. *FDA Consumer Magazine*. Accessed on March 4, 2009 at: http://www.fda/gov/Fdac/features/2003/203_dtc.html

4. N.M. Hadler, *The Last Well Person: How to Stay Well Despite the Health-Care System* (Ithaca: McGill-Queen's University Press, 2004).

5. N.M. Hadler, *Worried Sick: A Prescription for health in an overtreated America* (Chapel Hill: The University of North Carolina Press, 2008).

6. The Practitioner's Guide to the Data Banks. Accessed on March 5, 2009 at: http://www.npdb-hipdb.hrsa.gov

7. B.E. Levine (March 3, 2009). "Eli Lilly and the Case for a Corporate Death Penalty." AlterNet. Accessed on March 4, 2009 at: http://www.alternet.org/module/printerversion/129709

8. No author. Corporate manslaughter (February 21, 2009). Wikipedia. Accessed on March 4, 2009 at: http://en.wikipedia.org/wiki/Corporate_manslaughter

9. The American Geriatrics Society. Pay for Performance (P4P) Primer. Accessed March 3, 2009 at: http://www.americangeriatrics.org/policy/2006p4p_primer.shtml

10. Martha Lagace (April 14, 2003). "Pay-for-Performance Doesn't Always Pay Off." Harvard Business School – Working Knowledge. Accessed on March 3, 2009 at: http://hbswk.hbs.edu/item/3424.html

About the Author

Dr. Grace E. Jackson is a board-certified psychiatrist who graduated summa cum laude from California Lutheran University with a Bachelor of Arts in Political Science and a Bachelor of Science in Biology, as well as a Masters Degree in Public Administration. She earned her medical degree from the University of Colorado Health Sciences Center, in 1996, then completed her internship and residency in the U.S. Navy.

Following her transition from military service to civilian status in the spring of 2002, Dr. Jackson has worked for the North Carolina Department of Corrections, the Veterans Administration, and as a clinician in private practice.

An internationally renowned lecturer, writer, and forensic consultant, she has submitted testimony to governmental agencies and authorities on behalf of patients' rights, medical ethics, and health care reform; and she has served as an expert witness for the Law Project for Psychiatric Rights (a non-profit organization based in Anchorage, Alaska).

Dr. Jackson's first book, *Rethinking Psychiatric Drugs: A Guide for Informed Consent*, underscored the urgent need for societies and health care systems to recognize the unnecessary harmfulness of psychiatric medications, and to protect the rights of those who desire drug-free care. Expanding upon this same theme, *Drug-Induced Dementia: A Perfect Crime* presents a methodical analysis of the scientific and epidemiological evidence which confirms psychopharmaceuticals as a cause of brain damage and premature death. Hopefully, these publications will be used by laypersons, clinicians, lawyers, and policy makers to improve the quality and integrity of health care, and to safeguard the fundamental right of all patients to avoid unwarranted bodily harm – particularly, when that harm occurs in the form of misinformed, fraudulent, and/or coercive (involuntary) medical care.

Printed in Great Britain
by Amazon.co.uk, Ltd.,
Marston Gate.